UNDERSTANDING TERRORISM AND MANAGING THE CONSEQUENCES

Paul M. Maniscalco MPA, PhD(c), EMT/P

Hank T. Christen MPA, EMT/B

Prentice
Hall

Upper Saddle River, New Jersey 07458

Library of Congress Cataloging-in-Publication Data

Maniscalco, Paul M.
    Understanding terrorism and managing the consequences / Paul M.
Maniscalco, Hank Christen.
    p.   cm.
    Includes bibliographical references and index.
    ISBN 0-13-021229-6
    1. Terrorism—United States.   2. Emergency management—United States—
Planning.   I. Christen, Hank T.   II. Title.

HV6432 .M36   2001
363.3'2—dc21   2001021122

Dosage Selection:

The authors and publisher have made every effort to ensure the accuracy of dosages cited herein. However, it is the responsibility of every practitioner to consult appropriate information sources to ascertain correct dosages for each clinical situation, especially for new or unfamiliar drugs and procedures.

Pre-hospital personnel are reminded that they are required to comply with local treatment protocols as established within the jurisdiction that they are rendering service. Any questions regarding administration route, action, or dosages of medications should be presented to the appropriate medical director of your jurisdiction for answer or clarification.

Use of Trade or Brand Names:

Use of trade or brand names in this publication is exclusively for illustrative purposes and does not constitute or imply endorsement by the authors, editors, or publisher.

Publisher: *Julie Alexander*
Executive Editor: *Greg Vis*
Acquisitions Editor: *Katrin Beacon*
Director of Production and Manufacturing: *Bruce Johnson*
Managing Editor: *Patrick Walsh*
Manufacturing Buyer: *Patrick Brown*
Design Coordinator: *Maria Guglielmo*
Cover Designer: *Gary J. Sella*
Full Service Production/Formatting: *BookMasters, Inc.*
Editorial Assistant: *Kierra Bloom*
Printing and Binding: *RR Donnelley, Harrisonburg, VA*

Pearson Education LTD.
Pearson Education Australia Pty, Limited
Pearson Education Singapore, Pte. Ltd.
Pearson Education North Asia Ltd
Pearson Education Canada, Ltd
Pearson Educación de Mexico, S.A. de C.V.
Pearson Education—Japan
Pearson Education Malaysia, Pte. Ltd
Pearson Education, Upper Saddle River, New Jersey

10 9 8 7 6 5 4 3
ISBN 0-13-021229-6

Dedication

This book is dedicated to the men and women of "America's Front-line Defense," the emergency responders of the United States.

Whether it was the bombing of the World Trade Center in New York City, the Murrah Building bombing in Oklahoma City, the Atlanta Centennial Park bombing, or the myriad of other terrorist-related events that have transpired in the United States, in each instance, we have witnessed extraordinary people rising to the occasion and answering the call for help. You have met the challenge head-on to mitigate and remediate—at times, at great personal risk and peril—the horrific aftermath of these cowardly attacks on noncombatants.

All too often people discount the importance that members of Emergency Medical Services, Law Enforcement, Fire Service, Public Health, Security, Emergency Management, and Medicine play in the preservation of civil society and national security. With the ever-changing tactics and strategies of terror-related organizations, we now witness emergency responders being targeted for attack (such as the use of secondary devices focused on wreaking havoc and perhaps injuring/killing our peers while operating at the scenes of these incidents). These acts of cowardice have upped the personal threat ante to our fellow professionals. We developed this text to bring understanding, clarity, and sensibility to our fellow professionals so that when you respond to an act of terrorism, you may be safer and more effective in helping those who need your assistance in their time of dire need.

Although it is never our intention to discount the critical and vital role our national defense forces play in protecting us here and abroad, it is our belief that it is our nation's emergency response force, almost 3.5 million strong, that mean the difference between life and death each and every day for our citizens. In our business we have often said, "A good day is coming home. . . . A great day is coming home in one piece!"

We pray and wish for each of you many days coming home in one piece!

Paul M. Maníscalco
Hank T. Christen

Contents

Foreword

There is a paradox in modern America's concern with terrorism in the new millennium. The most powerful nation in human history finds itself preoccupied with the asymmetric tactics of much weaker states and a few terrorist groups and radical individuals. These weaker actors cannot directly

challenge the United States, so they are turning increasingly to terrorist tactics and unconventional weapons. Some states have no choice but to take the strongest possible measures against terrorism. For a state standing singly at the pinnacle of global power, however, methodical and comprehensive preparedness for terrorism is something of a luxury—but one which it is perfectly able to afford if it wishes.

This is not to suggest that the United States has no significant problem with terrorism. The United States has experienced terrible terrorist attacks in its recent history, both at home and abroad. The worst of these attacks were the bombings of World Trade Center (1993), the Murrah federal building in Oklahoma City (1995), Khobar Towers in Saudi Arabia (1996), two U.S. embassies in East Africa (1998), and the USS Cole (2000). These attacks have exposed high vulnerability of the American homeland as well as U.S. interests outside of the continental United States to terrorist attacks. The attacks also point to the tactical adaptation forced upon America's enemies by the preponderant power of the U.S. government.

The threat of terrorism is thus a byproduct of America's power: We should expect to see more of it in the future, and we should expect this terrorism to grow more destructive and more difficult to combat.

Especially worrisome is the possibility of a terrorist attack against the United States involving a chemical, biological, nuclear, or radiation weapon—a category known as weapons of mass destruction (WMD). Such an attack is less likely than a conventional terrorist attack, such as a bombing, but would have far more consequences. Although the United States has yet to experience an episode of WMD terrorism, over the last five years America's diverse public safety community has gradually been preparing itself for "consequence management" operations in the aftermath of a possible WMD attack at home. The two authors of this text have been at the forefront of this movement since its inception.

Historically, U.S. counterterrorism policy has focused on preventing acts of terrorism and defending against such acts should prevention fail. Recently, the U.S. government has established a "domestic preparedness" program that seeks to broadly reduce America's vulnerability to large, destructive acts of terrorism by improving a wide range of operational response capabilities across the country, at all levels of government. As a result, the U.S. government is now actively engaged in preparing the nation for highly destructive acts of terrorism. This effort involves multiple federal agencies and a wide variety of programs. The federal budget for WMD preparedness programs has grown from effectively zero in 1995 to approximately $1.5 billion in fiscal year 2000. This made the U.S. domestic preparedness program one of the fastest growing federal programs of the late 1990s. No other nation has embarked on a comparable counterterrorism preparedness program.

A unique aspect of this program is its focus on improving the ability of state and local agencies to manage the consequences of terrorist attacks, especially those involving weapons of mass destruction. The program emphasizes preparedness for an attack with a chemical or biological weapon because damage-limiting opportunities, such as medical intervention, are greatest at the early stages of such incidents. The efficacy of consequence management operations, therefore, will largely be determined by the abilities of the "first responders." These first responders will not come from federal agencies. There is simply no way the federal government can maintain the operational capabilities needed for a speedy response across the breadth of the United States. Instead, response capabilities will come from local and state agencies—such as law enforcement and fire departments, hazardous material teams, emergency medical services, emergency management, and public health agencies—which already possess many of the operational systems upon which the nation's emergency preparedness rests. Thus for reasons of efficiency and practicality, the U.S. domestic preparedness program has no choice but to leverage existing state and local resources and work closely with first responders across the nation.

There have been problems with this effort, and there undoubtedly will be more in the future. Even though the U.S. domestic preparedness program is in its early stages, the great difficulty of designing and implementing the program has already become apparent. There is disagreement over the actual threat of terrorism, especially WMD terrorism, on American soil. Because the preparation and training have taken place on an ad hoc basis, there is no clear overarching strategy or goal to serve as a benchmark to assess readiness, nor is there a consensus on what readiness is, and how to sustain it once it is achieved.

The lack of a clear, coordinated mission at the federal level means that state and local agencies must assume some of the initiative in establishing standards and integrating those standards into their operational training. For the public safety community, this is a significant challenge. Day-to-day operations frequently utilize the full capacity of public safety organizations. Budget constraints prevent these organizations from expanding their capacity. This is particularly true of the public health community and the large volunteer-based emergency response community.

This book is an important and insightful contribution to America's ongoing national domestic preparedness effort. It is a highly pragmatic volume designed for first responders and other operational public servants who may one day be called upon to respond to a terrorist incident in the line of duty. This makes the book especially valuable, because this community of operators has only recently been drawn into America's counter terrorism program. The first responders' contribution to domestic preparedness, however, is increasingly understood and appreciated.

Paul Maniscalco and Hank Christen are uniquely positioned to write this book. Through their current and past public service, they have experience in the classroom, in the policy-making arena, and on the street. Their book bridges the gap between these disciplines. Both men have been at the forefront not only of educating first responders about the terrorism threat and the operational requirements of effective consequence management operations, but of educating policy makers and experts on the capabilities and needs of the first responders. Their new book represents one of the first books on terrorism response written by people who have real field experience in emergency management. As such, policy experts on terrorism and counterterrorism can learn much from this book, and emergency response personnel and other first responders will find it invaluable as a primer on the myriad issues associated with the consequence management of terrorist incidents.

Richard A. Falkenrath
John F. Kennedy School of Government
Harvard University

Richard A. Falkenrath

Richard A. Falkenrath is an assistant professor of public policy at the Kennedy School of Government, Harvard University. He completed a three-year term as executive director of the Kennedy School's Belfer Center for Science and International Affairs (BCSIA) in 1998. He is principal investigator of the Executive Session of Domestic Preparedness (a Department of Justice-funded joint project of BCSIA and the Kennedy School Taubman Center for State and Local Government) and of the Jeddah Forum project funded out of Jeddah, Saudi Arabia. He is the author or co-author of *Shaping Europe's Military Order: The Origins and Consequences of the CFE Treaty* (1995), *Avoiding Nuclear Anarchy: Containing the Threat of Loose Russian Nuclear Weapons and Fissile Material* (1996), *America's Achilles' Heel: Nuclear, Biological, Chemical Terrorism and Covert Attack* (1998), and numerous journal articles and chapters of edited volumes. Dr. Falkenrath has been a visiting research fellow at the German Society of Foreign Affairs (DGAP) in Bonn, as well as a consultant the U.S. Department of Defense, the intelligence community, several congressional offices, the RAND Corporation, and a range of private companies in the defense sector. He is a member of the Nonproliferation Advisory Panel (Central Intelligence Agency), International Institute for Strategic Studies, the Council on Foreign Relations, the American Council on Germany, and the American Economic Association. He holds a doctorate from the Department of War Studies, King's College, London, where he was a British Marshall Scholar, and is a *summa cum laude* graduate of Occidental College, Los Angeles, with degrees in economics and international relations.

Preface

Terrorism is not a new phenomenon. It has, and always will remain a weapon of the weak. It is a low-cost, high-leverage method and tactic that enables small nations, sub-national groups and even individuals to circumvent the conventional projections of national strength (political, economic, or conventional military might). This is especially the case since the United States' swift and decisive victory in Desert Storm. Few, if any, nations would attempt to confront the United States tank for tank/plane for plane in a conventional war on the traditional battlefield, recognizing that terrorism and unconventional warfare is a more effective—and perhaps even unaccountable—means of offsetting a superpower's strength and striking at its weakness. Although the threat of terrorism is not new to the world stage, the face of terrorism is changing.

In our current climate of increased ethnic, tribal, religious, and nationalist conflict, terrorism is increasingly becoming a strategic tactic and weapon for non-state actors with interests inimical to our own. Subnational and non-state groups, many with a global reach, such as Usama Bin Laden's organization, Al Qaida, the Islamic Jihad, the Tamil Tigers of Eelam (LTTE) and a host of loosely affiliated networks of *fellow travelers* are difficult to monitor for a variety of logistical problems and legal constraints. Even when a conflict coalesces into a mass movement, terrorism tends to germinate within small cells, difficult to discern against the backdrop of the larger movement. Moreover, terrorists' motivations are multifaceted, and differ from group to group. Of unique concern is the irrationality of many non-state groups. Often feeling that they answer to a higher spiritual authority, groups such as the Aum Shinrikyo are not bound by traditional political ideology, nor do they strive for popular support and acceptance. These motivations translate into a myriad of dangerous intentions, including the propensity for mass casualties. In the past, politically motivated terrorists often tried to strike a casualty balance in their attacks: inflict enough to be heard on the global stage, but not too many where a backlash hurts the cause. Modern terrorists are no longer motivated or constrained by popular appeal. Some religious extremists don't seek a seat at the negotiating table. They want to blow up the table and replace it with their own.

Perhaps most troubling is that terrorists today can avail themselves of advanced technology. The access and know-how to such technology, particularly chemical, biological, radiological, or nuclear (CBRN) weapons, truly empowers a new class of adversaries.

What makes a major CBRN terrorist incident unique is that it can be a transforming event. A terrorist attack involving such weapons of mass destruction would have catastrophic effects on American society beyond the deaths it might cause. Although the probability of a significant CBRN attack may be low in the near future, the consequences are too severe to ignore. Aside from the actual physical effects and casualties resulting from a CBRN event, the psychological impact cannot be underestimated, possibly shaking the nation's trust and confidence in its government and institutions to the core. However, the longer the window that it does not occur, the more robust opportunity we have to undertake and implement the mechanisms to prevent and better manage the consequences of such an event.

To fully appreciate the considerable challenges the United States is now facing, it is important to put the current fears regarding the threat into perspective. For decades, terrorism experts have argued about the likelihood of a major terrorist incident occurring on U.S. soil. They also argued over the possibility of terrorists using weapons of mass destruction. The debating ended abruptly with the February 26, 1993, World Trade Center bombing and the May 20, 1995, sarin gas attack of the Tokyo subway. Threat calibrations did a 180-degree turn, and our nation's planners have been running ever since to catch up with the change and backfill shortfalls that had been allowed to grow during the debating years.

Two critical thresholds have now been crossed, forcing policy makers to plan for terrorism that may well be within our home territory and involve weapons more dangerous than ever before. This sea change in terrorism demands a vigorous and flexible response policy to counter the threat of terrorism—in both its novel and familiar form.

The terror threat has changed but what remains constant is the crucial influence of the emergency responders: the EMTs, paramedics, firefighters, and state and local law enforcement officials and in the case of bioterrorism, primary care physicians and public health and medical communities. A terrorist event has international implications, but the immediate consequences are local. The actions taken in the first moments after an event are critical to shaping and determining the outcome of a major terrorist event. The actions taken by these men and women will decide whether or not victims can be turned into patients and ultimately whether the battle will be won or lost.

Much has been written by inside-the-Beltway think tanks and policy wonks. The terrorist threat has been thoroughly studied and recommendations have been offered by the dozen. The next step is to generate awareness and effectively educate the professionals tasked with responding to these heinous acts.

There is currently a dearth of literature aimed at educating and training the real heroes in the fight against terrorism: the emergency responders. However, education and training also require equipment and exercises—which, in turn, must be underpinned by resources. The big mistakes must be made on the practice field, not on Main Street, U.S.A.

Resources, however, are limited. Our goal, therefore, must be optimal transition from the *ordinary* (coping with heart attacks or a HAZMAT incident) to the *extraordinary*. Standardized training, cross-jurisdictional/interoperable equipment and incident command structures, and communication of lessons learned from exercises so as to encourage (common) best practices—all are key to our common cause.

This text, *Understanding Terrorism and Managing Its Consequences*, is the first of its kind and fills a critical void. The authors, Paul M. Maniscalco and Hank T. Christen, provide the first road map for the practitioner for dealing with a mass-casualty terrorist event. It represents an invaluable reference to arm those at the tip of the spear [those at the forefront] with knowledge and awareness. This book will help improve the odds in favor of the emergency responders and in turn the American people.

Frank J. Cilluffo
Deputy Director and Senior Policy Analyst
Center for Strategic and International Studies
Washington, D.C.

Frank J. Cilluffo

Frank Cilluffo directs the Global Organized Crime Project's seven multiagency and multidisciplined task forces on information warfare and information assurance, terrorism, Russian organized crime, Asian organized crime, the narcotics industry, financial crimes, and the nuclear black market. These task forces comprise over 175 senior officials and experts from the academic, defense, diplomatic, intelligence, law enforcement, and corporate communities. Cilluffo regularly advises policy makers and practitioners in developing strategic plans, policies, and tactical procedures to better prepare for, deter, and respond to a range of emerging and interrelated threats such as terrorism, transnational crime, and information warfare. He has lectured to dozens of government agencies, corporate groups, colleges and universities, and has testified before the U.S. Congress and presidential, defense, and congressional commissions. In addition to publishing extensively in professional journals, newspapers, and magazines, Cilluffo coauthored and edited *Global Organized Crime: The New Empire of Evil* (CSIS, 1994), *Russian Organized Crime* (CSIS, 1997), *Cybercrime, Cyberterrorism, Cyberwarfare* (CSIS, 1998), and several forthcoming task force reports. He regularly appears on major television and radio networks worldwide, commenting on a variety of national security matters. Cilluffo is currently a World Economic Forum fellow, a Council on Foreign Relations term member, a member of the Office of the Secretary of Defense's Highlands Forum, and has served as a member of the Defense Science Board's Summer Study on Transnational Threats.

Acknowledgments

Paul M. Maniscalco PhD (c), MPA, EMT/P

To my wife Connie, whose generous love and encouragement is a constant source of inspiration. She continues to be the best thing that has ever happened in my life! To my father and role model, Anthony S. Maniscalco, whose positive and lifelong influence has played a critical role in my public service career. To my mother, Patricia O'Brien Maniscalco, her guidance and constant reinforcement of assisting others as a noble calling has brought me to appreciate my chosen career even more. To my brothers Peter, Mark, John, and Matthew whose support is constant and invaluable.

To Dr. Roy Sparrow, New York University–Wagner School of Public Service; Dr. Bruce Hoffman, RAND; Frank Ciluffo, Center for Strategic and International Studies; Dr. Brad Roberts, Institute for Defense Analysis; Dr. Richard Falkenrath, Harvard University–Kennedy School of Government; Dr. Jeff Isaacson, RAND; Gov. James Gilmore (Virginia); Congressman Curt Weldon (Pennsylvania); Lt. Gen. (U.S. Air Force retired) James Clapper; Lt. Gen. (U.S. Army retired) Bill Garrison; Lt. Gen. (U.S. Army retired) Bill Reno; FBI Asst. Director (retired) James Greenleaf; George Foresman, deputy state coordinator, Virginia Emergency Services; Ambassador L. Paul Bremer, U.S. State Department; Secretary of the Army (retired) and Congressman (Virginia retired) John O. Marsh; Giandomenico Picco, U.N. chief hostage negotiator (retired); Michael Wermuth, RAND; and Dr. Allan Braslow—the valuable guidance and assistance each of you have provided me is greatly appreciated and applied daily.

For the immeasurable longtime support and assistance provided to me by my many peers I would like to thank: Jeff Abraham, Doug Arbour, Alan Brunacini, MacNeil C. Cross, James Denney, Gerald Dickens, Neal Dolan, Walter Drivet, John Fitzsimmons, Judd Fuller, Erik Gaull, Daniel Gerard, Marc Griswold, David Handschuh, Donald Hiett, Dr. Keith Holtermann, Robert Iannarelli, Wade Jones, Dr. Paul Kim, Daniel Kaniewski, Don Lee, Maj. Michael Malone (USMC), Bob Morrone, Michael Newburger, Eugene J. O'Neill, Dennis Rubin, John Sinclair, Clark Staten, Mark Steffens, Matthew

Streger, Robert Sudol, and Dean Wilkinson; each of you has been a great friend and confidant.

To Hank Christen without whom my writing would be laborious and boring, thanks for being a great friend, partner, and contributor.

Hank T. Christen MPA EMT/B

This book was possible because of support from my best friend, my wife, Lynne. I am also proud to acknowledge my three sons, Hank, Eric, and Ryan. My career and professional influences began with my parents, District Chief (retired) Henry T. Christen Sr., Miami Fire Department, and the world's greatest mom, Louisa Worth Christen. I will always remember my career in the Atlanta Fire Department, where the adventure started. I especially will never forget the night of May 29, 1971, when Howard Beck, Charles Fernander, Lewis Grady, and Virlon Crider were killed in the line of duty. I want their names to be remembered.

I am proud of my association with Mike Hopmeier, Dr. Chuck Linden, Dean Preston, and the outstanding team at Unconventional Concepts, Inc. Other professionals that have taught me well throughout the years and continue to set an example are Lou Cuneo, David Chamberlin, Claude Lemke, Mac McDonnell, Phil Chovan, George Collins, Dr. David Goetsch, Wallace Moorehand, Mike Malone, Jim Denney, Dr. Ron Weed, Alan Brunacini, and Dennis Rubin.

My greatest personal reward has been my continued association with my professional partner and friend, Paul Maniscalco.

Contributors and Reviewers

The authors would like to thank the following contributors to our project. Their invaluable assistance and expertise enhanced the overall quality of our end product. They are:

Frank J. Cilluffo
Director, Terrorism Task Force
Chairman, Chemical, Biological, Radiological, Nuclear (CBRN) Task Force
Center for Strategic and International Studies
Washington, D.C.

David Cid
Asst. Special Agent in Charge
Federal Bureau of Investigation
Oklahoma City, Oklahoma

James P. Denney MA, EMT/P
Executive Director
Global Emergency Management Services Association
Alta Loma, CA
Principal
Organizational Strategic Solutions (OSS) Group

Gerald F. Dickens EMT-D (retired)
Hazardous Materials Technician
Logistics Coordinator
Special Operations Division
New York City Emergency Medical Service

Neal J. Dolan MCJ, NREMT-P
Special Agent in Charge
United States Secret Service

Michael J. Hopmeier MSME
Chief, Innovative and Unconventional Concepts
Unconventional Concepts, Incorporated
Eglin Air Force Base, Florida

Dennis R. Krebs CRT
Captain, Baltimore County Fire Department
International Survival Systems
Baltimore, Maryland

Michael V. Malone
Major, United States Marine Corps.

Susan S. McElrath MS(c), BSHP
McElrath and Associates
Powder Springs, Georgia

Eugene J. O'Neill NREMT-B
Rescue Specialist
WMD/Domestic Preparedness Lead Instructor
Special Operations Group
University Hospital EMS, Newark, New Jersey

Frederick R. Sidell MD
Former Chief, Chemical Casualty Care Office
Former Director, Medical Management of Chemical Casualties Course
U.S. Army Medical Research Institute of Chemical Defense
Aberdeen Proving Ground
Chemical Casualty Consultant
Bel Air, Maryland

Charles E. Stewart MD, FACEP
Emergency Physician
Colorado Springs, Colorado

Robert L. Walker
Explosive Ordinance Disposal
Subject Matter Expert
Ft. Walton Beach, Florida

We would also like to thank the following reviewers. They are:

Frank J. Cilluffo
Director, Terrorism Task Force
Chairman, Chemical, Biological, Radiological, Nuclear Task Force
Center for Strategic and International Studies

MacNeil C. Cross
Chief of Operations (retired)
New York City Emergency Medical Service

Richard Falkenrath
John F. Kennedy School of Government
Harvard University

Marc S. Griswold NREMT-P
Special Agent
United States Secret Service

Donald H. Hiett, Jr. EMT/CT
Assistant Fire Chief (retired)
Olympic Planning and Operations
Atlanta (Georgia) Fire Department
Principal
Organizational Strategic Solutions (OSS) Group

Keith A. Holtermann, DrPH, MBA, MPH, RN
Assistant Dean for Health Sciences
School of Medicine and Health Sciences
Chief of Research & Health Policy Analysis
Ronald Reagan Institute of Emergency Medicine
The George Washington University
Principal
Organizational Strategic Solutions (OSS) Group

Paul D. Kim MD
Area Emergency Manager
Stratton V.A. Medical Center
Albany, New York

Dennis L. Rubin MPA(c), BS, EMT-D
Fire Chief
Norfolk Fire and Paramedical Services
Norfolk, Virginia

About the Authors

Paul M. Maniscalco PhD (c), MPA, EMT/P is a deputy chief paramedic for Fire Department, City of New York EMS Command; an adjunct assistant professor with George Washington University, School of Medicine and Health Sciences; and a past president of the National Association of Emergency Medical Technicians. His previous commands in New York City have

Authors Paul M. Maniscalco and Hank T. Christen (L to R) were members of the ESF-8 overhead team for the 1996 Olympics.

included Manhattan, the Bronx, Brooklyn and Staten Island, Communications, Division of Training, Office of the Chief of the Operations, and Commanding Officer of the Special Operations Division. Chief Maniscalco has over 25 years of EMS and fire service response, supervisory, and management experience.

Paul has lectured extensively internationally and is widely published in academic and professional journals on emergency medical service, fire service, special operations, public safety, and crisis and consequence response to acts of terrorism and national security issues. He is the co-author of the Brady textbook, *The EMS Incident Management System: EMS Operations for Mass Casualty and High Impact Incidents*, a contributing author to the Chemical and Biological Arms Control Institutes' *Hype or Reality? The "New Terrorism" and Mass Casualty Attacks*, the U.S. Fire Administration's *Guide to Developing and Managing an Emergency Service Infection Control Program*, National Fire Service Incident Management System Consortium's *Model Procedures Guide for Emergency Medical Incidents, First Edition* and a contributing author for the International Federation of Red Cross and Red Crescent *IFRC Disaster Management Training Manual*.

Chief Maniscalco is a member of the U.S. Department of Defense, Defense Science Board, Transnational Threat Study; the Center for Strategic and International Studies—CBRN Terrorism Task Force; the Department of Defense, Interagency Board (IAB) EMS/Medical Working Group; an advisor to the Defense Advanced Research Project Agency; and is an appointee to the congressionally mandated National Panel to Assess Domestic Preparedness (Gilmore Commission). Paul also holds an appointment to the Harvard University, John F. Kennedy School of Government and Department of Justice Executive Session on Domestic Preparedness, a three-year panel examining the issues of responding to acts of terrorism.

Paul M. Maniscalco earned his baccalaureate degree in public administration—public health and safety from the City University of New York, a master of public administration—foreign policy and national security from the New York University Wagner Graduate School of Public Service and is presently a candidate for a doctoral degree in organizational behavior with a specialization in disaster management.

Hank Christen, MPA, EMT, is the director of emergency response operations for Unconventional Concepts Inc. He was previously a battalion chief for the Atlanta Fire Department, and director of emergency services, Okaloosa County, Florida.

He was the unit commander for the Gulf Coast Disaster Medical Assistance Team and responded to national level disasters including multiple hurricanes, wildfires, and the 1996 Olympics.

He is a member of the U.S. Department of Defense, Defense Science Board, Transnational Threat Study and has served on the Department of Defense, Interagency Board (IAB) EMS/Medical Working Group and as an advisor to the Defense Advanced Research Project Agency. He is currently a member of the Executive Session on Domestic Preparedness, John F. Kennedy School of Government, Harvard University.

Hank Christen has been a speaker at international level disaster conferences since the 1970s. He is a contributing editor to *Firehouse Magazine* and has published over 50 technical articles. He is the co-author of *The EMS Incident Management System—A Guide to Mass Casualty and High Impact Events* (Brady, 1998) and a contributing author for the International Federation of Red Cross and Red Crescent *IFRC Disaster Management Training Manual*.

Introduction

Terrorism is about violence. It is the tactical use of efficient high lethality weapons to accomplish a strategic purpose. Terrorism is designed to violently accomplish an objective.

With the advent of the twenty-first century the definition of terrorism by our government is expanding. In the traditional twentieth century model, terrorism was a volatile act committed by foreign operators or governments to promote a political agenda. This definition was applied by some federal

Source: Courtesy D. Handschuh

agencies for statistical purposes. (With this definition, the number of "official" terrorist acts remained small.) The authors attended a U.S. Department of Defense briefing that focused on the World Trade Center incident, Kobar Towers (Saudi Arabia), and the American embassy bombings in Africa, but excluded Oklahoma City because it was considered a domestic incident.

Modern terrorism has blurred the formal definition because the motives are changing and the players are changing. In recent times the motives include religion, revenge, and the furthering of a criminal enterprise. The players now include domestic terrorists that want to drastically change the American government or the American way of life. They also include criminal elements that vary from individuals, street gangs, single-issue zealot groups, and highly organized national/international cartels. There are also religious groups that choose to violently promote their beliefs by attacking people or institutions perceived as religious enemies.

The weapons of terrorism vary from traditional to cutting edge. Improvised explosive devices (IED) and firearms (especially automatic assault weapons) are low tech and inexpensive, yet very effective and more common in their utilization. There exists a threat for weapons of mass effect (WME). The breakup of the former Soviet Union has resulted in black market supplies of small and portable nuclear weapons based on 1997 statements by Russian officials. There are also chemicals available in many forms that could also be co-opted for hostile use. Recently we are witnessing heightened concerns about biological substances being used as weapons against unprotected civilian populations including pathogens and toxins that can cause widespread death or disease. These pathogens are often difficult, if not impossible to detect if released, and the result of an effective bioterrorism incident would severely stress medical treatment capabilities.

Lastly, information operations (sometimes called information warfare) has become prominent in today's modern cyber-world. Information operations are not violent, however, cyber-subterfuge can severely affect the ability of response agencies to process accurate information. It can also cripple critical infrastructures such as communications, 911 systems (as witnessed in Florida), transportation, petrol-chemical, electrical, and financial systems.

Convergent responders, emergency responders, emergency medical facilities, and the public health system inherit the effects of terrorism. We are the consequence managers and must act accordingly. The definition of a given act may blur between terrorism and violence, but the effects are the same. If there is a mass-shooting event at a shopping mall involving a foreign perpetrator, it is classified as terrorism. If the shooter is an American militia member, it is domestic terrorism. If the intent is to rob the mall, it is tactical ultraviolence.

In each case, the definition/classification of the event varies, but the consequences are mass medical casualties. The outcome is a high-impact EMS incident that threatens the safety of convergent responders and emergency responders, and taxes the capability of the 911, public health, and hospital systems.

In many ways, this book is about the unthinkable: A consequence response mode where people and systems are forced to confront mammoth tasks that stress systems like never before. However, through awareness and sustainable training, we can be effective at protecting ourselves and reducing the death and injury from the tactical violence that we call terrorism.

1
Terrorism: Meeting the Challenge

James P. Denney
Paul M. Maniscalco
Hank T. Christen
Frank J. Cilluffo

Source: Courtesy AP/Wide World Photos

Chapter Objectives

After reading this chapter, you will be able to accomplish the following objectives:

1. Understand the critical concept of local preparedness for terrorism/ tactical violence response.
2. Have an overview of the federal agencies in the national response spectrum.
3. Recognize the changing history and threat evolution of terrorism.
4. Discuss the capabilities and limitations of local response systems.
5. Understand the risk to responders in the modern terrorism environment.
6. Recognize the differences between advanced trauma life support protocols and casualty care in the urban combat environment.
7. Outline the key concepts in a mass fatality management plan.

Organized Terrorism

In the past, what was referred to as an "international terrorist network" was attributed to the Tricontinental Congress held in Havana, Cuba, in January 1966. At that congress, 500 delegates proposed close cooperation among socialist countries and national liberation movements in order to forge a global strategy to counter "American imperialism."

Meetings among terrorist organizations increased in the years that followed. Examples of networking include the American Black Panther leaders who toured North Korea, Vietnam, and China. In China, they had an opportunity to meet with Shirley Graham, wife of W.E.B. Du Bois, who was visiting with Zhou Enlai.[1] The network advanced terrorism on global, national, and regional levels by providing economic support, training, specialty personnel, and advanced weaponry. For example, in 1978 Argentine terrorists received Soviet-made rockets from Palestinian terrorists and the Baader-Meinhof group supplied three sets of American night vision binoculars to the IRA and American-made hand grenades to the Japanese Red Army. Terrorist training camps sprang up in Cuba, Algeria, Iraq, Jordan, Lebanon, Libya, and South Yemen, while Syria provided sites for advanced training.

Acts of terrorism had prompted the Tokyo and Montreal conventions of 1963 and 1971 on hijacking and sabotage of civilian aircraft. This was followed by the 1973 convention on crimes against diplomats, and the 1979 convention on hostage taking sponsored by The Hague. These conventions established categories of international crimes that are punishable by any state regardless of the nationality of the criminal, victim, or the location of the offense.

Following President Reagan's ordering of U.S. military forces to attack "terrorist-related" targets in Libya in 1986, seven Western industrial democracies pledged themselves to take joint action against terrorism. Those nations include the United States, Germany, Great Britain, Italy, Canada, France, and Japan. These nations agreed to deny terrorist suspects entry into their countries, to bring about better cooperation between police and security forces in their countries, to place restrictions on diplomatic missions suspected of being involved in terrorism, and to cooperate in a number of other ways.

The Cold War Ends, Threat Shifting Begins

The breakup of the former Soviet Union and its subsequent economic collapse significantly altered the global security landscape. At the time of the breakup, the Soviet Union was a superpower with well-developed nuclear, chemical, and biological warfare capabilities.[2] Unfortunately, a newer, more sinister threat has begun to emerge from the former Soviet Union, a thriving international black market in arms and weapons technology.

The economic collapse of the Soviet Union is manifesting in the nations' inability to meet their fiscal obligations, including military and other payrolls. As a result, Russia is suffering a tumultuous period with an ever present threat of coup d'état and organized crime emerging as a significant influencing factor within their domestic and international affairs.

With available weapons ranging from "suitcase" nuclear devices to submarines and MIGs to missiles, the former controlled threat has been transformed into an uncontrolled threat market strongly influenced by the Russian Mafia, a market where unpaid government employees have already attempted to sell limited amounts of fissionable material and other articles or substances that are of value to terrorist organizations.

In addition to the black market in weapons, there also exists a potential for weapon thefts, including those weapons stored at poorly guarded chemical and biological storage facilities located throughout the former Soviet Union and its satellite nations.

A New Threat Emerges

The Middle East region has emerged as the primary contributor of personnel, funding, training, and arms for the current terrorist movement. Unlike the Marxist-Leninist of the 1960s and 1970s, who were motivated by political and social conditions,[3] today's international terrorism is an amalgam consisting of anarchist-nihilist and Middle Eastern religious-based fundamentalists who despise American hegemony and the resulting influ-ence on world policy. Bent specifically on the destruction of American val-ues, influence, and economy, they have targeted U.S. interest worldwide.[4]

Globalization

Globalization[5] is changing the context in which terrorists operate. A transnational cast of characters that, in most cases, cannot be directly con-trolled by governments, either individually or collectively, increasingly af-fects even so-called domestic terrorism. Information technology has effectively removed the ability of countries to isolate themselves. Informa-tion and communication control is difficult, if not impossible, to achieve be-cause the information revolution has resulted in a democratic access to technology. Two important results are that the terrorists' prospect base has expanded exponentially and free speech and civil liberty have been given an inexpensive international medium with which to voice discontent.

The notional concept of a centrally controlled international terrorist network, previously investigated during the 1960s and 1970s, was deemed to be unlikely due to conflicting ideologies, motivating factors, funding, and arming and training among global practitioners. However, networks are now quite possible with the advent of public access to the Internet, the ability to transfer funds and conduct banking electronically, the interna-tional arms market, encrypted digital communication technology, and the emergence of "state-less" terrorism. An important result is that instant global communication between offensive action cells and their controllers are now possible. Controllers now have global reach and can run multiple independent cells from a single location with no interaction between the cells. They can also contract terrorism services utilizing mercenaries com-posed of local indigenous practitioners in a given target community.

Technology has also resulted in a reduced requirement for infrastruc-ture, security, and detection avoidance and has resulted in an asymmetry between cause and effect. The complexity of weapons acquisition, produc-tion, transportation, lethality, and delivery platform has been diminished.

Critical Factor: The historical, organizational, operational and behavioral complexi-ties posed by terrorist organizations are numerous. Although the previous section was designed to provide minimal clarity and foundation for the background material, it is only a snowflake on the tip of an iceberg. We encourage continued acquaintance with these issues by conducting in-depth research via any of the numerous comprehensive texts that were developed to address these topics.

FIGURE 1-1   Walt Disney helped design this gas mask for American children during WWII. *Source:* Courtesy Soldier Biological Chemical Command (SBCCOM)

Preparing for Threats Involving Weapons of Mass Destruction

The 1993 bombing of the New York City World Trade Center and the 1995 bombing of the Alfred P. Murrah Federal Building in Oklahoma City clearly illustrated America's vulnerability to terrorism—not only abroad, but to the homeland as well. These acts demonstrate the willingness and ability of determined terrorists to carry out attacks against high-impact targets. The March 1995 sarin gas incident in the Tokyo subway demonstrated their willingness to kill several thousand in a single event.

Twenty-first century threat evolution will require public safety and emergency service organizations to reevaluate their priorities and consider their ability to address multiple new tasks.[6] These tasks include support for new and emerging technologies; the increasing significance of resource linking and specialization; and a better understanding of issues such as the threat environment including terrorism, drugs, gangs, and crime. These changes, coupled with severe budgetary pressures to reduce costs and find more efficient ways to carry out their responsibilities, force these organizations to exercise greater care in matching available resources to priorities and to integrate political power with economic requirements.

To meet these challenges effectively, emergency service organizations must address flaws in their procedures for allocating personnel and financial resources to meet developing priorities and objectives, including training problems, staffing, and the inadequate integration of multifunction components. Unfortunately, these tasks will be made more difficult by past management reforms such as decentralization and the devolution of decision making to lower level administrative units that have resulted in the development of new multilevel policy systems.

Emergency Medical Services

Pre-hospital health care no longer consists of a simple emergency transportation service. Since the early 1970s, new levels of training have added to what was basically a first aid and transportation system. It is now known as the emergency medical service (EMS) system. The EMS system provides a sequential process that includes patient problem recognition, system access, basic and advanced intervention by first responders and bystanders, and transportation to the hospital, if necessary, by ambulance. The advent of this system completed the public safety management triad of law enforcement and crime prevention, fire prevention and suppression, and emergency medical care and transportation.

Pre-hospital health care has changed significantly with the advent of specialized emergency medical technician (EMT) and EMT-paramedic (EMT-P) training. Capabilities now include responding to the public need with mobile intensive care units, and providing specialized advance life-support measures when necessary.

Public safety EMS is available nationwide and is provided by private and public agencies. The management and provision of the service provided by them represents a sizable investment in training, human resources, supplies, and equipment. The ability to utilize various system delivery models has made it possible for agencies and communities of all sizes to participate in this vital public service.

Complex Organizational Capability

In the fire service, distribution of the workload is not characterized solely by volume nor is it completely configuration dependent. In order for the fire service to maintain its robust capability, it vertically integrates multiple competencies and each individual resource is configured for multitasking. Within the fire service, where consequence management is the fundamental imperative, "high-performance" is measured by organizational effectiveness in the areas of preparedness levels, community loss prevention and reduction, and decreased mortality and morbidity.[7]

Through integrated stability, the American fire service provides both high reliability and high trust to its constituency. High reliability organizations are complex, and complexity mitigates risk. High reliability organizations use organizational structure to adapt and then mitigate events through processes such as risk awareness, decision migration, and process auditing.

As a result of budget corrections, operational adjustments and technological advances that began in the early 1990s, the twenty-first century American fire service will emerge as a complex, precision force that possesses the nation's dominant body of knowledge regarding emergency service consequence management.

Complex Interface Development

Interface complexities involving multiple interdependent agencies, external resources, and federal assets can be mitigated through the implementation of a standard, multipurpose event management system that conforms to the scope and type of event and the capabilities of responding agencies. Although variations of this event management system exist, it has not yet been fully implemented within the inclusive public safety-emergency service amalgam.

UNREALISTIC EXPECTATION

Notional concepts, including the all-hazards capability, are unrealistic, unproven, and dangerous. The assumption that a local emergency resource is capable of managing an event equivalent to that found in military warfare could potentially result in the loss of significant operational assets and personnel and place the community at catastrophic risk.

Contemporary training is invariably conducted in daylight, in good weather, and consistently results in a positive outcome. The ability to survive a given operation is not addressed and casualties, if any, are attributed to the civilian population at risk. Training must now include an ability-to-survive-operation (ATSO) component and measures to implement when conventional practices fail.

Threat Issues

Since 1992, direct physical threats aimed at America's EMS, police, and fire responders have begun to emerge with increasing frequency. These threats specifically target first responders and public safety personnel. These tactics appear similar to those espoused in *The Minimanual of the Urban Guerilla,* written by Carlos Marighella in 1969, which states in part:

> *Thus, the armed struggle of the urban guerrilla points towards two essential objectives:*
> 1. *The physical elimination of the leaders and assistants of the armed forces and of the police;*
> 2. *The expropriation of government resources and the wealth belonging to the rich businessmen, the large landowners and the imperialists, with small expropriations used for the sustenance of the individual guerrillas and large ones for the maintenance of the revolutionary organization itself.*
>
> *It is clear that the armed struggle of the urban guerrilla also has other objectives. But here we are referring to the two basic objectives, above all expropriation. It is necessary for every urban guerrilla to always keep in mind that he can only maintain his existence if he is able to kill the police and those dedicated to repression, and if he is determined—truly determined—to expropriate the wealth of the rich businessmen, landowners and imperialists.*

The threats are tactical violence[8] and 911 target acquisition.[9] These threats have four applications:

1. To control the criminal environment
2. To divert law enforcement attention
3. To kill or injure law enforcement personnel
4. To compel law enforcement-assisted suicide

Tactical Violence

Since the 1992 Los Angeles civil disturbance, a disturbing phenomenon that directly affects emergency service providers has surfaced. The phenomenon has been identified as tactical violence.

Tactical violence is primarily employed by criminals. It is defined as the predetermined use of maximum violence in order to achieve one's criminal goals, regardless of victim cooperation, level of environmental threat to the perpetrator, or the need to evade law enforcement or capture. This method also usually results in physical or psychological injury or death to the victim. The rationale for employing tactical violence is predatory control of the immediate criminal environment through the creation of chaos and the infliction of terror, trauma, and death on presenting targets.

911 Target Acquisition

One tactic utilized by the criminal and extremist elements (including but not limited to militias, white supremacists, environmentalists, animal rights activists, and anti-abortionists) is 911 target acquisition. This tactic has been employed for the following purposes:[10]

- to divert public safety resources prior to, during, or after the commission of criminal acts
- to draw resources into an ambush situation
- to draw resources into environments laced with improvised explosive antipersonnel devices
- to compel law enforcement-assisted suicide

Although these threats primarily target law enforcement organizations (crisis managers), fire and EMS first responders (consequence managers) have the highest exposure by virtue of their response configuration and resource staffing levels.[11] Recovering from these events can take up to one year, during which time the community's vulnerability to all hazards is significantly increased.

Covert Intimidation

Local and federal law enforcement agencies are not the same as military units and rely to some extent on the intimidation factor. The primary goal of law enforcement is crime prevention/control. There are many subtle intimidation strategies incorporated into this function that have a psychological impact on the community, including vehicle color schemes and emergency lighting configuration, uniform design and color, the visible wearing of side arms, and officer bearing and demeanor.

Overt Intimidation

Law enforcement special teams include special weapons and tactics (SWAT), hostage response team (HRT). These units are semi-militarized and affect quasi-military status through the wearing of military-like/special operations/commando garb. They carry special weaponry and devices not seen in routine law enforcement operations. Their presence at an incident is an implied threat to targeted perpetrators and demonstrates the capacity to employ overwhelming force and special aggressive tactics if necessary.

Overt intimidation tactics have been adopted in other areas as well, such as the serving of high-risk warrants or drug raids. It is common to see various law enforcement agencies entering a suspect dwelling en masse creating confusion, shouting commands, and displaying drawn weapons. Although intimidation may play a roll in deterring terrorist activity, it will not be a factor once crisis management transitions into consequence management.

Outcome Training

Local law enforcement is not trained to participate against sustained armed resistance with an opposing force without external support.

In all cases, law enforcement personnel are trained in a manner that results in a positive outcome for them on any given incident. Although vulnerability, danger, and personal risk are discussed, it is presented in general terms suggesting that, although they exist, it is unlikely that they will occur to a particular individual if training is followed.

Special aggressor teams such as SWAT and HRT are trained to conduct assaults if necessary in order to affect closure of a particular incident. However, this training implies that an assault is both inclusive and conclusive relative to a given criminal problem. This means inclusive of all threats and conclusive once the threat is neutralized. An example is to gain entry and subdue/arrest/kill perpetrators and/or rescue hostages.

Unlike the military, law enforcement agencies are not prepared to conduct an assault in the face of withering automatic weapons fire, rocket-propelled grenades (RPGs), hand grenades, claymores, mortars, or other similar type weaponry. This is especially true if they incur substantial casualties among their personnel. They must rely instead on special teams for this type of activity (England discovered this during an IRA assault against Parliament, as well as the Los Angeles police department during the North Hollywood bank robbery). Nor are law enforcement agencies, whose primary objective is the safety of its personnel, prepared to conduct operations against a sustained, widespread, or diffuse hostile faction operating from several venues simultaneously. These groups have no acceptable loss provisions.

Acts of terrorism perpetrated against society present many problems to those charged with crisis and consequence management response. An act of terrorism is similar to other man-made disasters in that the main characteristic is sudden onset and the resultant effect is significant human injury and/or death. Because of the nature of the incident, in many cases, there is no opportunity for crisis management or intervention.

These incidents, by their nature, are not telegraphed in advance to the authorities and therefore, the appropriate response may consist solely of consequence management and criminal investigation. To the consequence manager, these responses may represent the greatest challenge that can be faced by an emergency agency.

To begin with, the physical impact of a "conventional" terrorist incident is characterized by rapid onset. Unlike a common conventional response, where initial units are dispatched and the event accelerates or expands based on the observation and assessment of first arriving companies, conventional terrorist incidents require an immediate maximal response of appropriate resources in order to optimize the survivability of victims.

This may indicate that a new standard or ad hoc response configuration is necessary for terrorism incidents—one that combines emergency medical care providers, hazardous materials management, and search and rescue capability with conventional resources on the initial dispatch to specific types of incidents.

Federal Asset, First Responder Integration

The Marine Corps Chemical Biological Incident Response Force (CBIRF) responds to incidents involving weapons of mass destruction (WMD) in support of local responders. Their mission is to deploy domestically or overseas to provide force protection and/or mitigation and to assist federal, state, and local responders in developing training programs to manage the consequences of a WMD event.[12]

The National Guard originally commissioned 10 WMD civil support teams (CST),[13] formerly known as rapid assessment and initial detection teams (RAID Teams),[14] consisting of 22 full-time members of the Army and Air National Guard. At the direction of the U.S. Congress, the Department of Defense commissioned an additional 17 teams. With this addition, the total number of states with WMD-CST capabilities is 27. The teams are stationed in Alaska, Arizona, Arkansas, California (two CSTs), Colorado, Florida, Georgia, Hawaii, Idaho, Illinois, Iowa, Kentucky, Louisiana, Maine, Massachusetts, Minnesota, Missouri, New Mexico, New York, Ohio, Oklahoma, Pennsylvania, South Carolina, Texas, Virginia, and Washington. The teams will deploy, on request or based upon mutual aid agreements, to assist local responders in determining the nature of an attack, provide medical and technical advice, and prepare the way for identification and the arrival of follow-on federal response assets.

Both of these response components will integrate with local responders through implementation of the incident management system. With the establishment of a unified command, a fully capable civil-federal management team will oversee management and operations of combined assets in weapons of mass destruction and other events impacting communities in the United States.

Although these external assets along with other federal agencies (other divisions of the Department of Defense, Department of Health and Human Services, Centers for Disease Control, Federal Emergency Management Agency, and Public Health Service) will respond to assist the primary impact, the response to the aftermath of an event remains the responsibility of local communities. It is critical that this facet is not lost in the planning process due to the reliance on federal response and capacity. Managers, chief officers, planners, and elected officials must remain cognizant of the critical factor that the arrival of these assets are time and distance limited. In many cases, depending on the proximity of federal resource stationing and/or the availability of aircraft with appropriate lift capacity, it can be many hours or perhaps days prior to arrival and full operational activity.

Physical Constraints

It is not unusual for terrorist incidents to involve structural collapse, multiple casualties, fire, and chemical release. Is there a difference between a terrorist act and a conventional incident? Aside from the possible psychological factors, not to the consequence manager, who must implement operations based on the present state of the event rather than the precipitating factors. Therefore, attribution is not a consequence priority. Regardless of motivation, both function as casualty generators that result in short-term resource commitment on a large scale. Both have the potential to cause collateral injury to responders and, by their nature, generally do not afford an opportunity for crisis management or intervention prior to the event.

However, because terrorism is a deliberate act, these incidents may present unique hazards to response personnel. For example, it would not be unusual for terrorists to plan secondary events that target emergency responders; events that can be triggered once emergency operations begin and responders are most vulnerable to attack. Therefore, although the results of a terrorist act may not be different than a conventional incident, the approach to them must be.

Aggressive response to these incidents must be curtailed. Courage, valor, honor, and integrity are not issues in these instances. Any response to suspected terrorist acts must be moderated and coupled with careful consideration of any potential secondary threats to responders and the general public.

Political Constraints

Terrorism is both a federal and a state crime. However, all incidents are manifested locally and initial reaction to them will be provided by local crisis and consequence organizations. As these incidents expand, they will involve multiple organizations. Therefore, both at the planning and response level, local crisis and consequence managers must interact with external entities with, in some cases, broader authority over the incident and in other cases, with near equal power but radically different missions, perceptions and value systems. They must also cooperate with political crisis managers and co-response organizations from within a unified command structure.

Political, strategic, operational, and tactical direction is necessary in every case of joint or combined force operations. Political direction will set

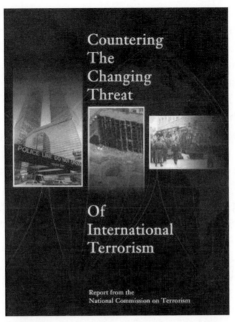

FIGURE 1-2 2000 National Commission on Terrorism Report. *Source:* Courtesy National Commission on Terrorism

political objectives, define basic strategy to achieve these objectives, and provide basic guidance for operations. Strategic direction will define desired operational target conditions and sequencing. Operational direction will coordinate the efforts of multiple organizations to achieve a successful outcome. Tactical direction will issue orders to the front-end operators actually engaged in problem resolution.

These unusual circumstances may result in the consequence manager's inability to manage the incident effectively, particularly if the responding organizations do not have a familiar, working relationship. It is therefore imperative that all responsible organizations plan, organize, and conduct training within a single incident management system/incident command system (IMS/ICS) and that individual organizational roles and responsibilities be clearly defined and agreed upon prior to an event and preferably during a comprehensive planning process.

Ethical Constraints

In unconventional terrorist incidents involving WMDs such as chemical, biological, or radiological agents, local agency limitations may preclude response into an impacted area. Therefore, the interval to intervention time will increase. This will be one of the more difficult concepts for emergency managers to instill in their personnel.

However, the current response capability of specialized federal resources is limited by geographic/regional time constraints and may not be available for the better part of a day. Additionally, the medical community currently has no capability to manage mass care for local populations.

Under these circumstances, the function of emergency response organizations may be to limit environmental expansion of the incident if possible, protect the unaffected population through evacuation, and establish shelters or safe havens capable of providing for basic human needs.

Because of potential delays in access to unconventional incident sites involving WMD, the critical interval to intervention time may be protracted. Therefore, the survivability potential for victims will diminish. Prolonged waiting times will increase mortality and the consequence manager may have to consider the implementation of disaster mortuary plans a priority once the site becomes accessible.

Unconventional incidents involving WMDs are likely to convert from EMS/fire/rescue incidents to body recovery, identification, and disposal operations in cooperation with criminal investigative organizations. This final fact, as observed in the Oklahoma City bombing, will have a tremendous psychological impact on emergency response personnel who are neither trained nor experienced in mass fatality management.

Evolution and Application of the IMS to Terrorism, Tactical Violence, and Civil-Military Coordination

The IMS began as a fire-fighting system for managing California wildfires. The system has now expanded into an all-hazard (all-risk) system and is employed at routine emergencies as well as natural disasters, technological disasters, and mass casualty incidents.

The IMS is designed to manage operations and coordinate resources. It is a functionally based system structured on the principles of common terminology, resource allocation, support functions, span of control, and chain of command. The IMS structure is based on a management staff that directs and supports four major sections. The management staff consists of an incident manager, a liaison officer, safety officer, and public information officer. The four sections are operations, logistics, planning, and administration.

The IMS is very similar to the military organization of NATO countries. The military staff positions of intelligence, logistics, and plans/operations closely parallel the four civilian IMS sections. The civilian system also utilizes a liaison officer and the concept of unified command to integrate with a military counterpart. This is an effective boilerplate for military support to civil authorities (MSCA) and coordination during a terrorism/tactical violence incident.

The mass casualty aspect of IMS makes it effective for managing terrorism/tactical violence incidents. For mass casualties, a medical branch is assigned to the operations section, and manages a triage unit, treatment unit, and a transportation unit. The system stresses early triage and transport of critical patients, and scene safety for the rescuers.

In the terrorism/tactical violence arena, the common weapons are still automatic weapons and improvised explosive devices (IED). A chemical attack is managed by an extension of the hazardous material IMS, with emphasis on detection, personal protective equipment, and patient decontamination. A biological attack has a delayed onset, and presents special problems relating to recognition, pathogen identification, treatment protocols, and personal protection.

Hospital Emergency Incident Command System

The Hospital Emergency Incident Command System (HEICS)[15] is a management tool modeled after the ICS. The core of the HEICS is comprised of two main elements: (1) an organizational chart with a clearly delineated chain of command and position function titles indicating scope of responsibility, and (2) a prioritized job action sheet (job description) that assists the designated individual in focusing upon their assignment.

The benefits of a medical facility using HEICS will be seen not only in a more organized response, but also in the ability of that institution to relate to other health care entities and public/private organizations in the event of an emergency incident. The value of the common communication language in HEICS will become apparent when mutual aid is requested of or for that facility.

The California Earthquake Preparedness Guidelines for Hospitals served as a cornerstone in the development of the HEICS. The HEICS has attempted to embody those same characteristics that make the ICS so appealing. Those attributes include the following:

1. Responsibility-oriented chain of command
2. Wide acceptance through commonality of mission and language
3. Applicability to varying types and magnitudes of emergency events
4. Expeditious transfer of resources within a system or between facilities
5. Flexibility in implementation of individual sections or branches of the HEICS
6. Minimal disruption to existing hospital departments

The HEICS includes an organizational chart showing a chain of command, which incorporates four sections under the overall leadership of an emergency incident commander (IC). Each of the four sections: Logistics,

Planning, Finance, and Operations, has a chief appointed by the IC responsible for the section. The chiefs in turn designate officers to subfunctions, with managers and coordinators filling other crucial roles.

Each of the 36 roles has prioritized job action sheets describing the important assignments of each person. Each job action sheet also includes a mission statement to define the position responsibility. The end product is a management system with personnel who know what they should do, when they should do it, and who to report it to during a time of emergency.

FEMA First Responder Nuclear, Biological, and Chemical Awareness Training

As a result of the changing nature of global politics and the increasing threat of terrorism, new information, training, and technology have begun to emerge. The ability of first responders to recognize and manage acts of terrorism has become essential. The First Responder Terrorism Program is an integral component of national preparedness. It is an introductory class developed for, and directed to, emergency service first responders, specifically fire, hazardous materials, and EMS providers. The secondary audience for this program would include members of law enforcement, emergency management, military, disaster response organizations, public health, and public works.

Metropolitan Medical Response System (MMRS)

Most nuclear, biological, and chemical events will present a relatively limited opportunity for the successful rescue of viable victims (usually no more than 3 hours). However, the more expeditious and aggressive the actions taken in the initial stages of the event, the more victims will be recovered that can be decontaminated, treated, and transported to medical facilities.

The MMRS was developed to provide support for, and to provide assistance to, local jurisdiction first responders in nuclear, biological, or chemical (NBC) terrorist incidents.[16] The MMRS has a strong emergency medical care focus and has the capability to provide rapid and comprehensive medical intervention to casualties of NBC events.

MMRS development was based upon providing a coordinated response to NBC incidents in a metropolitan environment. Special emphasis is placed on the ability to identify the specific agent involved and provide the earliest possible correct medical intervention for victims of these situations.

The MMRS is a specially trained and locally available NBC incident response team. These teams consist of 43 persons comprising five major functional elements: medical information and research, field medical operations, hospital operations, law enforcement, and logistics.

The MMRS is organized, staffed, and equipped to provide the best possible pre-hospital and emergency medical care throughout the course of an incident, and especially while on the scene. Medical personnel are responsible for providing the earliest possible medical intervention for the responders and civilian victims of NBC incidents through early identification of the agent type and proper administration of the appropriate antidote(s) and other pharmaceuticals as necessary.

It is not the intent of the field medical operations unit or hospital operations unit to be freestanding medical resources at the incident. However, they are part of the first line of intervention in a chain that stretches from the field to the local hospital medical system. It may be necessary for the hospital operations unit to contact local hospitals by telephone to give advice on decontamination procedures for convergent victims, for agent treatment protocols, and other information as requested by medical facilities.

Casualty Care in the Combat Environment

Unlike the military, civil first responders are not trained with consideration for the ability to survive a given operation. Civil mission priority does not prospectively accept personnel losses, and no first responder organization includes an acceptable loss ratio as part of their mission goal or planning efforts.

First responders are currently trained to manage trauma based on the principles of trauma care taught in the advanced trauma life support (ATLS) course, a standardized approach to trauma care. However, there are issues inherent in the ATLS training program relative to its appropriateness and application in the combat environment:

- It assumes hospital diagnostics and therapeutics can be accessed rapidly.
- It does not presuppose delayed transportation.
- It has no tactical context.

ATLS does not take into consideration the issues of care under fire, in complete darkness, various environmental factors, and delays to definitive care or command decisions.[17]

Tactical care objectives include turning casualties into patients, preventing additional casualties, and preventing response personnel from becoming casualties. Under these circumstances, basic protocols should be considered a starting point; ad hoc protocol modification may have to occur in response to a specific situation. In tactical situations, medics must be prepared and permitted to adapt and improvise within their scope of practice as conditions dictate.

The following new terminology and definitions appropriate to tactical field care (TFC) must be added to the ATLS program and incorporated into training in order to reflect the care rendered in the tactical environment:

I. **Care Under Fire**
   A. The care rendered by a medic at the scene of an injury, in a hostile environment, while at risk.
   B. Available medical equipment and supplies are limited to that which is carried by the individual medic.
   C. Aseptic technique is not a consideration.
   D. Control of hemorrhage is the top priority because exsanguination from extremity wounds is the number one cause of preventable death in the combat environment.
   E. Patient extraction is delayed due to the threat potential.

II. **Tactical Field Care**
   A. The care rendered by a medic at the scene of an injury, in a hostile environment, when not at risk.
   B. Available medical equipment and supplies are limited to that which is carried in the field.
   C. There is more time to render care.
   D. Care is rendered under non-sterile conditions.
   E. Patient extraction is possible.

Mass Fatality Management

When planning for nuclear, biological, chemical, or other acts of terrorism, the plan must include provisions for the management of mass fatalities. Evolving terrorist capabilities, coupled with the availability of WMDs, lead to consideration of the potential for mass fatalities. Many chemical agents result in fatal injuries to those exposed within a short period of time. In a similar manner, large improvised explosive devices placed strategically within a high-rise building, apartment, or other large public gathering can result in hundreds, if not thousands of deaths. Unlike the military, civil authority rarely finds itself in the position of managing, processing, and disposing of contaminated human remains.

A mass fatality management plan that provides organization, mobilization, and coordination of all provider agencies for emergency mortuary services is an imperative. It is important to delineate the authority, responsibility, functions, and operations of providers by agreement prior to an event of magnitude.[18]

Mass fatality plans should be regional and come under the auspices and management of the local coroner or medical examiner. The operational concept of the mass fatality management system is exclusive of cultural, religious, and ethnic beliefs and practices but may be modified as conditions permit. It is a utility system based on three levels of response:

- **Level I**—A minor to moderate incident wherein local resources appear to be adequate and available. A local emergency may or may not be proclaimed and fatalities may range from 50 to 100.
- **Level II**—A moderate to severe incident wherein local resources are not adequate and assistance is requested from other jurisdictions or regions. A local emergency declaration is imminent. Fatalities may exceed 100.
- **Level III**—A major incident wherein resources in or near the impacted area are overwhelmed and extensive state and/or federal resources are required. A local and state emergency will be proclaimed and a principal declaration of an emergency or major disaster will be requested. Fatalities may range in the thousands.

Mass Casualty Target Management

Many population centers include high-rise building complexes that commonly have transient censuses exceeding 10,000 people. The complexities involved in planning a response to these edifices can be a daunting task. These buildings exist globally and because of their accessibility represent attractive targets to terrorist.

For example, the New York City World Trade Center represented a target of colossal proportions, and an unprecedented opportunity to strike at the symbol of Western international commerce. In a single act of terrorism in 1993, over one hundred and fifty thousand individuals were placed at risk when an improvised explosive device detonated in a subterranean complex.

Fortunately, high reliability organizations such as EMS, fire, and law enforcement agencies use organizational structure to adapt to, and then mitigate the uncommon yet catastrophic event. This emergency was mitigated by a diverse public safety system, a convergent volunteer effort, and a systematic approach to disaster service operations. The adherence to an IMS provided the impetus for a unified command structure and an unprecedented level of cooperation between local and regional resources and state and federal assets. Through the application of this management process, a catastrophe was averted.

Mitigation of Individual Performance Degradation in Real Time As a Terrorist Event Evolves

Domestic NBC terrorist events are rare and catastrophic events. The unknowns are when and where these events will occur. The type and material used is also an unknown.

The basis of system function is performance of the individual, yet how citizens and first responders will perform is unknown. Algorithms developed to assist in responding are based on a reductionist approach that most problems can be identified, separated from the larger response, and prepared for. Algorithms for response are best guesses in a totally unknown arena where human and environmental pathology merge.

Latent system error, human performance error due to slips and mistakes, and normal human physiology interfere with well-planned algorithms, policies, and procedures. Organizational structure and decision-making methods for emergency events that can mitigate performance degradation already exist. Use of high reliability organizational techniques and decision-making methodology adapted from the military can mitigate NBC terrorist events in real time as the event evolves.

Latent error occurs when the system sets up the error, but the individual commits the act. On later review a long train of events tied together by ineffective policies and procedures can be identified as the actual cause of the error. Though latent error can sometimes be identified or prevented by review of the system as a whole, the more deeply enmeshed latent errors cannot. They will appear only when the system is strained under the conditions of a catastrophe. The perceptions and actions of the individual can identify and mitigate latent error early in the period of disorder or dysfunction.

It has been pointed out that error is a part of human performance, despite a commonly held view that error cannot exist without negligence. Some consider complications as errors when they are actually expected but unwanted results of interventions. Errors can be unconscious slips or conscious mistakes. Slips are rule-based or skill-based errors that can easily occur when an event evolves in unexpected ways or human performance cannot or does not match environmental demands. Mistakes are more complex, knowledge-based errors where the decision-making process is at fault. Interference with effective decision making comes from a number of heuristics and biases such as representativeness, availability, confirmation bias, reversion under stress, or coning of attention.

Fear, as fight or flight, is a well-known, expected, and natural physiologic response to life threats. Less well known is how these responses will present themselves during the threat. Fight manifests itself as anger, particularly when focused on an individual or tool. It is made plausible by the presentation that an individual's performance is inadequate causing the anger when, in actuality, it is the angry individual's response that leads to performance decrements. Flight manifests as avoidance and is made plausible by redirection of attention to a seemingly more important but safer task. Freeze is an often-unrecognized fear response mediated by cortisol and leads to confusion, inaction, or even paralysis (as in paralysis by analysis).

Decision-making techniques modified from the military, such as the closed-loop decision cycles, allow individuals to learn what works through action. They permit response when the nature of the threat is fuzzy and ill defined or ill identified. In fact, closed-loop cycles allow identification of the structure of the threat and success of the response. They provide a margin of safety by self-monitoring responses to actions for early warning of dangers created by actions of the response team. They are rapid and adaptable to those situations where unintended or unexpected outcomes occur.

Roberts and Libuser's high reliability theory, also developed from military studies, states that complexity will mitigate rare but catastrophic events. This apparent paradox occurs because disasters and terrorist events are nonlinear systems. Linear, reductionist approaches such as response algorithms and pre-arranged incident command systems do not apply. High reliability organizations can learn and have a culture of both safety and reliability. They allow decision migration and diminish authority gradients that will permit the necessary free flow of information during the terrorist event. They allow effective action before implementation of algorithms, policies, and procedures can be instituted in a rapidly evolving terrorist threat. These organizations will respond to a catastrophic event in the same manner they respond to routine events.

Chapter Questions

1. List and discuss at least three major response agencies at the local level. What are the limitations of these agencies in terrorism response?
2. Discuss the evolving threats of 911 target acquisition and tactical violence.
3. What are the basic functions of the following special units?
   - U.S. Marine Corps Chemical Biological Incident Response Force (CBIRF)
   - National Guard Civilian Support Teams (CST)
   - Metropolitan Medical Response System (MMRS)
4. Discuss how mass casualty care in the combat environment differs from standard advanced trauma life support protocols.
5. Outline the major elements in a mass fatality management plan.

Notes

[1]Rittenberg, Sidney, and Amanda Bennet (1993). *The Man Who Stayed Behind*. New York: Simon & Schuster.

[2]Sudoplatov, Pavel, and Anatoli P. Sudoplatov (1994). *Special Tasks*. New York: Little, Brown and Company.

[3]Ibid.

[4]Watson, Francis M. (1976). *Political Terrorism: The Threat and the Response*. Fairfield, CT: Robert B. Luce Co.

[5]bin-Laden, Osama, "Fatwa" *Al-Quds Al-Arabi News,* London, 1998.

[6]Strategic Management and Policy-Making, Globalisation: What Challenges and Opportunities for Government, Internet, September 1997.

[7]Office of the Inspector General, U.S. Department of State: Major Management Challenges, 1997.

[8]Denney, James, and Donald Lee (1997). *Millennium Issues Confronting the Fire Service.* Los Angeles: The MJS Group.

[9]Denney, James, and Donald Lee, "The Emergence and Employment of Tactical Ultra-Violence." Internet; ERRI. http://www.emergency.com/stratvio.htm, 1997.

[10]Denney, James, "Emerging First Responder Threats: 911 Target Acquisition." Internet; ERRI. http://www.emergency.com/911target.htm, 1998.

[11]Malone, Major Mike. "Chemical Biological Incident Response Force." USMC-ASPD, 1998.

[12]Cohen, William S. "National Guard Rapid Assessment Elements." Washington, D.C.: Armed Forces News Service, October 1998.

[13]Asst. Secretary of Defense, Weapons of Mass Destruction Response Team Locations Announced, Washington, D.C.: Armed Forces News Service, No. 512-98, October 1998.

[14]Hospital Emergency Incident Command System, Orange County Health Care Agency, Emergency Medical Services, 1991 DHHS Orange County, California.

[15]U.S. Public Health Service (1997). *Metropolitan Medical Strike Teams: Field Operations Guide,* 1997.

[16]Butler, Frank, M.D. (1996). *Combat Casualty Care.* Department of Defense briefing.

[17]Department of Coroner. Emergency Mortuary Response Plan: County of Los Angeles, Los Angeles County Department of Health Services, Los Angeles, 1993.

[18]Minihan, Lt. General Kenneth (1998). *Information System Security.* Washington, D.C.: National Security Agency.

[19]Molander, R. C., A. S. Riddile, and P. Wilson (1996). *Strategic Information Warfare: A New Face of War.* DocNo: MR-661-OSD, Rand Corporation.

[20]Hughes, Lt. General Patrick M. (1998). "Global Threats and Challenges: The Decade Ahead." Senate Select Committee on Intelligence; High Reliability Theory (Karlene H. Roberts, Managing High Reliability Organizations: California Management Review, Volume 32, Number 4, Summer 1990), also developed from military studies, states that complexity will mitigate rare but catastrophic events; Decision-making techniques modified from the military, such as Boyd's closed-loop decision cycles, allow individuals to learn what works through action.

2
The Basics of the Incident Management System

Paul M. Maniscalco
Hank T. Christen

Source: Courtesy P. Maniscalco

Chapter Objectives

After reading this chapter, you will be able to accomplish the following objectives:

1. Have an awareness of the history of the incident management system.
2. Define crisis management and consequence management and recognize the distinction between the two terms.
3. Diagram the basic structure of the EMS incident management system (EMS/IMS), fire IMS, and law enforcement IMS.
4. Recognize the importance of prescribing who should be in charge (lead agency).
5. Understand unified management (command).
6. Understand the importance of Emergency Management in supporting IMS.
7. Recognize the application of IMS as a planning tool.
8. Understand the use of IMS for integration with military units.

The History and Development of IMS (ICS)

In the 1970s, after a disastrous wildfire season, California fire managers recognized the need for change. In incident after incident, the same problem emerged: lack of interagency coordination. Specific problems were:

1. Uncommon radio codes—people could not talk to each other.
2. No command system—each agency operated on the personality of its leaders; in some cases, it depended on who was working that day.
3. No common terminology—when agencies did talk, they often misunderstood each other.
4. No method of effectively assigning resources—logistics depended on who got lucky.
5. No clear definition of functions, and how each function related to other elements.

Fortunately, a group of downsized Boeing engineers was assigned to develop the modern ICS. Using a system approach learned from the defense and aerospace industries, they implemented an ingenious management boilerplate that effectively integrated the myriad of agencies needed to combat a wildfire. In the 1980s, other emergency response disciplines recognized the universality of the system. ICS could literally be applied to any type of problem. In the 1990s, minor changes in terminology resulted in the system becoming a management system and referred to as the incident management system (IMS). Please note that throughout this text we use the terms *incident management system (IMS)* and *incident manager.* The terms *incident command system (ICS)* and *incident commander (IC)* are also frequently used and are acceptable terms.

Today the U. S. Coast Guard uses IMS for spill control operations, and law enforcement uses the system in tactical operations. EMS uses the EMS incident management system on mass casualty incidents. These response agencies have learned that IMS is a solution for effective coordination and liaison with their military counterparts in terrorism/tactical violence incidents.

Crisis Management/Consequence Management

Two new terms that have emerged from the federal level in terrorism response are the concepts of crisis management and consequence management.

Crisis management involves measures to identify, acquire, and plan the use of resources needed to anticipate, prevent, and/or resolve a terrorist threat or incident.

Consequence management involves measures to alleviate the damage, loss, hardship, or suffering caused by emergencies. Consequence management includes measures to protect public health and safety, restore essential government services, and provide emergency relief to affected governments, businesses, and individuals.

Crisis management is an important element and is under the direction of the FBI. Intelligence information (called "intel") from the FBI is sometimes shared with local law enforcement agencies, but rarely reaches EMS, fire, or emergency management agencies. Intel on potential terrorists is frequently classified information. This is understandable, as intel agencies must be careful about compromising sources. However, it is practical that managers of response agencies have a secret clearance and receive crisis management information on a need-to-know basis. This intel is not compromised if the source and detail are kept confidential from operations personnel. Managers can order protective actions without sharing the facts.

Consequence management is the action-oriented response that we are used to. On a bad day, when crisis efforts fail, we get the 911 call that drives us quickly into the consequence mode. The consequence mode is synonymous with incident management since consequence operations must have a system or game plan. Without IMS, consequence management is nothing more than well-meaning chaos. Our customers (the victims) deserve something better. In the next section, we will examine the basic IMS structure that allows us to leap into the consequence world and somehow come out alive.

The Basic Structure of the IMS

The system begins with someone running the show (more about that later). That special person is the incident manager (commander) and has a support staff on large and complex incidents. The management staff consists of a safety officer (always critical), a liaison officer (nice to have), and a public information officer (always important).

The heart and soul of the IMS are the four sections of operations, logistics, planning, and administration. Any event requires these four basic functions. These functions also apply to nonemergency activities. You could manage a business, run a nuclear submarine, or raise a family using the principles of IMS.

Operations cover the most obvious aspects of a terrorism/tactical violence response, including visible, dynamic, tactical hands-on functions. Operations is where the "rubber meets the road," where people get their hands dirty, and involves mass casualty EMS response, fire/rescue activities, public health operations, tactical law enforcement, and hazardous materials actions. If things go right, operations people get medals; if things go wrong, operations people get killed.

If operations types are the stars of the show (and rightfully so), then the logistics types are the unsung heroes. Nobody appreciates logistics (read support) until they run out of things. Operations units become dead in the water when they have no equipment, supplies, fuel, or communications. We all take our radios for granted until they go dead.

By nature, terrorism and tactical violence events consume resources. Throughout history military experience has shown that for every soldier, there are multiple support personnel. The same is true for consequence responders. To accomplish this support, the logistics section is divided into a support branch and a service branch. The service branch consists of a communications unit and a food unit. The support branch consists of a supply unit, a facilities unit, and a ground support unit.

In an average response, needed resources are available. In complex and protracted events, the logistics needs grow exponentially. Radio networks fail or require expansion and vehicles run out of fuel. EMS needs medical supplies, and everyone needs lights when it gets dark. The event requires more people and the people need to eat.

Along with resources, operations needs ideas; that is where the planning section enters the picture. The purpose of the planning section is to coordinate with the incident manager and other section chiefs in the development of an incident action plan (IAP). The IAP is a flexible plan that must respond to the dynamics of the incident, and is literally planning "on the fly" by reacting to rapidly changing information. Planning involves gathering data, assembling it into meaningful information, displaying the information in an easily readable form, and finally using the information to make decisions. If this sounds difficult, you get the point.

To accomplish these varied tasks, the planning section utilizes a resource status unit, a situation status unit, a documentation unit, and technical specialists. Early in the game, the incident manager better hit the bricks with a quick and dirty plan. As the incident increases in duration or complexity, gathering data and assembling it into key bits of information are necessary. Hence, the needs for a planning section. There is an old aviation expression that says, "never run out of air speed, altitude, and ideas at the same time." In our business, the planning section provides us with the ideas.

The administration section was designed as an element of the wildfire command system and for prolonged deployments. In terrorism and tactical violence incidents, administration units such as finance, time keeping, and worker's compensation are in place, but will not be discussed in this text.

Important aspects of the IMS that bear emphasis are:

1. IMS identifies key functions; it is not based on rank or hierarchy.
2. IMS delineates a relationship between four key sections: operations, logistics, planning, and administration.
3. IMS specifies a chain of command and a workable span of control (a leader does not supervise more than three to five subordinate units). The system emphasizes support (logistics) and decision making (planning) as elements complimentary to operations.
4. IMS applies to any agency or incident, event, or disaster. (It works on everything!)
5. IMS is a system of common terminology. (We can talk and understand each other.)
6. IMS is flexible, and is expanded or contracted depending on the demands of the incident.

The EMS Incident Management System

The primary objective of a terrorism/tactical violence incident is to create casualties and fatalities. The death and mayhem demonstrate a cause or help the perpetrators accomplish an immediate goal. For this reason, EMS operations (mass casualty operations) are critical objectives of the first response forces.

Emergency medical operations become a branch in the operations section called the EMS branch. The three functions of the EMS branch are triage, treatment, and transport. These functions are performed by one or two units in a minor incident or expanded to fully staffed sectors in a major disaster.

For example, consider a mass shooting with one fatality and two critical patients. Two units and a supervisor can handle the incident. The supervisor becomes the EMS branch director and coordinates his or her units with other IMS elements. The two units triage the patients, provide stabilization and treatment, and transport to an appropriate facility. In this case, the treatment is minimal; the "load-and-go" mode is in order.

If the same mass shooting resulted in fifteen casualties, the EMS/IMS must be expanded. There is still an EMS branch director, but the elements of triage, treatment, and transport become full sectors. Triage personnel under a sector supervisor triage and tag patients. Several units must be committed to patient treatment. Transport will also require several units (ground and/or air ambulances). The transport sector will have to coordinate with medical control to determine which medical facilities will be appropriate to receive patients.

All of the EMS functions just described do not operate in a vacuum. Simultaneously, a fire rescue branch and law enforcement branch will be working full tilt. Other types of events may require a hazardous materials branch, public health branch, or a public works branch.

Other sections such as logistics and planning are established if needed. Logistics is a busy player as the operational heroes begin to "suck up" supplies, equipment, and communications.[1]

FIGURE 2-1 Use of command vests allows sector officers to be readily identifiable. *Source:* Courtesy P. Maniscalco

Fire Incident Management Systems

The fire incident management system uses the same structure (boilerplate) as the EMS/IMS. There is the same management staff, with logistics and planning sections, fire/rescue, hazardous materials, and mass casualty (EMS/IMS). The wildland fire IMS is branched into ground operations and air operations. A wildland incident management team (IMT or sometimes called an overhead team) can easily restructure for a terrorism incident by utilizing the urban fire structure of fire/rescue, hazardous materials (haz mat), and mass casualty (EMS/IMS). The point is that both systems have the tools and flexibility to respond to the demands of a terrorism event.

The fire/rescue branch provides search, rescue, and fire fighting operations after law enforcement and haz mat personnel clear the area. The fire/rescue branch may need an urban search and rescue team(s) if the incident involves a structural collapse such as Oklahoma City. The fire/rescue branch also closely supports the haz mat branch and the EMS branch. In the haz mat scenario, fire/rescue assists in decontamination operations and provides the management structure for haz mat operations. In a terrorism mass casualty incident, fire/rescue assists in triage, basic medical care, and movement of patients to treatment areas (i.e., the Olympics bombing in Atlanta).

The haz mat branch is responsible for donning chemical personal protective equipment (PPE) and entering a chemical/biological/radiological hot zone (example: the sarin gas attack in Tokyo.) The haz mat branch also performs decontamination (decon). Decon expands to major proportions when there are large numbers of contaminated victims (more on decon in Chapter 11).

In a biological incident, the public health branch would be the lead operations branch. This branch is closely supported by a medical surveillance system, technical advisors (especially epidemiologists), and a medical laboratory system.

A major terrorism incident will require a large commitment of fire resources for a long duration. Operations of this nature place very high demands on logistics and planning sections. On the logistics side, there is a need for fuel, ground support, supplies, communications, food and water, auxiliary power, and personnel. On the planning side, incident action plan development, planning briefings, resource tracking, status boards, and technical specialists are needed.

Law Enforcement Incident Management

A senior law enforcement official once remarked to us, "Law enforcement is involved before the incident, during the incident, and after the incident; long after EMS and fire units have ceased operations." His statement is accurate, for law enforcement (local, state, or federal) is involved in crisis management, consequence management response, and finally long-term investigations and legal proceedings that may last for several years. Law enforcement crisis management and criminal investigation are beyond the scope of this text; however, it is important to elaborate on consequence response by law enforcement agencies.

In recent years, the law enforcement community has begun to adapt the tools of incident management for major tactical deployments. In the law enforcement incident management system (LEIMS) the principles of planning and support still apply. As in other systems, the major differences occur in the operations sections.

The LEIMS divides operations into groups or divisions that can include units, strike teams, or tactical teams. The primary mission groups are special emergency response teams (SERT, also called SWAT), hostage negotiation, bomb disposal, evacuation, air operations, traffic control, and perimeter security. Major terrorism incidents require most of these mission groups to work in liaison with fire and EMS operations. In major scenarios the FBI will establish a Joint Operations Center (JOC) that coordinates federal agencies above the local incident management level and provides appropriate intelligence information to the local incident management overhead team.

The logistics support of law enforcement IMS is similar to the EMS and fire models. Law enforcement requires an additional unit in the planning section called an intel unit. The purpose of the intel unit is to present crisis management intel to law enforcement operations commanders, and gather real-time intel about the incident or the perpetrators as the terrorist scenario progresses.

Who Is in Charge?

If an event meets the federal definition of terrorism, federal agencies will have jurisdiction. In accordance with Presidential Decision Directive 39 (PDD 39) the FBI (Department of Justice) has jurisdiction in crisis management and criminal investigation and FEMA coordinates consequence management. The key word here is *coordinate*, for FEMA does not have a response capability. FEMA can only coordinate federal teams with local teams. Even if everything goes perfectly, it will be many hours or longer before federal support is available.

The local community must be able to establish scene management immediately and maintain command throughout the incident. The critical step is to determine who is in charge. In most local governments the responsibility for scene management in terrorism/tactical violence events rests with the ranking law enforcement official. However, terrorism/ tactical violence incidents are complex and require a number of agencies. As previously explained, the incident will involve a large EMS and fire/rescue commitment with the appropriate command responsibilities.

There are two solutions. The first solution is a single agency incident manager, such as a sheriff or police chief, with EMS and fire/rescue operating at branch levels. The second solution is unified management (unified command). A unified management team is utilized when multiple agencies have jurisdiction in a complex incident. Consider a bombing with a building collapse and mass casualties; clearly, there are major EMS, fire/rescue, and law enforcement operations that must effectively coordinate. In this case, a command post with a joint EMS/fire/law enforcement management team is the solution.

Unified management requires that all managers be located in the same area (a command vehicle or an emergency operations center) and share information. The planning section makes this happen by effective status display and comprehensive interagency action plans. Separate command posts for each major agency never work. Each agency tends to freelance and coordination does not happen.

Unified command does not begin in the heat of battle. It starts with effective planning well before a terrorism incident. The responsibility of each agency must be specifically defined and written. Emergency management facilitates the formal development of unified management through the local comprehensive emergency management plan.

In many ways, the informal development of unified management is more significant than the formal process. Managers must get to know each other. As Phoenix Fire Chief Alan Brunacini says, "Unified command begins at lunch." He means that various agency managers initiate mutual support by breaking bread and sharing coffee. A community disaster committee is an excellent vehicle to schedule quarterly meetings where mutual incident management issues are discussed and agency heads get to know each other. In summary, managers cannot meet each other for the first time in a major incident, and expect to coordinate effectively.

Logistics

Throughout this chapter, we have emphasized that terrorism and tactical violence incidents consume resources quickly. Unfortunately, no IMS system can deliver resources, equipment, and supplies that are not immediately available.

Terrorism/tactical violence requires a push logistics system. In push logistics, equipment and supplies are estimated and stored in caches that can

FIGURE 2-2 Command Board. *Source:* Courtesy J. McMahon, M.D.

be quickly "pushed" to the scene. (Note: Pull logistics involves the ordering of equipment and supplies after their need is determined. Pull logistics is a necessary evil in long duration events, but totally ineffective in the early minutes/hours of a fast developing terrorism incident.)

The first step in push logistics is to determine supplies and equipment that will be needed immediately. There is no all-inclusive list; as future events occur, the list will grow based on lessons learned. A generalized list is:

1. PPE—protective suits and respiratory equipment for all response personnel, bulletproof vests for law enforcement.
2. Decon equipment—Tyvek suits for patients, red bags, cleaning equipment, privacy shelters, run-off control materials, and decon showers.
3. Medications—drugs such as atropine and antidotes.
4. Morgue materials—body bags, collection bags, and tags.
5. Medical supplies—oxygen, IVs, airway adjuncts, basic life support (BLS) trauma supplies, and disposable litters.
6. Crime scene supplies—evidence bags, film, tags, forms, perimeter control.
7. Detection equipment—air-sampling equipment, detection kits, radiological detection.

The primary cost of push logistics is the equipment and supplies. The secondary cost is the storage and maintenance. Some supplies must be rotated from storage to line service before expiration occurs such as medications.

The Safety Template

Safety is one aspect of the IMS that may not be fully appreciated. The most important duty of the incident manager is safety. The guideline for establishing a safety officer is simple: As soon as the incident manager cannot directly supervise all the safety aspects of an incident, the safety function should be assigned to a safety officer. This means appointing a safety officer early in the process. Few terrorism incidents are small enough to allow one person to be the incident manager and still maintain a grip on safety.

The safety officer has the authority to temporarily suspend any plan, procedure, or tactical operation that is unsafe to his or her people or customers. In addition, we need to make sure the terrorists are not harmed because the law enforcement people will badly want to talk to them later.

Safety becomes more important when you consider that the intent of terrorist events is to create a harmful scenario and hurt people. In some incidents, the sole objective of the terrorist is to hurt us, the emergency personnel. Examples abound including:

1. A secondary explosive device designed to detonate after first responders arrive at the scene.
2. A biological/chemical/radiological hot zone.
3. Armed terrorists still on the scene when early responders arrive.
4. Unstable structures that are partially collapsed.
5. An incident where emergency responders are intentionally harmed as a diversion for a crime or another terrorism incident.

Complex incidents require a safety group with several safety personnel under the command of a safety officer. These incidents create high safety demands because they are geographically dispersed and involve several diverse tactical operations. As an example, a subway chemical attack could affect several teams at two or three different stations. During the event, diverse operations would include safety issues such as law enforcement weapons safety, fire zones, and perimeter security; chemical hot zone entry, suit selection, and decon safety; and chemical safety for medical teams. In the Oklahoma City bombing there were dozens of areas with mass casualties, each having site-specific safety requirements.

On a long-duration incident, the safety officer closely coordinates with the incident manager, the planning section, and various operations branches. As incident action plans are developed and evolve, they are

FIGURE 2-3 In many instances face to face briefings at the scenes of major incidents are the most effective means of communication. *Source:* Courtesy L. Tan

evaluated for safety. The best place to stop unsafe ideas is in the planning stage before aggressive operations officers get creative and start to do dangerous things. Every incident action plan needs a safety officer's signature, and every incident manager's briefing must include a safety briefing.

Special teams have their own safety officers who provide relief for the scene safety officer. SWAT urban search and rescue teams and haz mat teams are all required to have safety officers. EMS must have an infection control officer (not an on-scene position). These team-specific safety officers share the workload and responsibility of incident safety.

A terrorism/tactical violence response means throwing our people into the most dangerous acreage on earth. The death, injury, or hostage taking of a responder greatly inhibits a successful outcome. We have to protect ourselves in order to protect others.

The Functions of Emergency Management

Throughout this chapter, emphasis has been on the importance of planning and logistics complementing operations. Many organizations have fine-tuned their operations because they get a lot of practice with routine events.

Likewise, their logistics and planning efforts are designed to support normal operations. For major terrorism/tactical violence events, logistics and planning may fall short because of lack of expertise. This is where an effective emergency management agency provides optimal support.

By definition, emergency management is a planning and support agency. Emergency management serves to coordinate response agencies, but does not directly provide response units or personnel. Emergency management is well suited for the planning role for two reasons. First, emergency management does not have a "turf protection" mind-set; the emergency response turf is owned by other agencies, namely EMS, fire, and law enforcement. Secondly, emergency management is often responsible for a county or region transcending many agencies. This places the emergency management agency in a good coordination and planning position.

An important emergency management function is training. Most agencies provide their own basic and recertification training. New issues are always on the horizon that require advanced training, beyond the scope of the response agency. These include terrorism response subjects like incident management, mass casualty operations, crime scene preservation, and chemical/biological response procedures. In many cases, emergency management can secure state and federal funding for countywide or regional training seminars. There are also new training technologies such as distance learning and virtual universities that emergency management can facilitate.

Disaster exercises closely complement a training program. Exercises allow emergency response agencies to test capabilities not frequently tested in day-to-day operations. Terrorism/tactical violence incidents certainly fall into this category. At least twice per year (quarterly is even better) response agencies should conduct joint exercises. Each exercise should test one or two functions of the interagency response system and include realistic scenarios. For example, key components include communications, logistics, unified management, mass casualty operations, medical surveillance (public health level) decontamination, operational readiness, air operations, mutual aid, medical transport, and media relations. Scenarios should include at least one terrorism/tactical violence event per year. Probable scenarios can be a workplace mass shooting, a chemical attack, an infectious disease outbreak from a biological weapon, or a bombing with a secondary device.

Emergency managers require certification in exercise planning, and are the pivotal people in exercise design and implementation. In fact, the disaster committee previously discussed is an excellent vehicle to promote exercise coordination. As an overview, there are several steps for effective exercise planning:

1. Determine critical components of the IMS to be tested, such as personnel recall and push logistics.

2. Create a realistic scenario, such as a pipe bomb incident or a mass shooting.
3. Coordinate the players through the disaster committee.
4. Use the IMS model to organize the exercises. The actual exercise must require operations, logistics, planning, and administration.
5. Establish an exercise controller and evaluators who are non-players and unbiased.
6. Adequately fund the exercise. Be realistic about overtime and supply costs.
7. Insist that safety is the key factor in the exercise. (People have been killed in exercises.)
8. Document the exercise; record it, videotape it, take pictures, and write about it.
9. Summarize the exercise results in a lessons-learned format.
10. Hold an after-exercise briefing. Candidly and tactfully point out shortcomings, and stress all the positive results.
11. Most importantly, establish realistic and measurable goals for implementing changes.

The last aspect of emergency management we will discuss is the emergency operations center (EOC) operation. EOC terminology is now evolving into the multiagency coordination center, or MACC. The EOC and MACC concepts have been tested over time in natural and technological disaster scenarios, including hurricanes, earthquakes, industrial explosions/fires, winter storms, airplane crashes, and many other disasters. Terrorism events such as the Oklahoma City bombing, the Olympics bombing, and the sarin gas attack in Tokyo are all examples of the effective use of an EOC or MACC in a terrorism incident.

In most communities, the emergency manager is the commander of the EOC/MACC during emergency operations. There are no specific guidelines to determine when an EOC/MACC should be fully or partially activated. A good rule is that the EOC/MACC is needed whenever the scope of an incident requires support and coordination that cannot be achieved at an on-scene command post. This is often the case when an incident is widespread or regional or there are multiple incidents in progress.

The EOC/MACC is structured like the IMS. All emergency support functions in the EOC are grouped under operations, logistics, planning, and administration. This system provides effective coordination and information flow from IMS elements on the scene with IMS elements at the EOC/MACC.

In summary, make sure emergency management is a major player in your community IMS. Emergency management performs diverse critical functions like planning, logistics, funding, interagency coordination, and liaison with state/federal agencies, training, exercise design, and EOC/MACC operations.

Military Coordination

The IMS is a model system for liaison with military special teams or national guard units. This is because IMS is based on the same principles as military command and control systems. Military organizations and special units have staff officers who are assigned command functions. The functions are identified as follows: (s-staff)

- S-1—administration
- S-2—intelligence
- S-3—plans/operations
- S-4—supplies/logistics

As an example, a military commander would refer a logistics issue to his or her S-4.

At the section level, the military system is almost identical to the civilian IMS. The difference is in terminology. If military officers fully understand the civilian terminology they can easily adapt to our system of operations. Likewise, by understanding military command/control terminology, we can integrate military units into the IMS.

The authors have heard numerous military officers ask, "How can I integrate with civilians? We are not used to going somewhere and not being in charge." The answer is IMS unified management (command), and branching. In unified management, the military commander becomes part of the management team along with the civilians. This ensures that the military commander is privy to all issues, communications, plans, etc. This system keeps civilian leaders in liaison with the military effort.

At the operations level, the military units are assigned to work with an existing branch or assigned as a separate branch. Support functions are carried out by the military staff or integrated with the civilian counterpart if practical. For example, military units have special communications and logistics plans that are executed internally. On the other hand, the civilian logistics section can furnish supplies, equipment, and facilities available at the local level.

Consider a hypothetical scenario of a chemical attack in a downtown urban area with mass patient contamination. The local first responders are on the scene for 10 hours when a military chem-bio team arrives. Patients have been decontaminated and transported, leaving a chemical hot zone. The military mission is to decontaminate buildings, equipment, and vehicles along with setting up decon sites for civilian teams.

Using IMS to integrate with the military command/control system, the following steps are taken:

1. The military commander integrates with the unified management team.

2. The military team is assigned to the haz mat branch.
3. The military team provides its own administration and planning (S-1 and S-2).
4. The military logistics officer (S-4) coordinates with the local logistics section chief for food and facilities and internally provides logistics for mass decon equipment and supplies.

Most military officers are not aware that the IMS or ICS exists in the civilian response community. We advocate a basic IMS awareness course for military officers who respond to civilian missions. This model has already been combat tested in the wildfire incidents in the western United States. Marines and soldiers were given basic fire fighting and IMS training and divided into crews with a civilian liaison. This system worked well and continues to be used during fire seasons.

CHAPTER SUMMARY

In summary, there are 10 major points in this chapter that are critical for an effective incident management system:

1. The IMS is a functionally based system. The emphasis is on common terminology, a chain of command with effective span of control, and the assignment of resources on a priority basis.
2. The IMS is an all risk system that applies to any technological or natural disaster incident, and terrorism/tactical violence incidents.
3. Consequence management involves measures taken to alleviate the damage, loss, hardship, or suffering caused by emergencies, and is the responsibility of local/regional emergency response agencies.
4. The four major sections of the IMS are operations, logistics, planning, and administration. These sections are coordinated and directed by an incident manager and a management staff.
5. An incident action plan is dynamic and "on the fly" in the early stages of an incident. The plan is written and formalized on complex and long duration incidents and developed by the planning section.
6. The EMS branch is responsible for mass casualty operations and is divided into the sectors of triage, treatment, and transport.
7. A local community disaster plan must specify who is in charge. There can be a single incident manager, or a unified management team if several disciplines or jurisdictions are involved.
8. Terrorism/tactical violence incidents consume supplies/equipment quickly. An effective solution is pre-staged disaster caches that are rapidly deployed (push logistics).
9. The IMS provides a template for safe operations. A safety officer supervises safety operations and coordinates with safety officers

on special teams. A terrorism/tactical violence incident is the most unsafe workplace on earth.

10. Local emergency management agencies provide critical functions including planning, support, interagency coordination, exercise planning, and EOC/MACC operations.

CHAPTER QUESTIONS

1. What were the significant problems that prevailed in major wildfires before the development of IMS?
2. Define consequence management and crisis management. What agencies are responsible for consequence management?
3. What are the positions on the management staff? Diagram and define the four sections of the incident management system.
4. What are the elements in the logistics section? Describe the critical logistics issues in terrorism/tactical violence incidents.
5. What is push logistics? Outline a system for placing disaster caches in your community including a supply list for the caches.
6. Describe the duties of the planning section. What is an IAP? What determines if an IAP is formal or informal?
7. Define and diagram the sectors of the EMS branch.
8. What other branches perform key operations in a terrorism/tactical violence incident? How do these branches coordinate and complement each other?
9. What determines who is in charge of a terrorism/tactical violence event? What is unified management and when should it be used?
10. How does the IMS provide a safer workplace? What are the major safety concerns at a terrorism/tactical violence incident?
11. What are the functions of emergency management? How can emergency management ensure coordination between diverse agencies?
12. What is an EOC? Define a MACC. What determines when an EOC/MACC is needed?
13. What are the staff functions in the military command and control system? How do these compare with the IMS? How can the IMS be used to coordinate with military or National Guard teams?

NOTES

[1]For detailed information about the EMS incident management system, study *The EMS Incident Management System—EMS Operations for Mass Casualty and High-Impact Incidents*, by Hank T. Christen and Paul M. Maniscalco, Brady, 1998.

3

Terrorism/Tactical Violence Incident Response Procedures

Paul M. Maniscalco
James P. Denny
Hank T. Christen

Source: Courtesy Reuters/KNBC-TV/Archive Photos

Chapter Objectives

After reading this chapter, you will be able to accomplish the following objectives:

1. Understand the definition of convergent responders and how convergent responder agencies relate to first responders.
2. Recognize the critical importance of scene awareness in a dynamic terrorism/tactical violence event.
3. Have an awareness of the on-scene complications present in a terrorism/tactical violence incident.
4. List the responsibilities of the first arriving unit in a terrorism/tactical violence event.
5. Understand the principles of "2 in 2 out" and LACES (lookout, awareness, communications, escape, safety).
6. Outline the key elements in a terrorism/tactical violence response protocol.
7. Recognize critical hospital response issues.

Introduction to Response

Response to an emergency is a routine function of public safety agencies. Thousands of times a day across the world, EMS, fire, rescue, and law enforcement agencies go "out of the chute" and effectively handle a myriad of incidents. However, no terrorism/tactical violence event is a garden-variety incident. Initial responders are confronted with an unfamiliar, unpredictable, and unsafe scene. It is often a combination mass casualty incident, rescue, haz mat incident, and crime scene scenario. No matter how global an event becomes, it starts with convergent responders trying to assist (more on this later) and first responders being overwhelmed upon arrival at the scene.

While many emergency scenarios share some similar characteristics and response demands, as stated previously, none are "garden variety" and all pose complex and confusing environments for the arriving emergency responders. Let's take a minute to review the expected sequence of activity at a terrorism/tactical violence event.

Event: Explosives detonated at a football stadium to punish the "Great Satan" nation of America.

Jurisdiction—Local EMS, fire, hospital, and law enforcement with eventual state and federal support

Criminals—Foreign nationals

Patients—150 fans including eight children; 25,000 terrorized

Resolution—Convergent responders and emergency responders; medical facilities and prolonged emergency service presence due to criminal investigation and structural instability

Event: A street gang attacks another street gang with automatic weapons.

Jurisdiction—Local EMS, fire, hospital, and law enforcement
Criminals—The winning gang and the losing gang
Patients—Eight gang members and two citizens
Resolution—Convergent responders and first responders; medical facilities and short-term emergency service presence

Event: A fired employee tosses a grenade into a warehouse and shoots four employees.

Jurisdiction—Local EMS, fire, hospital, and law enforcement
Criminal—Ruckus Ralph who was fired for making threats to his boss
Patients—Eleven injured workers; 40 terrorized workers
Resolution—Convergent responders and first responders; medical facilities

In each scenario the common denominator in the event is a *local* incident that draws down upon *local* assets and requires an effective and robust emergency response plan to manage successfully. It does not matter about the jurisdictions, or whether the event is official terrorism; the locals have to get there first, and hold the fort for hours or a day before state and federal assistance arrives and supports the local effort. Reliance upon state or federal assets to assure a successful outcome of your response is unwise. The amount of time that it will take for these resources to arrive at most emergencies creates a vast disadvantage that may result in great detriment to the patients and your community. Clearly, it is in the best interests of all concerned to ensure that our personnel and organizations are capable of mounting an effective sustainable response to high impact/high yield events.

Convergent Responders

Convergent responders are citizens or individuals from nonemergency agencies that witness a terrorism/tactical violence event and converge on the scene. We call ourselves first responders, but we almost never arrive first (we are the first professionally trained responders); convergent responders get there first, we arrive second.

Many response professionals perceive convergent responders in a negative light. They are viewed as undisciplined and in the way. Reality is somewhat different than this myth. In many disaster and terrorism incidents, we have seen video footage of convergent responders digging victims out of rubble, manning hose lines or rendering first aid to victims. Convergent responders are the people who call 911 and report that a terrorism/tactical violence incident has occurred and provide vital incident information to the 911 operator or dispatcher. This information is often conflicting and sometimes hysterical, but multiple calls paint an initial picture that is extremely important in determining the nature of an incident and the level of response.

Realistically, convergent responders become victims, create a crowd, and may hinder initial response efforts. On the surface, it appears that the good and the bad aspects of convergent responders are beyond control. This is not the case. Response agencies are unaware that many convergent responders are from organized and disciplined agencies. At any moment, a local street may have convergent agencies, companies, or individuals performing work assignments. These organizations include:

1. Utility crews such as power, gas, telephone and cable
2. Postal workers and express delivery services (like FedEx or UPS)
3. Meter readers and inspectors
4. Transit authority, school buses and taxis
5. Public works crews
6. Social workers or probation officers
7. Private security agencies
8. Real estate agents
9. Crime watch and neighborhood watch volunteers

Many of these people are in radio-equipped vehicles or have cellular telephones and can provide early and accurate reports of a suspicious scene or a terrorism/tactical violence event in progress. Festivals, events, and public assembly buildings have security guards, ushers, ticket takers, and concession workers who are employees of formal organizations.

With training, convergent agencies are a positive initial response component. The level of training is similar to the hazardous materials awareness model. Members are exposed to basic awareness material in the following subject areas:

• A brief overview of terrorism/tactical violence history and local threats
• Recognition of potential threats such as suspicious people, weapons, or devices
• Procedures for reporting an incident and the critical information needed by 911 communications
• Safe scene control and personal safety
• Basic first aid techniques

While it may not be practical to conduct awareness training for most of the public, adoption of the Community Emergency Response Team (CERT) initiative, which is a familiar program in communities vulnerable to earthquakes, hurricanes, and tornados, may be a mechanism to afford these members of the community an awareness of emergency actions and the role that they may (or should) play at an incident.

It is also feasible to identify agencies that are possible convergent responders, solicit their participation, and assist them with basic awareness training. This is no different in concept than the CERT program or the law enforcement agency training programs for developing volunteer crime/ neighborhood watch organizations. In summary, convergent responders are a viable force that the emergency service community needs to tap, for they will always get there first.

Scene Awareness

A terrorism/tactical violence event is a scene that resembles a boiling cauldron. The scene is a hot zone that can include chemical/biological/radiological hazards; individuals firing weapons (on both sides), partially exploded devices, secondary devices or booby traps, fires, collapsed structures, and multiple injured victims—many screaming for your assistance. On a time scale, first responders arrive when the event is getting started (we stumble into a can of worms), when it is already over (we're seldom that lucky), or somewhere in between (usually the case).

Scene awareness principles begin before the response in any incident (routine or terrorism). Being familiar with the neighborhood, the surroundings, and the violence history of the area is important. Does the area have a gang history? Are special events being held? Is the stadium full or empty? Has there been recent unrest such as political protests or union/ labor issues? Are special religious ceremonies being held? These questions only scratch the surface of information that response agencies must consider before the 911 call. In fact, this type of information (intel) should be obtained pre-event if possible and incorporated into your organization's pre-incident response plans.

During an emergency response, scene awareness jumps into high gear. A terrorism/tactical violence 911 call seldom starts with an accurate description of the event. What sounds like a single shooting turns into a mass shooting with automatic weapons. A suspicious package report becomes a pipe bomb explosion, and a report of respiratory distress at a stadium evolves into a chemical release with contamination and mass casualties.

While responding, mentally explore the possibility that the scene may be far worse than the initial information. Replay your pre-response knowledge. For example, if you are en route to an abortion clinic for an "unknown medical," expect more than a simple gyn patient. Monitor the radio traffic; turn up the volume and really *listen.* You cannot afford to miss any additional information about scene conditions. Local protocol must emphasize that new information be immediately forwarded to all responding units. Radio traffic from convergent responder agencies and other emergency response agencies is also critical. Know the wind direction.

When you get close enough for the scene to appear in the distance, start really looking. Visually scan the entire scene periphery (this may be limited at night or in bad weather). On arrival, there is a tendency to get tunnel vision. EMS focuses on patients, firefighters focus on fires, and cops focus on a crime scene. These are habits based on years of practice that are detrimental in a complex terrorism/tactical violence incident. Force yourself to look around the incident. At this point you are still in the vehicle but slowing down. If you see threats to your safety, or if you have a sixth sense that something is wrong, stop and even retreat. Always trust your intuition on the street.

When exiting your vehicle, you are especially vulnerable. Look for indications of an unsafe scene. Start looking above, behind you, and for bad guys on roofs. Your head should be on a swivel. Look for any indication of people with weapons, explosive devices, or evidence of a chemical agent. The forward bodyline (FBL) may unfortunately define a hot zone. The FBL is a military term that describes the boundary, or forward line of bodies in a mass casualty zone. The FBL may be an indicator of a chemical hot zone. Remember that unsafe scene entry may result in emergency responders becoming the FBL.

Critical Factor: Do not become blinded by tunnel vision. Survey the entire scene, including the area above you, for threats to your safety.

In the structural fire service, there is a new safety guideline referred to as "2 in 2 out." This Occupational Health and Safety Administration (OHSA) guideline specifies that two members should enter a fire hot zone, remain in personal contact, and exit together. Two others are outside. This is the buddy principle that kids are taught in swimming lessons. The 2 in 2 out principle is a rule that applies to all first responders (EMS, fire, and law enforcement) in a terrorism/tactical violence incident when the scene is unstable. *Don't leave your partner!*

Another effective scene principle is one adopted from the wildland fire community, and suggested for terrorism by Assistant Chief Phil Chovan, Marietta (GA) fire/rescue. The principle is called LACES, and stands for lookout, awareness, communications, escape, and safety. These principles were developed after hard-learned lessons where wildland fires made sudden and unexpected changes in direction and/or intensity and trapped fire suppression crews. The same scene dynamics occur in a terrorism/tactical violence hot zone; especially when there is a chemical agent or automatic weapons fire.

Let's analyze the LACES principle:

L—Lookout: Someone is responsible for watching the overall scene from a safe distance and warning crews of danger.

A—Awareness: All members on the scene have situational awareness and are ready for unpleasant surprises.

C—Communications: Exposed crews must have effective communications which include direct voice or hand signals as well as portable radios.

E—Escape: Plan an escape route and an alternate route from any unstable scene.

S—Safety Zones: Escape to a safe area that provides distance, shielding, or upwind protection.

Consider a mass-shooting example where a 911 call describes shots fired with one possible victim in a warehouse. A first due EMS unit finds a single victim by the doorway. A law enforcement unit declares the scene secured. A second EMS unit and fire company arrives and enters a large open floor space with six victims. There is dead quiet (no pun intended). Things do not look or feel right. Fire crews work in pairs using the 2 in 2 out principle. An EMS commander arrives and places himself as a **lookout** watching the street and the crime scene. Crews in the area are looking around the periphery and above (**awareness**). The lookout suddenly spots three men with weapons moving on the far side of the building. He broadcasts an evacuation message (**communication**) to the fire and medical crews who run down an escape route and retreat behind an industrial dumpster (**escape** and **safety**). Heavy gunfire breaks out, but our people are safe because they used the LACES principle of lookout, awareness, communications, and escape and safety.

New Scenes—New Surprises

The twenty-first-century terrorism/tactical violence scene offers new problems previously unseen by civilian responders. It is important to understand that we are not heroes to everyone. In fact, the purpose of the event may be to harm the emergency response troops.

In an Atlanta area abortion clinic bombing, there was a secondary device (binary weapon) designed to detonate after EMS, fire, and law enforcement personnel arrived. In the video *Surviving the Secondary Device— The Rules Have Changed* (Georgia Emergency Management Agency), Assistant Chief Phil Chovan makes several important points:

1. On a bombing, suspect a partially exploded device or a second device.
2. If a second device is found or suspected, evacuate 1,000 feet in every direction (including above).
3. Patients should be removed as if they were in a burning vehicle; use minimal spinal precautions and omit invasive procedures until patients are in a safe area. A terrorism/tactical violence incident may be a diversion for a major crime or terrorist attack.

In Gainesville, Florida, a bank bombing resulted in a high-impact incident. Another bank was robbed while responders were at the original incident. In Texas, criminals placed explosive devices on a hazardous materials pipeline to create a diversion for an armored car robbery. Fortunately, the explosives did not detonate.

Remember that any terrorism/tactical violence incident is a high-impact event; there is a detrimental affect on a community's ability to deliver other 911 services. If possible, utilize the military principle of "uncommitted reserves" for 911 services and/or another terrorism/tactical violence incident. In small communities this principle requires early initiation of mutual aid and support. In summary, be aware of diversions; there's no rule against having two simultaneous terrorism/tactical violence events on a very bad day.

Another trend is tactical ultra-violence. Terrorists, criminals, political militias, and street gangs carry the latest automatic weapons and wear body armor. Often law enforcement is out-gunned. Responders are threatened with projectiles and splintering from concrete, glass, wood, and metals. There is also a danger of being caught in a crossfire between law enforcement and the criminals.

The best defense is avoiding entrance into a ballistic hot zone by observing the scene awareness principles. If you are exposed to shooting, distance and shielding are your only choices (run for cover). If possible, hide behind solid objects instead of vehicles. If you are in an area that requires body armor, you are in the wrong place. Obviously, law enforcement is the exception.

Along with the bombings and shootings, consider the brutal reality of a chemical/biological/radiological (CBR) hot zone. Biological mass casualties on a single scene are unlikely because biological effects take days or

FIGURE 3-1 Always remain aware of the environment to remain safe. *Source:* Courtesy Jamie Francis/AP/Wide World Photos

weeks to manifest themselves. Patients will be spread in ones and twos throughout a region, state, or the nation. Chemical attacks and radiological weapons are another story. First, the dispatch information is unlikely to paint a clear picture. The event may begin with a report of a single patient having respiratory distress in a stadium, auditorium, or airport. Second, the usual indicators of a haz mat incident will not be present. The location will be benign of haz mat storage or transportation facilities, and not have identifiable containers, placards, material safety data sheets (MSDS), or shipping papers. First responders may not immediately recognize a chemical incident; it will take longer (maybe days) to accurately identify the substance. The CBR threat is covered in more academic detail in Chapters 6, 7, and 9.

First responders must look for indicators and remember several important points:

1. Suspect a chemical agent if you are presented with several non-trauma patients with like symptoms.
2. Check for patients that may be scattered throughout a crowd or facility.

3. Look for convergent responders that are showing symptoms; beware of direct exposure or transfer of mechanisms of injury from patients to responders.
4. Does the patient area smell funny or unusual?
5. In extreme cases, an FBL indicates the hot zone (no man's land).
6. Listen for radio traffic from other units indicating multiple patients.
7. Establish a hot zone fast; get the walking patients out.
8. Call for special teams quickly; control hot zone entry.

Set up a decon area for mass casualty patients (see Chapter 11).

In summary, your scene awareness in a chemical attack can save your life and the lives of your patients. Act fast, establish a hot zone, and observe safe hot zone entry principles. In all terrorism/tactical violence events prepare before arrival. Use all of your senses; be suspicious of anything that does not sound, look, feel, or smell right. Use the safety principles of 2 in 2 out and LACES. Lastly, remember that the purpose of the event may be to kill you. Don't join the FBL!

The First Arriving Unit

No matter where you fit in the management hierarchy, someone has to get there first. The first arriving unit has a drastic effect (positive or negative) on the progress of the incident. Remember the key principle of IMS: Scene management builds from the bottom up. This means the first arriving unit is the incident manager, and is responsible for operations, logistics, and planning. Obviously, this is an impossible task without prioritizing. Getting help is the first priority. Make a quick scene survey and transmit a radio report. The first report is almost always inaccurate; that is okay, give the details later. In the initial report give your command post location (remember you are the incident manager at this point) and a basic description of the event such as "mass shooting victims" or "multiple non-trauma patients." Make sure the report is received by your communications center. Don't leave your vehicle until you get feedback confirming reception.

If the event is a nonviolent mass casualty incident, your partner or crew members can begin triage efforts. If the event is terrorism/tactical violence, stay with other crew members (2 in 2 out). Try to determine the scope of the hot zone, an approximate number of patients, and a mechanism of injury. Talk to convergent responders and make every effort to gain relevant information from the usual scene hysteria.

It is critical that you transmit scene threats over the dispatch radio system. Make sure dispatchers relay this information to other units and agencies. This includes information such as:

1. Shooters on the scene or perpetrators with weapons
2. Suspicious device(s)
3. A possible CBR hot zone

As help arrives, the first arriving unit assigns operational functions. These functions must be conducted with scene awareness and safety in mind.

Finally, command is relinquished by the first due unit when a senior officer, supervisor, or manager arrives. This process is done on a face-to-face basis and announced over the dispatch radio channel.

The following example demonstrates the responsibilities of the first arriving unit. A call is dispatched for respiratory difficulty at a religious institution conducting worship services. Rescue 7 arrives at the front entrance; already things don't look right. There are two patients on the building steps; respiratory difficulty with no apparent trauma. Several convergent responders are removing other patients from the building and everyone is coughing.

The unit transmits the following report: "Rescue 7 on scene; we have multiple respiratory patients; send two additional rescues and a chief officer; command post is in front of 5530 58th Street."

The paramedic sends her EMT partner into the building to begin triage and obtain a patient count. The EMT stops at the doorway after discovering at least ten patients inside gasping for breath. The building interior is too "hot" to enter.

The paramedic orders the EMT out of the area, and transmits the following report: "Rescue 7, priority traffic; we have multiple respiratory patients, at least ten in the building; declaring a hot zone in the interior and front entrance for unknown chemical agent; request a full first alarm fire response and the haz mat team; request additional law enforcement units; notify emergency management."

Critical Factor: Proper actions by the first responding units are critical for an effective outcome.

Scene Control

In any emergency incident, scene control is a difficult issue. Terrorism/tactical violence incidents present new and challenging scene control problems. The objective of scene control is to establish a secure perimeter around the scene/hot zone for the purpose of controlling entry and exit from the incident area. Entry control prevents civilians or media from converging on the area. Effective scene control also establishes entry points where personnel and units are logged in for accountability purposes. Unfortunately, terrorism/tactical violence hot zones are very dynamic (the scene can grow in many directions). Shooters on the move or rooftop snipers expand an incident by several blocks. The discovery of an unexploded device or a secondary device requires an evacuation area of 1,000 feet in every direction including up and maybe down.

Keeping people in the hot zone from leaving the scene is a new scene control issue unfamiliar to most of us. In a chemical, biological, or radiological incident, it is important to keep victims in an evacuation area. We do not want contaminated victims spreading the mechanism of injury. When patients leave in a hundred different directions, they contaminate vehicles, people, and buildings. In urban areas, commuters will get on trains or return home in their cars, only to later present at suburban emergency rooms. (Back to the scenario of contaminated people arriving at ERs everywhere.)

The solution to this problem is difficult. First, there's no corporate memory of a large civilian CBR event. However, if large-scale industrial acci-

FIGURE 3-2 Students from Columbine High School are extricated from the hot zone during the tragic 1999 tactical violence incident. *Source:* Courtesy Hal Stoelzle/Denver Rocky Mountain News/Corbis/Sygma

dents are indicators, there will be great difficulty in controlling patient exits. Many patients will leave before we get there, especially if untrained convergent responders are eager to transport them. Other victims will heed warnings to stay in place, but later grow impatient with a long line waiting for decontamination. Lastly, some patients will refuse the indignity of stripping down to their underwear to have themselves and their loved ones scrubbed by men or women in funny suits.

The answer is to do the best you can. Realize that many patients will initially escape. (That is why it is crucial, really crucial, for ERs to have an aggressive decontamination plan). Secondary patients have to be decontaminated as dictated by practicality. In most cases (not all), basic steps of clothing removal and wash down, done quickly, will suffice. If you lose control, at least settle for the clothing removal. Some alarmists raise the issue of using force to contain grossly contaminated patients to prevent further injury to rescuers or civilians. In practicality, such force will not be accepted, because the person exerting the force must be fully suited for protection from the patients (this is an opinion based on educated guesses and a few case histories). In the Tokyo sarin gas attack, patients were not effectively confined. They scattered from the hot zone to get breathable air and affected rescuers.

The following example demonstrates some practical scene control issues:

> *Event:* A crime syndicate plans a multimillion dollar armored car robbery. To create a diversion, a street gang is subcontracted to explode a chlorine tank car on a downtown railroad siding. At 12:07 P.M., the tank car has a violent pressure rupture caused by a few pounds of strategically placed plastic explosives.
>
> *Casualties:* In a second, 40 people are killed. Within minutes 160 people are injured from acute chlorine exposure. Thirty convergent responders and three police officers approach the scene and are immediately overcome by chlorine gas.
>
> *Response:* The incident generates an immediate second alarm fire response and the arrival of two EMS units. Units approaching from the north see a large green fog accumulating in the low-lying area around ground zero. A chlorine cloud is seen drifting south. Units approaching from the south begin driving into the chlorine cloud and have to divert their response.
>
> *Victims:* Mass casualties are reported with an FBL. Many patients are staggering in all directions from the scene. Some patients are leaving in taxis and private vehicles. Victims receiving minimal exposure are moving farther away or back to their offices.

Scene Control Procedures:
- There is a static hot zone at ground zero.
- There is a moving hot zone traveling to the south.
- Patients are diverted to three staging areas (west, north, east) for decontamination and medical treatment.
- An evacuation is conducted south of the incident.

Scene Control Issues:
- Many patients have left before scene control is established. Within 10 minutes the two nearest hospitals are inundated with contaminated patients arriving in private vehicles.
- In spite of a heavy chlorine odor, media, bystanders, and friends or relatives of the victims begin trying to enter the patient treatment areas.
- Because of a slow decontamination process, many "walking wounded" patients are becoming adamant about leaving to go home or get personal medical care from their private physicians.
- Within one hour, multiple 911 calls are being received for patients in offices, transit stations, and residences that were originally on the incident scene.
- For the next 48 hours, reports are continually received about patients presenting at suburban emergency rooms as far as 50 miles from the event.
- The scene requires entry by hundreds of EMS, fire, and law enforcement agencies.
- CNN breaks the story, opening the floodgates for every conceivable type of media response. Media pressure to get closer and closer to ground zero escalates.

Scene Control Solutions:
- Hospitals throughout the region are alerted. The importance of decontamination before admission is stressed.
- On-site monitoring is used to determine the extent of the hot zone. The perimeter is reduced accordingly.
- A strong law enforcement perimeter is established in the crime area. All evidence is properly preserved, photographed, cataloged, and removed.
- Scene entry points are established for public safety personnel to control warm zone and hot zone access. A local personnel accountability system is effectively utilized.
- All receiving hospitals initiate entry control procedures. Contaminated vehicles are diverted to a remote and secured parking lot (not the ER driveway). Arriving patients are diverted to a decon area outside the building for clothing removal and initial wash down. No patient is permitted entry into any hospital area without being decontaminated.

- 911 calls outside the incident area are handled by a beefed-up response force of mutual aid units and recall personnel. Records from all chlorine exposure patients are separately maintained for inclusion into a final after-action report. All cases are also reported to law enforcement for future investigation.
- The media is instructed to establish a media pool. The media pool is given a closely supervised escorted tour of areas cleared by the safety officer and law enforcement.

Effective control of an incident perimeter places a high drain on law enforcement resources. Consider controlling a four square block area. A chemical hot zone can easily get this big. This hypothetical area requires a minimum of 16 units for control. To place a unit halfway down each block on the perimeter requires 32 law enforcement units. This is a commitment of resources that severely taxes even the largest urban police or sheriff's department.

Another issue is maintaining law enforcement response times for 911 services unrelated to the major event. A third consideration is that many units will be engaged within the perimeter in a mass shooting or bombing scenario.

In summary, major terrorism/tactical violence incidents that require perimeter security are very high-impact law enforcement events. Effective sealing of a large urban area is very difficult and resource intensive under the best of circumstances. If possible, the scene area should be reduced as soon as practical. In the previous example, if the four square block area was reduced to two square blocks, the number of perimeter control units would be reduced from 32 to 16 units.

An uncontrollable factor in a chem-bio event is wind change. A 90 degree wind change, or a change in wind speed that optimizes a chemical plume footprint greatly changes the size and/or location of the hot zone. Sudden changes make effective perimeter control impossible in the short term because of a shifting and dynamic perimeter. Such changes require altering the evacuation zone; a step that cannot be done well on the run.

The last significant factor is public reaction—the "CNN factor." Informal observation by many peers suggests that a media evacuation announcement draws people to an area. These are not the helpful convergent responders that we discussed earlier. Amazingly, some people actually say, "Gee whiz, Martha, I've never seen a chemical attack before. Let's load up the kids and go watch." Other entry issues include friends or relatives rushing to a scene or parents entering an area to retrieve their kids. This situation will never change; onlookers and parents will always present scene control problems.

Response Training

In most emergency response agencies, the survival skills for living through a terrorism/tactical violence incident require development and training. The most important component is the implementation and daily use of the IMS. IMS must be the "gold standard" for the operation of response and support agencies (see Chapter 2, "The Basics of the Incident Management System").

Every agency must be trained in basic IMS as a starter. The IMS must pervade all areas of agency operations, not just emergencies. Daily business should be conducted using IMS forms and jargon. This means that IMS is a way of life. For example, Tad Stone, the previous public safety director for Citrus County, Florida, used the IMS boilerplate for meeting agendas. Items were listed under four columns entitled operations, logistics, planning, and administration. He also produced a biweekly plan using the official ICS incident action plan form. The point: Use IMS every day in all facets of your agency. When a major terrorism/tactical violence event unfolds, the teams will know how to play.

Convergent responder awareness is another important training objective. Trained convergent responders provide valuable crisis management information (intel) and assist early in the consequence response.

The first step in convergent responder training is to identify the government agencies and utilities. Identify the private business agencies on a separate list. Mandate training for government agencies and solicit cooperation from utilities and private businesses.

Using a traditional "brick and mortar" classroom to train a myriad of agencies in a convergent responders program is time consuming and expensive for all parties. Virtual training is one answer. Using the Internet, e-mail, and teleconference calling, information and basic awareness programs can be disseminated to any organization. Virtual technology is also being used to monitor certification and new employee orientation. By using the Internet, response agencies become a training source with minimal expenditure of personnel. Private businesses can determine when to utilize the material and schedule employee awareness sessions. At the very least, have a training video and supplement it with handouts.

On major high-threat events, specialized convergent responder awareness training should be conducted for security guards, staffers, ushers, maintenance personnel, concession workers, tour guides, and others. This includes events such as a papal visit, political conventions, presidential visits, political issue rallies, (pro-life, pro-choice, animal rights, etc.) and major sporting events. Before the 1996 Olympics, thousands of workers received terrorism/tactical violence awareness training. Such a program is time consuming and expensive, but a single truck driver reporting a "suspicious incident" can keep an attack in the crisis mode and prevent a consequence tragedy.

Special teams have established standards and training certification levels that must be maintained. This sounds like a given, but in some locales, special team activity is infrequent, and the specialists lose their edge. These teams also need terrorism/tactical violence awareness programs, including SWAT units, SWAT medics, search and rescue teams, and haz mat teams.

All emergency responders must be trained in scene awareness. The principles of 2 in 2 out and LACES (lookout, awareness, communication, escape, and rescue) should be in a written protocol and part of a scene awareness program. All terrorism/tactical violence incidents, especially the small ones, should be reviewed for scene awareness issues. An effective post incident analysis is also a good training aid.

Medical Facility Response

Medical facilities do not respond in the same sense as emergency responders. However, a terrorism/tactical violence incident requires facilities to alter their general mode of operation and respond to the demands of the event. Granted, hospitals do not respond by driving to a scene, but they do respond by preparing for unusual patients in mass numbers.

The initial element in hospital preparation that drives the rest of the hospital response system is communications. Medical receiving facilities must be alerted early in the event chain. This is the same process used in standard mass casualty procedures. Early warning gives facilities time to prepare for a patient onslaught. Information about the type of incident and numbers of patients is critical. For example, 10 mass shooting patients, 13 bombing victims, or 20 chemical attack victims all trigger different preparedness responses.

Communications are assured (or have a high potential for success) through the use of correct protocol and technology. Your local protocol must specify that terrorism/tactical violence events initiate immediate notification of appropriate receiving facilities. As EMS communications transmits scene information, medical facilities must be in the receiving loop. Medical control is a dispatch center or control medical facility that monitors local/regional bed status and specialties available and is a critical element.

A protocol for hospital communications must specify feedback from critical medical facilities. In many systems, EMS and/or fire communications are monitored in the ER. There are a million reasons why initial information about a terrorism/tactical violence event may not be heard on a busy night (okay, a million is an exaggeration, but you get the point). Don't depend on one-way monitoring. Require by protocol that medical receiving facilities acknowledge receiving notification of an unusual event.

Medical facilities should have several layers of technology to ensure receipt of communications. There must be auxiliary power and back-up systems for failure of private systems and electrical power. (A detailed discussion of communications technology is not within the scope of this text.) At the least, a secondary system must back up the telephone system.

Hospital security is a key nonmedical element in hospital response. Workplace violence and gang-related mass shootings might migrate from the scene to the ER. The number one cause of workplace death for physicians and nurses is gun violence. The assailants in several case histories have gone to the ER attempting to finish off their victims. In a 1998 incident in Toledo, Ohio, an assailant was killed near the ER. He was attempting to kill two children in the hospital after killing two women and wounding two fire department paramedics several miles away.

Hospital security should be immediately alerted when mass shooting victims are being transported to the facility. Armed officers should secure entry to the ER and related treatment areas.

Contaminated patients en route to an ER require a different type of security plan. Security should initiate a complete facility lock down. Every effort must be made to protect the facility from contaminated patients bypassing decon procedures. Erik Auf der Heide, in his excellent book *Disaster Response* (Mosby, 1989), points out that in mass casualty incidents, over 50 percent of the patients arrive at the hospital in private vehicles or by other means. Additionally, the Hazardous Substances Emergency Event Surveillance Annual Report indicates that only 18.5 percent of contaminated victims are treated at the scene of an exposure. The remainder seek out contamination treatment at medical facilities/hospitals. You cannot expect chemically exposed patients to arrive via EMS in a "clean" condition.

Critical Factor: Patients may also present at physician offices or emergicare centers hours or days later.

If the decon area is separate from the ER entrance, security must be positioned to direct arriving vehicles (EMS and private cars) to the decon corridor. In any terrorism/tactical violence incident, the CNN factor goes into hyper-drive. Victims' friends and relatives follow the media swarm. Both situations require security control. All private vehicles are assumed to be contaminated and must be diverted from driveways to a vehicle staging area. The vehicles can be decontaminated later after patient removal.

Unfortunately, some hospitals use contract security that is not permanently assigned to the facility. A guard assigned to the hospital one day may have been at a park on the previous day. This is especially true on weekends. For this reason, security checklists and protocols for each guard should be written in a pocket guide format and issued to each security officer, especially guards on rotation. The pocket guide procedures should be brief and consist of critical factors only. Forget the usual two-inch thick notebook. Remember that if your facility has contracted security that frequently rotates, the traditional methods of orientation and training will not work.

Critical Factor: Hospital/medical facilities *must* adopt terrorism incident protocols within their disaster plans as well as secure the necessary equipment to support actions related to these types of events.

Reliance upon local emergency response agencies to augment hospital capacity to manage patient load or decontamination is unwise. We discourage this dependency based upon the understanding that most, if not all, emergency assets will be either committed to the management of the incident (at the site) or sustaining 911 operations for calls unrelated to the terrorism incident.

CHAPTER SUMMARY

The key response issues relating to a terrorism/tactical violence incident are:

1. Local agencies will always be the cornerstone of an effective terrorism/tactical violence response. Plan to be self-sustaining for 12 to 24 hours before federal help arrives.
2. Convergent responders arrive before first responders. Identify convergent responder agencies and develop an awareness training program for them.
3. A terrorism/tactical violence scene is definitely a hostile workplace environment. Focusing on victims instead of overall scene awareness can get you killed. Use your senses; look around before charging in. Beware of a chemical hot zone, shooters in the area, and primary or secondary explosive devices.
4. Use the buddy system when entering an unsecured scene by applying the 2 in 2 out principle. Also apply the wildland fire fighting principle of LACES: lookout, awareness, communications, escape, and safety zone.
5. A trend of mass shootings and tactical ultra-violence means more exposure to ballistic hot zones. Examine the peripheral areas of a scene, including roofs or other high places. Use the principle of distance and cover for self-protection. If you're debating about ducking and running you should have already been doing it.

6. The IMS is the boilerplate for management of terrorism/tactical violence incidents. Learn it, implement it, train on it, exercise it, and use it.

7. Scene control is a critical factor in a terrorism/tactical violence event and has high impact on law enforcement. Victims, witnesses, and possible assailants must be kept on the scene. Other groups will attempt to enter the scene such as media, onlookers, and friends and families of the victims.

8. All emergency responders must be specially trained for terrorism/tactical violence incidents. This includes IMS proficiency, scene awareness, and special team training.

9. Medical facilities must initiate response protocols for processing and treating mass casualty terrorism/tactical violence patients. Facility response begins with effective communications to provide early scene information.

10. Hospital security is a critical factor in medical facility preparedness. In a mass shooting, the ER must be physically protected from possible perpetrators. In a chemical attack, contaminated patients must be diverted to the decon area and the facility should be locked down.

CHAPTER QUESTIONS

1. What are convergent responders? What are their strengths and weaknesses?

2. List the public and private convergent responder agencies in your community.

3. What are the unsafe conditions that may be present at a terrorism/tactical violence incident?

4. What is the 2 in 2 out principle? Discuss the LACES principle.

5. What is the importance of IMS in terrorism/tactical violence incidents?

6. List and discuss the major scene control issues in a terrorism/tactical violence event.

7. What are some of the security responsibilities of a medical facility when a terrorism/tactical violence incident is in progress?

8. What procedures should be initiated by a medical facility when contaminated patients are en route?

Simulation

Research the response procedures of your local fire, EMS, and law enforcement agencies. Do these procedures address tactical violence scenes? Develop a comprehensive terrorism/tactical violence operational procedure based on the concepts and key issues discussed in this chapter.

4
Planning for Terrorism/Tactical Violence

Paul M. Maniscalco
Hank T. Christen

Source: Courtesy H. Christen

Chapter Objectives

After reading this chapter, you will be able to accomplish the following objectives:

1. Understand the importance of planning as a front-end concept.
2. Recognize the importance of incorporating terrorism/tactical violence in an emergency operations plan (EOP).
3. Outline the key (IMS) elements in a consequence plan.
4. Distinguish between an EOP and a real-time (IAP).
5. Understand the importance of logistics in an emergency operations plan.
6. Understand the importance of liaison with other agencies in effective emergency planning.

Planning—A Front-End Concept

Effective consequence response to a terrorism/tactical violence event is perceived as action oriented. We picture quick and aggressive operations and tactics. This picture is hopefully accurate in a well-organized event. However, the use of tactical units never occurs without effective planning.

Advanced planning is a front-end concept. This means that for a positive outcome (efficient operations) to occur at the back end, we must heavily load the front end. No one is lucky enough to dynamically deploy in a terrorism/tactical violence incident and be successful without planning.

Critical Factor: Plan on the front end to ensure success on the back end.

A front-end plan is different than a reference guide. Many disaster plans are a two inch thick binder that includes radio frequencies, names and addresses, lists of vehicles, units, and descriptions of resources. These guides are important for reference information, but are not documents that yield "quick and dirty" guidelines in the heat of battle. There is nothing wrong with reference information; it's important, but it's not a plan that guides you through the critical steps of a disaster or terrorism/tactical violence incident.

An effective plan is an easy-to-read document that guides you through the critical factors of a dynamic event; the plan identifies operations and support functions that complement operations. In other words, the plan is an incident management system (IMS) guide.

IMS—The Boilerplate for Planning

IMS is not only a response system but also a template for planning. An effective plan guides you through the IMS functions. As discussed in Chapter 2, ("The Basics of the Incident Management System"), all emergency functions consist of a management staff, operations, logistics, planning, and administration. It makes sense to use the IMS model when developing effective operational plans.

An emergency operations plan (EOP) for terrorism/tactical violence incidents should identify key elements such as:

- Management staff
- Operations
- Logistics
- Administration
- Planning

Who Is in Charge?

Command turf wars have no place in a terrorism/tactical violence incident. Emergency response agencies have to sit down ahead of time and work out who will manage what type of event. This may not be easy, but it must be done. Federal and state agencies have elaborate plans and statutes that identify who (or what) will manage the event, but they are at least hours away and maybe days away from your community. If you are local, you will be there first, and you are on your own.

If an event meets the federal definition of terrorism, federal agencies will have jurisdiction. In accordance with PDD 39 (Presidential Decision Directive) the FBI (Department of Justice) has jurisdiction in crisis management and criminal investigation and the FEMA coordinates consequence management. The key word here is "coordinate," for FEMA does not have a response capability. FEMA can only coordinate federal teams with local teams. Even if everything goes perfectly, it will be many hours or longer before federal support is available.

The local community must be able to establish scene management immediately and maintain command throughout the incident. The critical step is to determine who is in charge. In most local governments the responsibility for scene management in terrorism/tactical violence events rests with the ranking law enforcement official. However, terrorism/tactical violence incidents are complex and require a number of agencies. As previously explained, the incident will involve a large EMS and fire/rescue commitment with the appropriate command responsibilities.

There are two solutions. The first solution is a single agency incident manager, such as a sheriff or police chief, with EMS and fire/rescue operating at

FIGURE 4-1 Identification of lead agencies prior to an incident, during the planning process, is good planning policy and avoids confusion. *Source: Courtesy M. Cross*

branch levels. The second solution is unified management (unified command). A unified management team is utilized when multiple agencies have jurisdiction in a complex incident. Consider a bombing with a building collapse and mass casualties; clearly, there are major EMS, fire/rescue, and law enforcement operations that must effectively coordinate. In this case, a command post with a joint EMS/fire/law enforcement management team is the solution.

Unified management requires that all managers be located in the same area (a command vehicle or an emergency operations center) and share information. The planning section makes this happen by effective status display and comprehensive interagency action plans. Separate command posts for each major agency never works. Each agency tends to freelance and coordination does not happen.

Unified management does not begin in the heat of battle. It starts with effective planning well before a terrorism incident. The responsibility of each agency must be specifically defined and written. Emergency management facilitates the formal development of unified management through the local comprehensive emergency management plan.

> **Critical Factor:** An essential step in consequence management planning is to determine who will be in charge.

In many ways, the informal development of unified management is more significant than the formal process. Managers must get to know each other. The corporate world uses the term "networking." It means that various agency managers initiate mutual support by breaking bread and sharing coffee. A community disaster committee is an excellent vehicle to schedule quarterly meetings where mutual incident management issues are discussed and agency heads get to know each other. In summary, managers cannot meet each other for the first time in a major incident, and expect to coordinate effectively.

Operations

The EOP must identify operations agencies in the local/regional area that are trained and equipped to handle specific operational problems. On the surface this may seem like an easy task. Obviously, EMS, fire, public health, and law enforcement have clearly defined and visible tasks. However, terrorism/tactical violence incidents present unique operational problems. Many of these issues require special teams not available in even large urban areas. Special terrorism/tactical violence operational problems include the following:

1. Heavily armed tactical teams to combat terrorists with automatic weapons.
2. Haz mat teams equipped to detect, penetrate, and decontaminate radiological incidents.
3. EMS teams that triage, treat, and transport large numbers of mass casualty patients.
4. A decontamination unit that can decon hundreds of patients in a chemical attack.
5. A public health unit that can detect and track mass disease patients in a biological attack (medical surveillance).
6. Search and rescue teams for structural collapse events.
7. Computer antiterrorist units.

The listed scenarios are not common incidents. At the time of this writing, there is not a single city in the United States that has the capability of managing all of these events. If special teams are not locally available, the plan must identify other sources. These sources include federal response teams, military units, state and regional agencies, and private response resources.

FIGURE 4-2 Pre-packaged caches of equipment transport ready are the most effective means for logistic support during a high-impact—high yield incident. *Source:* Courtesy H. Christen

In reality, no single local government can afford all the anticipated resources. However, an effective threat assessment and planning process reveals areas of vulnerability and identifies possible operational units capable of threat response.

Logistics

Terrorism/tactical violence events are often destructive because equipment and supplies are consumed faster than they are replaced. We can manage a small bombing, but in an Oklahoma City type event, we run out of radio channels, people, vehicles, supplies, medications, and hospital beds. Examples of critical equipment needs in terrorism/tactical violence incidents are:

- People
- Weapons/crime scene supplies
- EMS units
- Protective equipment
- Medical supplies/ventilators
- ER/surgical facilities
- Detection and decontamination equipment
- Vehicles/aircraft

- Medical laboratories
- Medications/antidotes (pharmaceutical cache)
- Communications

In long duration incidents there is an added demand for crew rehabilitation, food/water, sleeping facilities, and sanitation measures.

An effective plan identifies logistics sources. The plan is based on the correct assumption that first response resources are immediately depleted. The availability of resources that address the previously listed problem areas are outlined in the plan. There are several solutions for critical logistics needs such as:

1. Logistics caches for regional support, such as a decontamination cache of washing supplies, protective equipment, and 100 Tyvek™ suits.
2. Mutual aid agreements, such as an agreement for a neighboring fire department to provide haz mat detection equipment.
3. Military support, such as a memorandum of understanding (MOU) with a local military installation for critical personnel or equipment.
4. Private contracting, such as an agreement with a chemical spill response company to provide chemical spill control or radiation equipment.
5. State/Federal support, such as procedures for requesting state and federal logistical support; nationwide pharmaceutical cache system.

Logistics caches must be packaged and stored for rapid deployment. Plastic carrying boxes, also called nesting boxes, are recommended. They are quickly loaded on a vehicle and rapidly transported. Avoid storing caches in a back room where they gather dust. Supplies should be accessible, packaged, and in good condition when needed for a "once in a career" incident.

Cost and quantities are other real world issues. How much should we buy and how much can we afford? These are tough questions. The history of the local area and the political climate greatly influence the financial commitment. A selling point is that equipment caches are needed for all types of disasters, not just terrorism. The question of quantities is just as complex. There is no formula based on population or threat assessment to determine how many Tyvek™ suits, body bags, or atropine injections a community needs. On the surface it seems obvious that high-threat urban areas should have the biggest caches, but what about a small college town that has a football stadium filled with 50,000 fans on a given Saturday? Again, there is no answer.

At a minimum, there should be enough medications and protective equipment for first responders and mutual aid personnel. This sounds self-serving, but the logic is irrefutable. We must protect ourselves and ensure

FIGURE 4-3 Remember in your planning to incorporate provisions to be self-sustaining for at least 48–72 hours. That is how long it will take for the federal special teams to arrive if you require their assistance. *Source:* Courtesy J. Dickens

our own survivability so we can take care of our customers. Key logistics points are:

1. Logistics takes planning.
2. Equipment and supplies must be pushed quickly to a scene.
3. Logistics caches must be accessible and maintained.
4. Budgets and political realities will always be logistics issues.
5. Estimates of logistics quantities are at best a semi-educated guess.
6. We need enough equipment and supplies to protect ourselves.

Administration

Administration functions are omitted from most emergency plans. In normal operations, individual agencies take care of the usual administrative duties such as finance/purchasing, workers compensation, and payroll records.

During major disasters, administrative requirements escalate. Normal administration is no longer suitable. Purchasing and contracting in the middle of the night, or tracking personnel from state and national agencies is a function that local agency administrative personnel have not experienced.

An EOP addresses administrative needs by establishing an administrative section chief who supervises the following:

1. Finance unit—performs purchasing/contracting duties, maintains records of all fiscal activities, develops post-incident financial reports for state/federal reimbursements.

2. Time unit—tracks all personnel and maintains hours and pay records; develops payroll report for state/federal reimbursement.
3. Compensation/claims unit—tracks and administrates all claims relating to the incident; tracks and handles submission of workers compensation claims.

The typical reader of this book is tactically oriented and may give little thought to administration until they have to account for expenditures approaching $1 million per day. If you are a mid-level or upper level manager, go back and read the administration paragraphs again. If an administrative section is not included in your EOP, your plan is inadequate.

Critical Factor: Don't allow your operations/logistics mind-set to blind you to the administrative needs of a major terrorism/tactical violence incident.

Planning Section

If there is no provision for planning, an EOP would not exist. The EOP is usually developed and maintained by emergency management, by a local interagency disaster committee, or both. An effective emergency manager plays a coordinating role in bringing competing agencies together and implementing the EOP. As many agency managers as possible should participate in the EOP development. The document should also be formally signed. When conflicts arise, agency heads are reminded of their responsibilities according to the EOP gospel.

The EOP should establish the position of planning section chief for disasters and terrorism/tactical violence incidents. The planning section chief is responsible for developing the IAP, conducting planning briefings, and supervising the following units:

- Resource unit—establishes an incident check-in procedure; maintains a master list of all incident resources; tracks location and status of all resources; assigns resources based on operational requests.
- Situation unit—collects and analyzes situation data; prepares incident map displays; displays critical incident information; prepares incident predictions at request of planning section chief; obtains and displays incident weather data.
- Demobilization unit—coordinates with resource unit to track incident resources; obtains incident manager's demobilization plans; establishes resource check-in; determines transportation and logistics needs related to demobilization.

- Documentation unit—responsible for maintaining accurate and complete incident files including duplication services, packing, and storing.
- Technical advisor(s)—specialists in any field relevant to the incident (e.g., a meteorologist, or an epidemiologist in a biological incident).

EOP Summary

The EOP is a comprehensive plan (sometimes called a comprehensive emergency management plan) that uses the IMS as a benchmark to establish an interagency plan of attack for terrorism/tactical violence incidents. The key steps in the implementation phase of an EOP are:

1. Coordinate with all convergent and local/regional response agencies.
2. Assign EOP responsibility to emergency management.
3. Conduct a terrorism/tactical violence threat assessment in your community.
4. Identify operational and logistics needs based on your threat assessment.
5. Produce a formal document signed by cooperating agency managers.
6. Perceive the EOP as an evolving document that is never done.

The EOP has the following key elements:

1. Specify who is in charge.
2. Make provisions for unified management in complex incidents. Base scene management on the overhead team concept. (An overhead team is the incident management team.)
3. Establish a management staff for public information, safety, and liaison.
4. Establish an operations section and identify operational agencies.
5. Establish a logistics section and identify logistics requirements.
6. Establish a planning section.
7. Establish an administration section.
8. Train section chiefs and unit leaders.
9. Conduct tabletop and operational exercises to test your plan.

Smaller communities are overwhelmed with the personnel demands of a full scale EOP implementation. Remember that in any disaster, overhead team positions are only filled as needed. The critical concept is that any function not assigned by the incident manager is his or her responsibility. It is very important to identify critical personnel from other support agencies or mutual aid agencies and train them to function as the overhead team. This is especially significant in locales that have inadequate staffing.

The IAP

An IAP is a plan of objectives for implementing an overall incident strategy. The IAP is a tactical plan for an operational period. Unlike the EOP, the incident action plan is related to a specific incident and a set of dynamic circumstances. The EOP is reference oriented whereas the IAP is action oriented.

On routine incidents or a small mass casualty event, a scene manager follows a protocol without a written plan. Planning is "on the fly" with the aid of a quick checklist. Major incidents need a formal IAP because an escalating event requires increased resources, increased variables, and an extended time period. The event also increases in geographical area requiring several divisions. Incidents of this nature present three major planning problems:

1. No individual can keep all the strategies in his or her head.
2. No single individual can plan all the strategies without consultation and help.
3. There is no way to convey the strategies, objectives, and assignments to others without a written document.

The solution is a formalized IAP. "When do you go formal?" is a valid question. In a short-duration incident, an informal plan communicated verbally will suffice. When an incident progresses beyond a 12-hour operational period, a written plan is prepared, briefed, and distributed to operational shifts. This is especially important if night operations are continued. If night operations are scaled down, a daily morning briefing is adequate.

The development of a 12-hour or daily IAP keeps a planning section chief busy. Fortunately, there is a nationally accepted format developed by the National Wildfire Coordinating Group. This format is fire oriented, but has been adapted for all risk type incidents, including terrorism/tactical violence events. There are several forms associated with the IAP that serve as checklists completed by the appropriate supervisor, and require information that greatly eases the IAP process.

The forms address critical planning areas:

1. ICS 202, Incident Objectives—includes basic strategy/objectives for the overall incident and includes critical areas such as safety concerns and forecasted weather.
2. ICS 203, Organizational Assignment List—includes operations sections with all branch directors and division/group supervisors; includes logistics section, planning section, and finance (administration) section with the respective units for each section.
3. ICS 204, Division Assignment List—a list for each branch (operations), divisions assigned to each branch, division managers, and units assigned to each division.

4. ICS 205, Incident Radio Communications Plan—lists administrative frequencies, operations frequencies, and tactical channels; lists telephone numbers for managers and key agencies; specifies an overall communications plan including secondary and tertiary communications networks.
5. ICS 206, Medical Plan—formalizes medical support for overhead team and on-scene responders; includes nearest hospitals, medical frequencies, and a roster of EMS units. *Note:* In mass casualty incidents, medical operations are delineated in ICS 204 relating to organizational assignments and division assignments.
6. ICS 220, Air Operations Summary Worksheet—lists all air units, air objectives, frequencies, and assignments.
7. Incident Map—a map, computer graphic, or aerial photo depicting the incident area.
8. Safety Plan—safety objectives and hazard warnings (prepared by the safety officer.) *Note:* In biological, chemical, and radiological events, safety procedures are technical and complex.

The formal IAP is not as overwhelming as it may appear. First, the forms serve as a checklist; you fill in the blanks. (Many agencies have their forms on notebook computers.) Second, the IAP is a joint effort. The incident manager, operations units, and specialized units such as medical, safety, weather, and communications contribute to the effort.

Critical Factor: The IAP forms serve as an effective planning checklist.

The IAP Briefing Cycle

In long duration incidents, the IAP is presented during a formal briefing every twelve hours. This is called a planning briefing cycle. The nature of the incident may dictate an eight-hour briefing cycle (at shift changes), or a 24-hour briefing cycle when there are diminished night operations.

The briefing is a very important formal process. All section level and division level commanders, along with appropriate unit leaders should attend the briefing. A 50-person audience or more is not unusual in a complex event.

The IAP is presented by the planning section chief and includes briefings by other specialists such as weather, safety, and technical experts who may include such specialties as medical, radiological, bio-weapons, etc.

Copies of the IAP are distributed to each attendee. An IAP can be 10 pages in length in a major disaster or terrorism/tactical violence incident.

FIGURE 4-4 Remember, if you have integrated military assets into the incident management they need to be included in your briefings. *Source:* Courtesy P. Maniscalco

The IAP copies become a reference for all managers, especially the sections on communications frequencies, unit assignments, and safety procedures.

The National Wildfire Coordinating Group publishes the following formal guidelines in the *Fireline Handbook.* In a 12-hour briefing cycle, IAP preparation progresses as follows:

1. Shift change—receive field observations; one hour.
2. Prepare for planning meeting; one hour.
3. Planning meeting with management staff, section chiefs, agency representatives; one hour.
4. Prepare incident action plan (IAP); four hours.
5. Finalize IAP; two hours.
6. Prepare for operations briefing; one hour.
7. Briefing of management staff, section chiefs, branch/division/unit supervisors; one hour.
8. Finalize reports; one hour.

The briefing cycle is a guideline, and is flexible. The cycle can be shortened or lengthened as appropriate.

Critical Factor: The IAP must be developed via a briefing cycle suitable for the incident.

Planning Scenario—Example

To demonstrate the use of an EOP and informal and formal IAPs, consider the following scenario:

It is 09:30 hours in Simulation City, California. There is an explosion originating from a parked van on Center Street. One person is killed, 14 people are injured, and windows are shattered throughout the block.

First due law enforcement units arrive at 09:32; fire/rescue and EMS units arrive at 09:35. At 09:41 there is a massive explosion in a complex of government buildings, immediately adjacent to the van explosion site. Many of the first responders are injured or killed. A five-story building collapses (estimated occupancy of 500 people). The major trauma center next to the building is heavily damaged; there are multiple injuries in the hospital from flying glass.

The EOP

The EOP has several sections directly related to an explosive attack on building complexes based on a previous emergency management threat assessment. Key response agencies immediately begin to run checklists based on elements of the EOP. Critical agency managers are immediately notified, as well as the state operations center. The EOP is based on the incident management system and identifies several critical areas:

1. Management—A unified management team is specified, with joint command between law enforcement and fire/rescue.
2. Operations—The three critical operations branches are law enforcement, fire rescue, and EMS. The EOP lists state/federal law enforcement mutual aid, procedures for fire mutual aid, and for requesting an urban search and rescue team (USAR), and procedures for medical mutual aid, including use of a disaster medical assistance team (DMAT).
3. Logistics—The logistics plan lists sources for scene control, vehicles, lighting, generators, back-up communications, fuel, emergency food, construction equipment, and lumber for shoring.
4. Administration—Specifies that expenditures, claims, worker's compensation issues be tracked and establishes the city purchasing director as the supervisor of expenditures.
5. Planning—Establishes the activation of the emergency operations center by emergency management and the establishment of a fully

staffed planning section if the incident becomes protracted, including a resource unit and situation unit.

The Informal IAP

Initial response managers begin arriving and consult checklists developed from the EOP. It is apparent that a major incident is in progress. The chief of police and fire chief establish a command post four blocks from the incident.

An informal IAP is developed in minutes by identifying critical factors. This plan is communicated verbally and by radio to response managers and units. The key elements are:

1. All units search their operational and staging areas for possible secondary devices.
2. Three branches are established: collapse branch, Center Street branch (including the van explosion), and trauma center branch (the EOP never addressed the loss of the trauma center).
3. A recon group is established to assess damage on all sides of the original explosion.
4. Immediate control of the scene perimeter is established.
5. Appropriate EMS, law enforcement, and fire units are assigned to groups/divisions.
6. Triage, treatment, and transport of MCI patients throughout the incident area are addressed.

As hours pass, new problems such as an area power failure and loss of the downtown communications repeater arise. There is also disturbing news that many patients appear to have chemical injuries and a combination explosive/chemical attack is suspected. As these problems evolve, the IAP is affected accordingly.

The unified management team becomes aware in the first hour that this will become a complex, long-duration incident. As a result, they request a pre-established incident team for full staffing of incident management positions. This overhead team is comprised of certified people from a multitude of local and state response agencies.

The scenario progresses to day two:

Day Two—The Formal IAP

The planning section, in coordination with the unified command team and the logistics and operations section chiefs, develops the IAP. Emergency management also establishes a multiagency coordination center (MACC).

The formal action plan is developed using a laptop computer and ICS forms 202–206 (see pages 77–90). Copies are made for each member of the overhead team and all branch directors and division supervisors.

Day Two also dawns with the arrival of federal assistance. The FBI has a full response team assigned to law enforcement operations. The U.S. Marine Corps CBIRF is assigned to the haz mat branch for detection and decontamination. CBIRF strike teams are also available for divisions requiring haz mat support. Lastly, a state National Guard unit is assigned to law enforcement operations for perimeter security.

The following pages demonstrate a completed IAP for an incident of this level. These pages include an incident radio communications plan (ICS 205), medical plan (ICS 206), and an air operations summary (ICS 220). See pages 77 to 90.

CHAPTER SUMMARY

Planning for terrorism/tactical violence is a front-end concept. For a successful tactical outcome on the back end, effective planning must occur at the front end.

The model of an effective emergency operations plan (EOP) is the incident management system (IMS). The plan addresses the key IMS functions of management staff, operations, logistics, planning, and administration. The management component identifies who is in charge and recognizes the concept of unified management. The operations section identifies operations agencies/teams based on an emergency management threat assessment.

Terrorism/tactical violence incidents and disasters consume logistics rapidly. The logistics section of an emergency operations plan must identify logistics needs and agencies or organizations that provide the appropriate people, supplies, and equipment. In a major event logistical support may be regional, state, or federal. Critical logistics requires logistics caches (push logistics) that are quickly deployed to a scene.

An EOP also identifies an administration section and a planning section. The administration section includes a time unit, procurement unit, and a compensation unit. The planning section consists of a resource unit, situation unit, and mobilization/demobilization unit.

An EOP is a reference tool, and not a tactical guide. Incidents are extremely dynamic; critical events change in seconds or minutes. In the "fog" of combat the incident manager develops an informal and verbal incident action plan (IAP).

A long-duration incident (longer than a day) requires a written IAP developed by the planning section chief in coordination with the incident manager and other section chiefs. The National Wildfire Coordinating Group has developed forms for the IAP. The forms serve as a checklist for organizing the event. All section groups and divisions are given a daily briefing along with copies of the IAP.

INCIDENT OBJECTIVES	1. INCIDENT NAME Simulation City explosion	2. DATE PREPARED 3/1/00	3. TIME PREPARED 05:37

4. OPERATIONAL PERIOD (DATE/TIME) 08:00 to 20:00

5. GENERAL CONTROL OBJECTIVES FOR THE INCIDENT (INCLUDE ALTERNATIVES)
Scene control entry/exit
Detailed secondary search
Evidence recovery and preservation
Chemical detection and decontamination
Heavy rescue
Equip and re-open trauma center

6. WEATHER FORECAST FOR OPERATIONAL PERIOD
 Heavy fog restricting visibility to 10:30
 Wind light from 270 deg.
 50% chance rain from 17:00 to 19:00

7. GENERAL/SAFETY MESSAGE
Type of chemical(s) not confirmed in collapse area; treat entire area as "hot zone."
Full ppe required in hot zones
Bio precautions for all divisions
Possible non-detonated explosives; possible secondary devices
Lightning in afternoon thunderstorms

8. ATTACHMENTS (X IF ATTACHED)

 - ORGANIZATION LIST (ICS 203) **XXX**   INCIDENT MAPS (2)   Safety Message **XXX**

 - DIVISION ASSIGNMENT LIST (ICS 204) **XXX**   TRAFFIC PLAN   Homestead Map

 - COMMUNICATIONS PLAN (ICS 205) **XXX**   - Air Operations   Structural Fire Plan
                                             Summary (ICS 220) **XXX**

 - MEDICAL PLAN (ICS 206) **XXX**

202 ICS 3/80	9. PREPARED BY (PLAN- NING SECTION CHIEF)	10. APPROVED BY (INCIDENT COMMANDER)

ORGANIZATION ASSIGNMENT LIST		1. INCIDENT NAME Simulation City explosion		2. DATE PREPARED 3/1/00	3. TIME PREPARED 06:01
5. INCIDENT COMMANDER AND STAFF POSITION — NAME		4. OPERATIONAL PERIOD (DATE/TIME) 08:00 to 20:00 — 3/1/00			
INCIDENT COMMANDER	Christen/Maniscalco	9. OPERATIONS SECTION			
SAFETY	P Chovan	CHIEF		C Lynne	
		DEPUTY			
INFORMATION OFFICER	T Blackmon			L Foley	
TECHNICAL SPECIALISTS:	J Denney	BRANCH DIRECTOR	I	R Carmine	
		GROUP	A	J Knowles	
7. PLANNING SECTION		GROUP	B	R Sampson	
CHIEF	D Chamberlin	GROUP			
SITUATION UNIT	R Gallagher	BRANCH DIRECTOR	II	M McGuire	
RESOURCES UNIT	D Chavis	GROUP	C	J Sanford	
DOCUMENTATION UNIT		GROUP	D	C Poliseno	
DEMOBILIZATION UNIT		GROUP			
		BRANCH DIRECTOR	III	T Weatherford	
STATUS/CHECK-IN RECORDER		GROUP	E	S Barney	
		GROUP	F	G Solomon	
TECHNICAL SPECIALISTS:	E Robbins				
		BRANCH DIRECTOR	IV	R McAllister	
		GROUP	G	H Tiffany	
		GROUP	H	R Colbert	
8. LOGISTICS SECTION		GROUP			
CHIEF	M Hopmeier	AIR OPERATIONS BRANCH			
SUPPORT BRANCH DIR	L Adams	AIR OPER.BR.DIRECTOR		J Hughes	
SUPPLY UNIT	J Hawkins	HELIBASE MANAGER			
FACILITIES UNIT	R Holden	AIR ATTACK SUPERVISOR			
GROUND SUPPORT	R Pitts	AIR SUPPORT SUPERVISOR			
SERVICE BRANCH DIR	S Katz	HELISPOT MANAGER			
COMMUNICATIONS UNIT	R Stancliff	10. FINANCE SECTION			
MEDICAL UNIT	K Burkholder	S Fallon			
FOOD UNIT	B Dennis	CLAIMS UNIT		R Cohen	
		CHIEF		TIME UNIT	
FOR FORMER BRANCH III		PROCUREMENT UNIT		G Collins	
COORDINATION CONTACT		COST UNIT		R Antonio	
		ADMIN PAYMENT TEAM			
		COST UNIT		R Antonio	
203 ICS 1/82					

1. BRANCH I Collapse	2. DIVISION/GROUP A	DIVISION ASSIGNMENT LIST ICS204

3. INCIDENT NAME Simulation City explosion	4. OPERATIONAL PERIOD DATE 3/1/00 TIME 08 to 20

5. OPERATIONS PERSONNEL

OPERATIONS CHIEF C Lynne   DIV/GROUP SUPERVISOR J Kowles
BRANCH DIRECTOR _ R Carmine   AIR ATTACK SUPERVISOR NO.

6. RESOURCES ASSIGNED THIS PERIOD

STRIKE TEAM/TASK FORCE/ RESOURCE DESIGNATOR	LEADER	NUMBER PERSONS		DESCRIPTION	
FBI	J Wilson	5		Evidence task force	
USAR	F Domino	35		Fl TF-1 urban search & rescue	
Fire/Rescue	J Redner	16		Eng16 eng4 eng6 trk4	
DMAT	G Moore	5		Type II Fl-2 medical task force	
TOTAL		61			

7. OPERATIONS

.Victim search/ collapse rescue; medical support

8. SPECIAL OPERATIONS   Evidence location/removal

9. MEDICAL UNIT   DMAT task force

10. FREQUENCIES   USAR tac 1   Sim City freq 2

PREPARED BY (RESOURCE UNIT LEADER)	APPROVED BY (PLANNING SECTION CHIEF)	DATE 3/1/00	TIME 07:24

1. BRANCH I Collapse	2. DIVISION/GROUP B	**DIVISION ASSIGNMENT LIST ICS204**
3. INCIDENT NAME Simulation City explosion		4. OPERATIONAL PERIOD DATE 3/1/00 TIME 08 to 20

5. OPERATIONS PERSONNEL

OPERATIONS CHIEF C Lynne
BRANCH DIRECTOR _ R Carmine

DIV/GROUP SUPERVISOR R Sampson
AIR ATTACK SUPERVISOR NO.

6. RESOURCES ASSIGNED THIS PERIOD

STRIKE TEAM/TASK FORCE/ RESOURCE DESIGNATOR	LEADER	NUMBER PERSONS		DESCRIPTION	
FBI	L Freeh	6		Crime scene task force	
USAR	D Salinger	32		Va TF-1 urban search & rescue	
Fire/Rescue	P Williams	17		Eng1 eng58 sq4 trk11	
DMAT	G Moore	5		Medical task force	
TOTAL		60			

7. OPERATIONS

.Victim search/ collapse rescue; medical support

8. SPECIAL OPERATIONS   Evidence removal

9. MEDICAL UNIT   DMAT task force

10. FREQUENCIES   USAR tac 2   Simulation City freq 3

PREPARED BY (RESOURCE UNIT LEADER)	APPROVED BY (PLANNING SECTION CHIEF)	DATE 3/1/00	TIME 07:24

1. BRANCH II Trauma center	2. DIVISION/GROUP C	DIVISION ASSIGNMENT LIST ICS204

3. INCIDENT NAME Simulation City explosion	4. OPERATIONAL PERIOD DATE 3/1/00 TIME 08 to 20

5. OPERATIONS PERSONNEL

OPERATIONS CHIEF C Lynne
BRANCH DIRECTOR _ M McGuire

DIV/GROUP SUPERVISOR J Sanford
AIR ATTACK SUPERVISOR NO.

6. RESOURCES ASSIGNED THIS PERIOD

STRIKE TEAM/TASK FORCE/ RESOURCE DESIGNATOR	LEADER	NUMBER PERSONS		DESCRIPTION	
DMAT	E Robbins	35		MA-1 primary care	
DMAT	K Allen	32		OH-1 emer room	
DMAT	C D'Angelo	34		CA-2 clinic	
TOTAL		101			

7. OPERATIONS

Maintain ER operations, support EMS, operate treatment clinic, support primary care

8. SPECIAL OPERATIONS

.

9. MEDICAL UNIT   N/A

10. FREQUENCIES   DMAT tac 4, Simulation City EMS Med 2

PREPARED BY (RESOURCE UNIT LEADER)	APPROVED BY (PLANNING SECTION CHIEF)	DATE 3/1/00	TIME 07:26

1. BRANCH II Trauma center	2. DIVISION/GROUP D	DIVISION ASSIGNMENT LIST ICS204

3. INCIDENT NAME Simulation City explosion	4. OPERATIONAL PERIOD DATE 3/1/00 TIME 08 to 20

5. OPERATIONS PERSONNEL

OPERATIONS CHIEF C Lynne   DIV/GROUP SUPERVISOR C Poliseno
BRANCH DIRECTOR _ M McGuire   AIR ATTACK SUPERVISOR NO.

6. RESOURCES ASSIGNED THIS PERIOD

STRIKE TEAM/TASK FORCE/ RESOURCE DESIGNATOR	LEADER	NUMBER PERSONS		DESCRIPTION	
EMS strike team	Chief Neal Dolan	10		Sim City EMS 6, 14, 8, 7, 2	
LAFD res 27	Chief Neil Cross	3		Heavy rescue	
Orange Co 62	Lt. Reza Golesorkhi	2		ALS unit	
Orange Co 24	EMT M. Griswold	2		BLS unit	
TOTAL		17			

7. OPERATIONS

City wide EMS response to support local 911 system

8. SPECIAL OPERATIONS
.

9. MEDICAL UNIT   N/A

10. FREQUENCIES   EMS med 4

PREPARED BY (RESOURCE UNIT LEADER)	APPROVED BY (PLANNING SECTION CHIEF)	DATE 3/1/00	TIME 07:26

1. BRANCH III Center St.	2. DIVISION/GROUP E	DIVISION ASSIGNMENT LIST ICS204	
3. INCIDENT NAME Simulation City explosion		4. OPERATIONAL PERIOD DATE 3/1/00 TIME 08 to 20	

5. OPERATIONS PERSONNEL	
OPERATIONS CHIEF C Lynne BRANCH DIRECTOR _ T Weatherford	DIV/GROUP SUPERVISOR S Barney AIR ATTACK SUPERVISOR NO.

6. RESOURCES ASSIGNED THIS PERIOD

STRIKE TEAM/TASK FORCE/ RESOURCE DESIGNATOR	LEADER	NUMBER PERSONS		DESCRIPTION	
FBI	SSA W. Drivet	10		Crime scene unit	
Simulation City PD	Chief Patricia O'Brien	6		Crime scene unit	
Medical Examiner	Dr. K.A. Holtermann	3		Body/remains recovery	
Fire/rescue	Chief Don Hiett	8		Strike team	
TOTAL		27			

7. OPERATIONS

Evidence and body recovery; scene control

8. SPECIAL OPERATIONS   Bio protection and bio waste disposal

9. MEDICAL UNIT   N/A

10. FREQUENCIES   Simulation City PD channel 1 Fire freq 3

PREPARED BY (RESOURCE UNIT LEADER)	APPROVED BY (PLANNING SECTION CHIEF)	DATE 3/1/00	TIME 07:27

1. BRANCH III Center St.	2. DIVISION/GROUP F	DIVISION ASSIGNMENT LIST ICS204
3. INCIDENT NAME Simulation City explosion		4. OPERATIONAL PERIOD DATE 3/1/00 TIME 08 to 20

5. OPERATIONS PERSONNEL	
OPERATIONS CHIEF C Lynne BRANCH DIRECTOR _ T Weatherford	DIV/GROUP SUPERVISOR G Solomon AIR ATTACK SUPERVISOR NO.

6. RESOURCES ASSIGNED THIS PERIOD

STRIKE TEAM/TASK FORCE/ RESOURCE DESIGNATOR	LEADER	NUMBER PERSONS		DESCRIPTION	
ATF	SA M. Maniscalco	4		Bomb evidence recovery	
FBI	SSA Connie Patton	4		Crime scene unit	
EMS rescue 23	PM Minga	2		ALS medical unit	
Simulation City PD	Capt. M. Newburger	12		PD strike team	
TOTAL		22			

7. OPERATIONS

Center Street damage assessment
Perimeter control

8. SPECIAL OPERATIONS   Evidence detection, photo evidence, and recovery
.

9. MEDICAL UNIT   EMS rescue 23

10. FREQUENCIES   Simulation City PD channel 2 Fire freq 4

PREPARED BY (RESOURCE UNIT LEADER)	APPROVED BY (PLANNING SECTION CHIEF)	DATE 3/1/00	TIME 07:27

1. BRANCH IV Haz mat	2. DIVISION/GROUP G	DIVISION ASSIGNMENT LIST ICS204
3. INCIDENT NAME Simulation City explosion		4. OPERATIONAL PERIOD DATE 3/1/00 TIME 08 to 20

5. OPERATIONS PERSONNEL

OPERATIONS CHIEF C Lynne   DIV/GROUP SUPERVISOR H Tiffany
BRANCH DIRECTOR _ R McAllister   AIR ATTACK SUPERVISOR NO.

6. RESOURCES ASSIGNED THIS PERIOD

STRIKE TEAM/TASK FORCE/ RESOURCE DESIGNATOR	LEADER	NUMBER PERSONS		DESCRIPTION	
Simulation City haz mat	Gene Chantler	12		Bomb evidence recovery	
Orange Co. haz mat	Bill Lewis	14		Crime scene unit	
USMC CBIRF	Bob Morrone	20		ALS medical unit	
Johnson Co. EMS medic 4	Matt Streger	4		Haz mat medical support	
TOTAL		50			

7. OPERATIONS

Chemical detection and identification
Decontamination for on-scene personnel

8. SPECIAL OPERATIONS
.

9. MEDICAL UNIT   Johnson Co medic 4

10. FREQUENCIES   Simulation City haz mat channel

Orange Co tac 1

USMC bravo channel

PREPARED BY (RESOURCE UNIT LEADER)	APPROVED BY (PLANNING SECTION CHIEF)	DATE 3/1/00	TIME 07:27

1. BRANCH IV Haz mat	2. DIVISION/GROUP H	DIVISION ASSIGNMENT LIST ICS204
3. INCIDENT NAME Simulation City explosion		4. OPERATIONAL PERIOD DATE 3/1/00 TIME 08 to 20

5. OPERATIONS PERSONNEL

OPERATIONS CHIEF C Lynne   DIV/GROUP SUPERVISOR R Colbert
BRANCH DIRECTOR R McAllister   AIR ATTACK SUPERVISOR NO.

6. RESOURCES ASSIGNED THIS PERIOD

STRIKE TEAM/TASK FORCE/ RESOURCE DESIGNATOR	LEADER	NUMBER PERSONS		DESCRIPTION	
USMC CBIRF	Sgt. Maj. A.S. Maniscalco	22		Chemical detection	
USMC CBIRF	Sgt. R. Christen	18		Patient decontamination	
Johnson Co Sheriff Ofc	Deputy P. Chovan	17		Perimeter/scene control	
TOTAL		57			

7. OPERATIONS

Chemical detection and decontamination in trauma center

8. SPECIAL OPERATIONS
.

9. MEDICAL UNIT   N/A

10. FREQUENCIES   Johnson Co S/O freq 4

                               USMC alpha channel

PREPARED BY (RESOURCE UNIT LEADER)	APPROVED BY (PLANNING SECTION CHIEF)	DATE 3/1/00	TIME 07:27

INCIDENT RADIO COMMUNICATIONS PLAN

1. INCIDENT NAME	2. DATE/TIME PREPARED	3. OPERATIONAL PERIOD (DATE/TIME)
Simulation City explosion	3/1/00 05:50	08:00 to 20:00

4. BASIC RADIO CHANNEL UTILIZATION

BRANCH/SYSTEM/CACHE	CHANNEL	FUNCTION	FREQUENCY/TONE	ASSIGNMENT	REMARKS
Simulation City haz mat	Tac 1	Haz mat teams	152.250	Haz mat branch	Haz mat ops only
Orange Co	Tac 1	Haz mat team	800.500	Haz mat branch	Haz mat ops only
Johnson Co Sheriff	Freq 4	Dispatch channel	457.400 tx 445.120 rx	Haz mat branch	

5. PREPARED BY (COMMUNICATIONS UNIT)

205 ICS 9/86

INCIDENT RADIO COMMUNICATIONS PLAN

1. INCIDENT NAME	2. DATE/TIME PREPARED	3. OPERATIONAL PERIOD (DATE/TIME)
Simulation City explosion	3/1/00 05:50	08:00 to 20:00

4. BASIC RADIO CHANNEL UTILIZATION

BRANCH/SYSTEM/CACHE	CHANNEL	FUNCTION	FREQUENCY/TONE	ASSIGNMENT	REMARKS
USAR	Tac 1	Rescue teams	172.250	Branch I A	
USAR	Tac 2	Rescue teams	177.500	Branch I B	
DMAT	Tac 4	Medical at trauma center	174.200	Branch II C	For internal team com
Simulation City fire	Freq 2	Collapse area tac & command	458.100 tx 453.600 rx	Branch I A	
Simulation City fire	Freq 3	Collapse area tac and command	457.265 tx 453.700 rx	Branch I B	
Simulation City EMS	Med 2	Med dispatch		Branch II C	
Simulation City EMS	Med 4	Med dispatch		Branch II D	
Simulation City PD	Ch 1	Tac and command	457.100 tx 453.100 rx	Branch III E	
Simulation City PD	Ch 2	Tac and command	456.950 tx 454.015 rx	Branch III F	

5. PREPARED BY (COMMUNICATIONS UNIT)

205 ICS 9/86

MEDICAL PLAN	1. INCIDENT NAME Simulation City explosion	2. DATE PREPARED 3/1/00	3. TIME PREPARED 06:30	4. OPERATIONAL PERIOD 08:00 to 20:00

5. INCIDENT MEDICAL AID STATIONS

MEDICAL AID STATIONS	LOCATION CONTACT THRU DISPATCH	PARAMEDICS	
		YES	NO
Trauma Center 3 DMATs with physicians	123 Pratt St	XXX	
Center St branch II	704 Center St	XXX	

6. TRANSPORTATION   A. AMBULANCE SERVICES

NAME	ADDRESS	PHONE	PARAMEDICS	
			YES	NO
Simulation City EMS & mutual aid	City-wide 754 Main Ave	555-2376	XXX	
Johnson Co EMS	County-wide 117 Wilson St	555-2311	XXX	

B. INCIDENT AMBULANCES

NAME	LOCATION	PARAMEDICS	
		YES	NO
Orange Co 62, 64	Trauma center	XXX	
Simulation City EMS 23	Center St	XXX	
Johnson Co medic 4	Haz mat operations – Center St	XXX	

7. HOSPITAL NAME	ADDRESS	TRAVEL TIME (Hr)		PHONE	HELIPAD		BURN CENTER	
		AIR	GRD		YES	NO	YES	NO
Simulation City Trauma Center	123 Pratt St	N/A	1 min	555-2624	XXX			
Baptist Hospital	2766 Mullen Dr	10	30	555-4555	XXX			
Mount Sinai	1749 125th Ave	5	15	555-4002	XXX		X	

8. MEDICAL EMERGENCY PROCEDURES

All pediatric patients to Baptist – burn patients to Mount Sinai
Trauma Center patients admit directly to DMAT tents
All divisions have EMS coverage – requests for additional EMS go to operations section chief

206 ICS 8/78	9. PREPARED BY (MEDICAL UNIT LEADER)	10. REVIEWED BY (SAFETY OFFICER)

AIR OPERATIONS SUMMARY

1. INCIDENT NAME	2. OPERATIONAL PERIOD	3. DISTRIBUTION
Simulation City explosion	08:00 to 20:00	PD air unit, Lifeflight, medical unit leader

4. PERSONNEL AND COMMUNICATIONS

	NAME	AIR/AIR FREQUENCY	AIR/GROUND FREQUENCY
AIR OPER. DIRECTOR	Chief E. O'Neill	122.50	123.75
AIR ATTACK SUPER.			
HELICOPTER COOR.			
AIR TANKER COOR. HELIBASE MANAGER			

5. REMARKS (Spec. Instructions, Safety Notes, Hazards, Priorities)

PD helicopter dedicated to incident site and perimeter

Lifeflight available on-call basis

6. LOCATION/ FUNCTION	7. ASSIGNMENT	8. FIXED WING		9. HELICOPTERS		10. TIME		11. AIRCRAFT ASSIGNED	12. OPERATING BASE
		NO.	TYPE	NO.	TYPE	AVAIL.	COM- MENCE		
Simulation City PD	Perimeter and traffic surveillance			1	4			PD 1	PD hdqtrs
Baptist Hosp	EMS response			1	3			Lifeflight	Baptist Hosp
13. TOTALS				2					

14. AIR OPERATIONS SUPPORT EQUIPMENT

15. PREPARED BY J Hughes (Include Date & Time) 3/1/00 07:01

220 ICS 3/82

NFFS 1351

CHAPTER QUESTIONS

1. Define the front-end concept of planning.
2. What are the key elements of the incident management system and how do they relate to planning?
3. Discuss at least four operational problems related to terrorism/tactical violence incidents.
4. Why is administration important in a long-duration terrorism incident?
5. List at least eight critical equipment needs in a terrorism incident.
6. What are the key steps in the implementation of an EOP?
7. What are the key elements of an EOP?
8. What determines if an IAP is formal or informal?
9. Define an IAP briefing cycle. What are the related steps in the cycle?

Simulation I

Obtain an after-action report for a major event in your community or state. Based on the incident report, develop an IAP based on the event by using copies of blank ICS forms on pages 77–90. Establish IMS sections, and use the branch/division concept.

Simulation II

You have been appointed as emergency manager in your community. Your elation is short-lived when you find out your first assignment is to develop an EOP. Develop a community EOP by observing the following principles:

- Obtain a realistic threat assessment or design a hypothetical threat assessment.
- Identify key operations agencies.
- Determine who will be the incident manager in various types of incidents.
- Identify logistics sources on a local and regional level.
- Identify sources of support personnel and support agencies.

5
The Federal Response Plan

Paul M. Maniscalco
Hank T. Christen

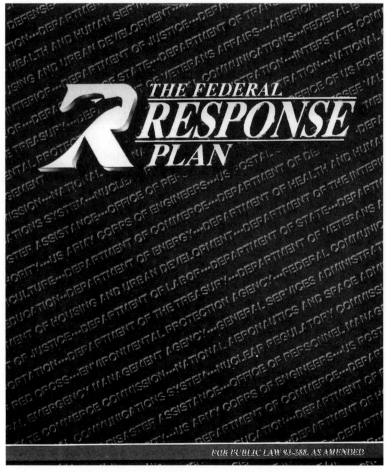

Source: Courtesy Emergency Mangement Agency

Chapter Objectives

After reading this chapter, you will be able to accomplish the following objectives:

1. Discuss Presidential Decision Directive 39 (PDD 39).
2. Understand the policies and assumptions of the Federal Response Plan (FRP).
3. Understand the concept of emergency support functions (ESF).
4. List primary support agencies in the FRP.
5. Have an overview of the Terrorism Incident Annex.
6. Recognize the difference between Emergency Support Functions (ESF) and the Incident Management System (IMS).

Introduction

The federal government, as previously discussed, is slow to respond to local consequence events due to communications delays and long distances. However, the federal government has a major strength; namely, enormous quantities of personnel and logistical assets.

A key document that describes federal support responsibilities is the Federal Response Plan (FRP). This plan addresses several important elements such as:

- Fundamental policies and assumptions.
- Federal resources for response, recovery, and mitigation.
- Twelve Emergency Support Functions (ESF) and their respective primary agencies.
- Linkages to the federal emergency operations plan for specific incidents.

The Scope of the FRP

The FRP applies to major disasters as defined in the Robert T. Stafford Disaster Relief Assistance Act. The act defines a disaster event as a natural catastrophe, fire, flood, explosion, or any instance where the president determines that federal assistance is needed on a local operation.

The FRP is far ranging in the federal government scope. The plan applies to all agencies that have signed the agreement, and to independent agencies that may offer response or support. Under the FRP, the American Red Cross functions as a federal agency in coordinating mass care operations. The plan applies to all states, as well as the District of Columbia, and the territories of Puerto Rico, Guam, Virgin Islands, American Samoa, and the Northern Mariana Islands.

FEMA is responsible for federal operations in the areas of preparedness, planning, management, and disaster assistance functions. FEMA also has the responsibility for federal disaster assistance policy, and as such has the lead role in maintaining and developing the FRP.

History of the FRP

The FRP was originally issued in 1992. The second edition of the FRP was published in 1999. The second edition integrates recovery and mitigation functions into the response structure, and describes relationships to the other emergency operations plans. It also includes four new annexes: community relations, donations management, logistics management, and occupational safety and health.

Lastly, the terrorism incident annex has been added. This annex establishes policy relating to terrorism and consequence management of weapons of mass destruction (WMD). The terrorism incident annex will be discussed later in this chapter.

FRP Organization

FEMA is the primary agency responsible for developing and maintaining the FRP. The Stafford Act, along with Executive Orders 12148 (Federal Emergency Management), and 12656, (Assignments of Emergency Preparedness Responsibilities), delegates the role of coordinating emergency preparedness, planning, management, and disaster assistance functions.

The FRP consists of the following sections:

- Basic plan—policies and concept of operations that guide how the federal government will assist state/local government; summarizes federal planning assumptions, response and recovery actions, and responsibilities.
- Emergency Support Functions (ESF) annexes—missions, policies, concept of operations, and the responsibility of primary and support agencies.
- Recovery Function Annex—planning and concepts of operations that guide assistance operations to help disaster victims and affected local governments to return to normal and minimize future risk; delivery systems include individual assistance, family assistance, business assistance, and assistance to local/state governments.
- Support Annexes—mission and policies of the activities required to conduct overall federal disaster operations; include community relations, congressional affairs, donations management, financial management, logistics management, and occupational safety and health.

- Incident Annexes—the mission and concept of operations in specific events that require an FRP response along with federal plans that implement authorities and functions outside of the Stafford Act; the terrorism incident annex is the first of a series of new incident annexes.

The Terrorism Incident Annex

Presidential Decision Directive 39 (PDD 39), a portion of the U.S. policy on counterterrorism, establishes policy to reduce the nation's vulnerability to terrorism. PDD 39 is the lead document that outlines our nation's policy of aggressively responding to terrorism. This policy includes relief to victims of terrorism (similar to natural disaster relief under the Stafford Act.)

The federal terrorism response provides crisis management and consequence management operations. (As emergency responders our thrust is consequence response.) It must be reemphasized that consequence response is the primary responsibility of local governments, with state governments providing additional resources. The federal government assists the state/local operations.

Policies of the Terrorism Incident Annex

There are responsibilities of lead federal agencies (LFA) in terrorism response efforts. The Department of Justice (DoJ) is the lead agency for crisis management and consequence management operations within the United States. The DoJ has assigned operational response to the Federal Bureau of Investigation (FBI). The FBI is the on-scene manager for the federal government. FEMA is the lead agency for consequence management and response in the United States. FEMA utilizes the FRP to coordinate all assistance to state/local governments.

To ensure chain of command, FEMA supports the DoJ in a major terrorism incident (the FBI is the LFA until the U.S. attorney general transfers the LFA role to FEMA). In a major terrorism event that occurs without warning, FEMA can implement the FRP. This support will be coordinated with the FBI as appropriate.

Planning Assumptions

The terrorism incident annex is based on several important planning assumptions, including:

- No single agency at any level of government (local, state, or federal) has the capability or authority to act unilaterally on the many facets of a threat or act of terrorism.
- A WMD event, especially in a large population center, will overwhelm the capabilities of local/state governments.

- Federal capabilities may also be overwhelmed, especially in a multiple location event.
- Local, state, and federal responders will have overlapping responsibilities. Adequate coordination must be established.
- Hot zone entry may be delayed until safe levels are achieved; responders may encounter secondary devices.
- Operations may involve multiple areas in one state, or incidents across state lines.
- Operations may cross U.S. boundaries requiring Department of State coordination with foreign governments.

Crisis Management

PDD 39 establishes the FBI responsibility for crisis management response. The FBI response is tiered for the following range of incidents:

- Credible threats (written, verbal, or intel based).
- An act of terrorism that exceeds the local FBI capability.
- Confirmation of an explosive device or WMD capable of significant destruction.
- Detonation of an explosive device, WMD, or other catastrophic event that results in substantial injury or death.

Consequence Management

The terrorism annex divides consequence response into pre-release and post-release modes. Critical elements of the pre-release mode include:

- Alerting of appropriate FEMA officials on receipt of initial notification of the FBI of a credible threat.
- Deployment of the domestic emergency support team (DEST) and staff provisions for the Joint Operations Center (JOC); FEMA will staff the JOC consequence management group.
- FEMA will consult (if appropriate) with governors' offices and the White House to determine if federal assistance is required.
- The establishment of an Emergency Support Team (EST). FEMA regions may also activate a Regional Operations Center (ROC).

Post-release functions include:

- Pre-release agencies at a JOC will transition to consequence responsibilities.
- If an incident occurs without warning, FEMA and the FBI will concurrently initiate crisis management and consequence management actions.

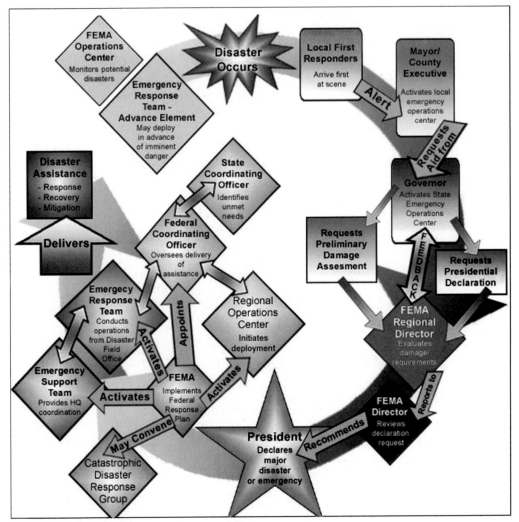

FIGURE 5-1 Federal Response Plan Actions After a Disaster Occurs. *Source:* Courtesy Federal Response Plan

- FEMA will consult with the governor's office and the White House to determine if federal assistance is required (federal resources will be assigned under the Stafford Act).
- The LFA (FBI or FEMA) will initiate a joint information center (JIC) to serve as a focal point for public and media information.
- The senior FEMA official and the FBI on-scene commander will provide resolution of conflicts between crisis management and consequence management agencies.
- The DoJ will advise the White House, establish a JOC, appoint an FBI on-scene commander, and initiate status tracking.

Agency Responsibilities

The responsibilities of lead agencies are specified in PDD 39. The DoJ is the LFA for threats or acts of terrorism that take place in the United States. The Attorney General can transfer the LFA role to FEMA.

FEMA is a critical agency for emergency responders, as it is the lead agency for consequence response. There are several federal agencies that support technical operations. These agencies include:

- Department of Defense—coordinates military operations in the United States with civilian agencies.
- Department of Energy—implements (with the FBI) the Federal Radiological Emergency Response Plan (FRERP), which includes on-site management, radiological monitoring, and development of protective actions.
- Department of Health and Human Services (DHHS)—activates and supports the Medical Services Support Plan. The DHHS plan may include threat assessment, agent identification, epidemiological investigation, decontamination, public health support, medical support, and pharmaceutical support.
- Environmental Protection Agency (EPA)—coordinates with agencies identified in the National Oil and Hazardous Substances Contingency Plan (NCP). EPA responses may include environmental monitoring, decontamination, and long-term site restoration.

Emergency Support Functions

The FRP is based on support functions that are likely to be needed at a state level. Other ESFs are designated to support federal operations such as communications and transportation. Each ESF is headed by a primary agency. The primary agency is selected based on resources, expertise, and capabilities.

Federal support requests are channeled from a local government through an appropriate state coordinating office (usually an emergency operations center). Federal ESFs then coordinate with their counterpart state agencies and downward to local agencies. Federal fire, rescue, and emergency medical agencies that may respond to a local area are integrated with similar local response agencies in the incident management system (IMS) structure.

Military support is also available for major disaster incidents. Military support is based on the premise that the disaster response will not compromise military contingencies. Military support is utilized when other federal resources are not available.

Emergency Support Functions (ESFs)

The FRP is a very detailed reference document. The previous discussion of the terrorism incident annex and the following discussion of ESFs are an overview at best. We recommend that emergency planners obtain a copy of the FRP (9230.1—PL) from FEMA or through the FEMA website at www.fema.gov. It is important to remember that the FRP is a reference guide for support operations and is not an incident management system, nor is it designed to serve as an operations guide or tactical document.

ESF 1—Transportation Annex

Primary agency: Department of Transportation

Secondary agencies: Department of Agriculture
                                   Department of Defense
                                   Department of State
                                   Department of the Treasury
                                   Federal Emergency Management Agency
                                   General Services Administration
                                   Tennessee Valley Authority
                                   U. S. Postal Service

The transportation annex assists local and state governments as well as other federal agencies in the transportation of people, resources, and equipment. Other responsibilities include the operation of field Movement Coordination Centers (MCC) and assessment of the transportation infrastructure. The transportation ESF also provides alternative transportation systems to temporarily replace mass transit systems.

The transportation ESF is based on the following planning assumptions:

- The local/regional transportation system will sustain damage that limits access to the disaster area.
- Transportation requirements will exceed state/local assets.
- Local transportation distribution systems may remain unusable for a long period after a disaster.
- Controls may be needed to coordinate movement of relief supplies in areas of congestion.

ESF 2—Communications Annex

Primary agency: National Communications System

Support agencies: Department of Agriculture
                              Department of Commerce
                              Department of Defense
                              Department of the Interior

Federal Communications Commission
Federal Emergency Management Agency
General Services Administration

The communications annex ensures telecommunications support to state and local governments during a disaster, and is a supplement to the National Plan for Telecommunications Support in Non-Wartime Emergencies (NTSP).

The NTSP is the foundation for planning efforts relating to the use of telecommunications assets in support of the Stafford act. The communications ESF is based on the following planning assumptions:

- Lifesaving efforts are the local focus in a disaster. Local officials will work with telecommunications industries to restore and reconstruct telecommunications facilities as the situation permits.
- Initial damage assessment of telecommunications facilities may be fragmented and inaccurate.
- Mobile and transportable communications equipment may have a delayed deployment into an area because of inclement weather.
- A disaster region may be isolated from the rest of the nation because of communications failures.
- The location of a disaster field office (DFO) will be dependent on the availability of telecommunications facilities.

ESF 3—Public Works and Engineering Annex

Primary agency: Department of Defense, U.S. Army Corps
of Engineers

Support agencies: Department of Agriculture
Department of Commerce
Department of Health and Human Services
Department of the Interior
Department of Labor
Department of Veterans Affairs
Environmental Protection Agency

The public works and engineering annex provides technical evaluation, engineering services, contracting for construction management, contracting for repair of water and wastewater treatment facilities, potable water and ice, emergency power, damage mitigation, and disaster recovery activities. The public works and engineering ESF is based on the following planning assumptions:

- Access to a disaster will be dependent on the reestablishment of ground and water transportation. Debris clearance and road repair will be given top priority to support lifesaving and response operations.

- Early damage assessments will be fragmented and incomplete. Rapid assessment of the disaster area is needed to determine critical response times and work loads.
- Legal clearance will be needed for the proper and safe disposal of disaster debris and materials.
- Large numbers of personnel with construction and engineering skills, along with equipment and materials, will be needed from outside the disaster area.
- Damaged structures will require inspection and reevaluation.

ESF 4—Firefighting Annex

Primary agency: Department of Agriculture, U.S. Forest Service

Support agencies: Department of Commerce
Department of Defense
Department of the Interior
Environmental Protection Agency
Federal Emergency Management Agency

The firefighting annex detects and suppresses urban and wildland fires resulting from disasters or major emergencies. It provides personnel, equipment, and support for local and state fire suppression operations. The firefighting ESF is based on the following planning assumptions:

- Many urban and wildland fires may result from major disasters.
- There may be widespread wildland fires elsewhere in the United States along with disaster related fires. The result may be a scarcity of firefighting resources.
- Radio communications may be necessary resulting in a need to order radio starter systems from the National Interagency Coordinating Center (NICC).
- Aerial delivery of fire retardant may be needed in disaster areas with water shortages.
- Local and state agencies may need ICS support from wildland fire overhead teams.

ESF 5—Information and Planning Annex

Primary agency: Federal Emergency Management Agency

Support agencies: Department of Agriculture
Department of Commerce
Department of Defense
Department of Education

Department of Energy
Department of Health and Human Services
Department of the Interior
Department of Justice
Department of Transportation
Department of the Treasury
American Red Cross
Environmental Protection Agency
General Services Administration
National Aeronautics and Space Administration
National Communications System
Nuclear Regulatory Commission
Small Business Administration

Support organizations: Civil Air Patrol
Voluntary organizations

The information and planning ESF collects, analyzes, disseminates, and processes information relating to a potential disaster or emergency event to facilitate the actions of the federal government on assisting affected states. ESF 5 becomes the information and planning section of the ROC, or the emergency response team at the disaster field office. Similar support can also be provided at FEMA headquarters.

The information and planning ESF is based on the following planning assumptions:

- In a disaster there is a need for central information collecting points where situation information can be compiled and prepared for decision makers.
- Regional data, using state and local government sources and federal input, becomes the primary information source for the ESF.
- Officials involved in disaster response operations require information on ongoing situations.
- Field observers or assessment personnel may need to be deployed to obtain information about resource needs for disaster victims.

ESF 6—Mass Care Annex

Primary agency: American Red Cross

Support agencies: Department of Agriculture
Department of Defense
Department of Health and Human Services
Department of Housing and Urban Development

Department of Veterans Affairs
Federal Emergency Management Agency
General Services Administration
U. S. Postal Service

The mass care ESF coordinates federal assistance in support of state and local operations for mass care needs of disaster victims. Mass care support includes feeding, shelter, and emergency first aid to disaster victims. Bulk distribution of emergency relief supplies and the collection of disaster welfare information are also included in this annex.

The mass care annex is based on the following planning assumptions:

- A major disaster event produces mass casualties and damage.
- Mass care facilities receive priority for structural inspections to ensure occupant safety.
- Mass care operations and related logistical support are priorities of federal agencies.

ESF 7—Resource Support Annex

Primary agency: General Services Administration

Support agencies: Department of Agriculture
Department of Commerce
Department of Defense
Department of Energy
Department of Labor
Department of Transportation
Department of the Treasury
Department of Veterans Affairs
Federal Emergency Management Agency
National Aeronautics and Space Administration
National Communications System
Office of Personnel Management

The resource support ESF provides operational assistance in a potential or declared disaster or emergency. The scope of the resource support annex includes relief supplies, office equipment and supplies, space, telecommunications, security services, federal law enforcement liaison, and personnel to support response activities.

The response support ESF is based on the following planning assumptions:

- Federal support requirements may need to be met by agencies and resources outside the disaster area.
- Transport of resources may require a mobilization center. National Guard or military bases will be available for use.
- Resource support will be required for immediate lifesaving and life-support operations.

ESF 8—Health and Medical Services Annex

Primary agency: Department of Health and Human Services

Support agencies: Department of Agriculture
Department of Defense
Department of Energy
Department of Justice
Department of Transportation
Department of Veterans Affairs
Agency for International Development
American Red Cross
EPA
FEMA
General Services Administration
National Communications System
U.S. Postal Service

The health and medical services annex coordinates federal assistance to state and local resources for public health and medical operations during a disaster or emergency. The Department of Health and Human Services, through the assistant secretary of health (ASH), directs the efforts of this annex. Resources are deployed when state/local assets are overwhelmed.

The health and medical services ESF is based on the following planning assumptions:

- Medical resources for casualty clearing and medical treatment within a disaster area will be overwhelmed.
- Additional federal resources and supplies will be drastically needed.
- A major WMD release or damage to sewer lines and water systems will result in environmental toxicity and public health hazards.
- A major event that causes mass death and injuries will overwhelm state and local mental health resources.
- Loss of power and mass sheltering increase the potential for disease and injury.
- The local medical infrastructure may be damaged or destroyed, necessitating assessment and restoration.

ESF 9—Urban Search and Rescue (USAR) Annex

Primary agency: Federal Emergency Management Agency

Support agencies: Department of Agriculture
Department of Defense
Department of Health and Human Services
Department of Justice
Department of Labor
Agency for International Development
National Aeronautics and Space Administration

The urban search and rescue annex provides for the deployment and operation of specialized lifesaving assistance to state and local governments. USAR activities include locating, extricating, and on-site medical services to victims trapped in collapsed structures.

The urban search and rescue ESF is based on the following planning assumptions:

- State and local rescue organizations will be overwhelmed in a disaster.
- Local rescue assets may lack specialized training and equipment.
- Convergent responders and volunteers will require direction and coordination.

ESF 10—Hazardous Materials Annex

Primary agency: Environmental Protection Agency

Support agencies: U. S. Coast Guard
Department of Agriculture
Department of Commerce
Department of Defense
Department of Energy
Department of Health and Human Services
Department of the Interior
Department of Justice
Department of Labor
Department of State
Department of Transportation
Nuclear Regulatory Commission

The hazardous materials annex provides federal support to state and local governments for a hazardous materials discharge or potential discharge.

FEMA determines, with consultation from the U. S. Coast Guard and EPA, if activation of the haz mat annex is needed. In a terrorism incident, ESF 10 is activated to provide assistance during crisis and consequence management phases.

The hazardous materials ESF is based on the following planning assumptions:

- State and local agencies will be overwhelmed by the requirements of assessment, mitigation, control, and cleanup of hazardous materials spills.
- A disaster will produce numerous coastal and inland hazardous materials incidents.
- Industrial and chemical facilities in the disaster area will require assessment and monitoring.
- WMD events will require additional coordination procedures and specialized response actions. Radiological events will be consistent with the FRERP.

ESF 11—Food Annex

Primary agency: Department of Agriculture, Food and Nutrition
                    Service

Support agencies: Department of Defense
                    Department of Health and Human Services
                    American Red Cross
                    Environmental Protection Agency
                    Federal Emergency Management Agency
                    General Services Administration

The food annex identifies, secures, and arranges for the transportation of food to disaster areas. These activities include the determination of needs, obtaining of supplies, and arrangements for transportation. This annex also authorizes disaster food stamp assistance.

The food ESF is based on the following planning assumptions:

- In a disaster, food processing facilities and the distribution infrastructure will be damaged.
- Water supplies will be unusable requiring juices and potable water.
- Disruption of energy sources will affect cold storage and freezer equipment.
- Schools and institutions on the disaster fringe will be capable of supplying large food quantities for up to three days.

ESF 12—Energy Annex

Primary agency: Department of Energy

Support agencies: Department of Agriculture
                   Department of Defense
                   Department of the Interior
                   Department of State
                   Department of Transportation
                   National Communications Commission
                   Tennessee Valley Authority

The energy annex assists in the restoration of the nation's energy system following a major event or disaster. ESF 12 also gathers, assesses, and shares information on energy system damage and coordinates with suppliers and energy officials in energy system restoration. Energy includes production, refining, transporting, generating, conserving, building, and maintaining energy systems.

The energy ESF is based on the following planning assumptions:

- A disaster may produce widespread power failures.
- There may be damage to the telecommunications and transportation infrastructures.
- Loss of power may result in delays in the production, refining, and delivery of petroleum products.

ESFs Compared with the IMS

The FRP and the related ESFs are designed to serve as support functions at the federal level (the key word is "support"). The ESFs were never intended to be a command system or IMS.

In some state and local EOCs the ESFs are being integrated into the command system. This plan is workable if the ESFs are integrated into the IMS structure and are assigned to the management staff or the operations, logistics, planning, or administration/finance sections. For example, ESF 8, health and medical services, is an operations function and assigned to the operations section chief; ESF 2, communications, is assigned to the service branch of the logistics section.

A system of ESFs without an IMS has several deficiencies including:

1. The FRP identifies all support functions as equals. IMS prioritizes resource allocation (support functions) through the incident manager and the operations, logistics, and planning sections.

2. There is no provision in the FRP for dividing support functions. In the IMS, there are methods of assigning assets through the principles of branching, divisions, and groups.
3. There is no provision in the FRP for information flow between the support functions. IMS provides for information flow through chain of command, and unity of command and communications.
4. The FRP makes no provision for tracking resource assignments or demobilizing support agencies. In the IMS there is a situation unit, resource unit, documentation unit, and demobilization unit for effective resource tracking and demobilization operations.

CHAPTER SUMMARY

The federal government has large quantities of resources and personnel that can support a state or local community after a terrorism event. The principal document that defines and outlines federal support procedures is the Federal Response Plan (FRP). The FRP identifies twelve Emergency Support Functions (ESF). The 1999 edition includes the four new support annexes of community relations, donations management, logistics management, and occupational safety and health.

Presidential Decision Directive 39 (PDD 39) is the U.S. policy on counter terrorism. This document defines crisis and consequence management and includes a policy on relief for victims of terrorism.

The policies in PDD 39 are detailed in the terrorism incident annex. This annex identifies the Department of Justice as the lead federal agency for crisis and consequence management. The terrorism incident annex is based on the primary assumption that no single agency at any level of government can respond to the many facets of a major terrorism incident.

The emergency support functions in the FRP are:

1. Transportation annex
2. Communications annex
3. Public works and engineering annex
4. Firefighting annex
5. Information and planning annex
6. Mass care annex
7. Resource support annex
8. Health and medical services annex
9. Urban search and rescue annex
10. Hazardous materials annex
11. Food annex
12. Energy annex

ESFs provide support operations in a disaster or terrorism incident, and although they are not intended to be a management system, they should be integrated into the IMS. The IMS makes effective use of support elements through the principles of chain of command, unity of command, planning, situation status, resource status, and demobilization.

CHAPTER QUESTIONS

1. Discuss the major provisions of PDD 39.
2. What is the lead federal agency for crisis management? Consequence management?
3. List and describe three of the five sections of the FRP.
4. What are the major planning assumptions of the terrorism incident annex?
5. What are emergency support functions? List at least five ESFs and the primary agency for each.
6. What is the difference between the ESFs and the IMS? Why is the FRP an ineffective system for managing a disaster or terrorism incident?

Simulation

List the twelve ESFs at the federal level. Select a lead agency and several support agencies (if appropriate) from your community or region. Outline how these agencies can be integrated into a local comprehensive emergency management plan and IMS.

6
Weapons of Mass Effect: Chemical Terrorism and Warfare Agents

Fred Sidell, M.D.
Paul M. Maniscalco
Hank T. Christen

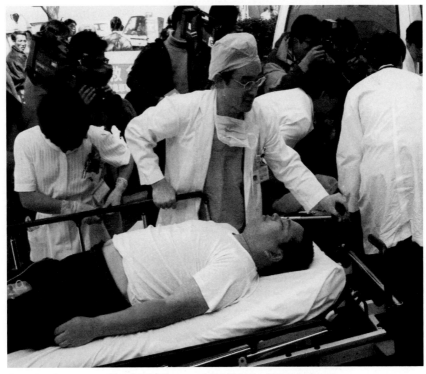

Source: Courtesy Chiaki Tsukumo/AP/Wide World Photos

Chapter Objectives

After reading this chapter, you will be able to accomplish the following objectives:

1. Understand the importance of community preparedness for a chemical attack.
2. Recognize the characteristics of nerve agents.
3. Outline patient treatment procedures for nerve agent exposure.
4. Outline treatment procedures for cyanide exposure.
5. Define vesicants and list the symptoms for exposure to specific vesicants.
6. Recognize the symptoms of exposure to pulmonary agents.
7. Define the common riot control agents.
8. Recognize the importance of triage in mass patient incidents.

Introduction

Chemical warfare agents are chemical substances that were developed for use on the battlefield to kill or injure. For 79 years, from their first use in 1915 in World War I until 1994, the intentional use of these chemicals to kill or injure was limited to battlefield use. In June 1994 the religious cult Aum Shinrikyo disseminated one of these agents, sarin, throughout an apartment complex in Matsumoto, Japan, with the intent of causing widespread harm or death to people. In March 1995 the same group released sarin on the Tokyo subways causing injuries in over 1,000 people and death in 12. Terrorists had a new weapon, a chemical weapon.

Rogue individuals and organizations continue to possess or have access to these weapons. There are probably several dozen countries with the capacity to manufacture these chemicals, and some of these countries are known to be sponsors of terrorist groups and acts of terrorism. In addition, instructions for the synthesis of these agents are widely available to terrorist groups and rogue individuals in books, on the Internet, and in other places such as militia newsletters.

Some very toxic chemicals are regularly manufactured in large amounts in this country and are transported daily on our highways and railways. Chemicals such as cyanide, phosgene, and chlorine, all of which were once military agents, are widely used in large amounts. Many commonly used pesticides are very similar to nerve agents.

With the ever-increasing number of toxic chemicals in the world and the existence of rogue organizations that are willing to use them to further their causes, it is essential that communities and emergency response organizations prepare to confront the challenges presented by a chemical terrorist event.

Critical Factor: The emergency response community must be prepared for a consequence response to a chemical attack.

Terrorist incidents involving military chemical agents are, from an individual patient treatment foundation level, not much different from a regular haz mat incident; patient care is the same. The critical differences are: (1) it is a deliberate release, (2) most likely it is a high impact/high yield incident with numerous patients, and (3) the identity of the substance released will be difficult to ascertain with standard haz mat equipment.

Overall, there are differences between an accidental spill of a toxic chemical and a deliberate release of the same chemical. Probably the most important difference is the fear and anxiety generated in the community and among the emergency responders who must deal with unknown factors. This fear and anxiety may be present in possible victims, who do not know and who cannot be immediately reassured that they have not been harmed, or it might be present in responders who fear a secondary hostile device at the scene designed to injure them. Many more agencies become involved in a deliberate release incident, including emergency medical service, fire/rescue, emergency management, law enforcement, and all levels of government, which will converge on the scene in an effort to assist. The site of a deliberate incident is a crime scene, a chemical hot zone, a biological hot zone, and in most cases a "high impact/high yield" multiple casualty incident. These events result in EMS and the medical community playing a significant lead role in consequence management.

Delivery/Dissemination

Chemical agents are disseminated in many ways. Bombs, rockets, mines, and other explosive devices are used by the military. When these explode some of the agent remains as liquid, some immediately evaporates to form vapor, and some will exist as small droplets of the agent suspended in air, or an aerosol. These small droplets eventually evaporate and become a vapor. The result is a hazard from the liquid agent, both as the original liquid and as the aerosol droplets (which neither remain long nor travel far) and a hazard from agent vapor. Agents can also be sprayed from airplanes during battlefield use, as some agents were during the Vietnam war.

Liquid chemical agents might be employed with an explosive device by nonmilitary users. The user would have to make the device, which is not without hazard, and then detonate it in the right place at the right time. Other means of disseminating the agent are more likely. Insecticides are sprayed from airplanes and helicopters in both crop dusting or for mosquito eradication. This is an obvious and an expensive way to disseminate an agent, but it is effective for spreading it over a large area providing the weather and winds are favorable. Vehicle-mounted spray tanks can be driven through the streets of a target area to disseminate an agent. (In Matsumoto the agent was spread from the back of a vehicle in which a container of agent was heated (to help it to evaporate), and the vapor was blown through the street by a fan.) Indoor areas might be attacked by putting an agent in the air system, and rooms could be thoroughly and quickly contaminated by the use of a common aerosol spray can. There are other methods of disseminating liquids. Remember that it does not take an explosive device to disseminate chemical agents.

Critical Factor: There are multiple methods of disseminating chemical agents over a wide area.

FIGURE 6-1   Personal protection equipment use is critical to the health and safety of the emergency responder. *Source:* Courtesy P. Maniscalco

Nerve Agents

Nerve agents are toxic materials that produce injury and death within seconds to minutes. The signs and symptoms caused by nerve agent vapor are characteristic of the agents and are not difficult to recognize with a high index of suspicion. Very good antidotes are available that will save lives and reduce injury if administered in time.

Nerve agents are a group of chemicals similar to, but more toxic than, commonly used insecticides such as Malathion. These substances were developed in pre-World War II Germany for wartime use, but they were not used in that war. They were used in the Iran–Iraq war and they were used by the religious cult Aum Shinrikyo in Japan on a number of occasions. The first use injured about 300 people and killed seven in Matsumoto in June 1994, and the second, more publicized incident, killed 12 and injured over 1,000 in the Tokyo subways in March 1995. The common nerve agents (with military designations) are Tabun (GA), Sarin (GB), Soman (GD), GF, and VX (GF and VX have no common names).

The nerve agents are liquids (not "nerve gases") that freeze at temperatures below 0°F and boil at temperatures above 200°F. Sarin (GB), the most volatile, evaporates at about the rate of water, and VX, the least volatile, is similar to light motor oil in its rate of evaporation. The rates of evaporation of the others lie in between. In the Tokyo subway attack, sarin leaked out of plastic bags and evaporated. Serious injury was minimized because the rate of evaporation of sarin is not rapid so the amount of vapor formed was not large. If the sarin had evaporated more rapidly (e.g., like gasoline or ether), much more vapor would have been present, and more serious injury in more people would have occurred.

Nerve agents produce physiological effects by interfering with the transmission between a nerve and the organ it innervates or stimulates, with the end result being excess stimulation of the organ. They do not act on nerves. Normally an electrical impulse travels down a nerve, but the impulse does not cross the small synaptic gap between the nerve and the organ. At the end of the nerve the electrical impulse causes the release of a chemical messenger, a neurotransmitter, which travels across the gap to stimulate the organ. The organ may be an exocrine gland, a smooth muscle, a skeletal muscle, or another nerve. The organ responds to the stimulus by secreting, by contracting, or by transmitting another message down a nerve. After the neurotransmitter stimulates the organ, it is immediately destroyed by an enzyme so that it cannot stimulate the organ again. Nerve agents inhibit or block the activity of this enzyme so that it cannot destroy the neurotransmitter or chemical messenger. As a result the neurotransmitter accumulates and continues to stimulate the organ. If the organ is a gland it continues to secrete, if it is a muscle it continues to contract, or if it is a nerve it continues to transmit impulses. There is hyperactivity throughout the body.

Nerve Agents

- Block the activity of an enzyme
- Cause too much neurotransmitter
- Cause too much activity in many organs, glands, muscles, skeletal muscles, smooth muscles (in internal organs), and other nerves

---

In the presence of nerve agent poisoning, exocrine glands secrete excessively. These glands include the tear glands (tearing), the nasal glands (rhinorrhea or runny nose), the salivary glands (hypersalivation), and the sweat glands (sweating). In addition, the glands in the airways (bronchorrhea) and in the gastrointestinal tract secrete excessively in the presence of nerve agents.

The clinically important smooth muscles that respond are those in the eye (to produce small pupils, or miosis), in the airways (to cause constriction), and in the gastrointestinal tract (to cause vomiting, diarrhea, and abdominal cramping).

Skeletal muscles respond initially with movement of muscle fibers (fasciculations, which look like rippling under the skin), then twitching of large muscles, and finally weakness and a flaccid paralysis as the muscles tire.

Nerve Agent Effects

Glands

- lachrymal (tearing)
- nose
- mouth (salivation)
- sweat
- bronchial (in airways)
- gastrointestinal

Skeletal muscles

- fasciculations, twitching, weakness, paralysis

Smooth muscles

- airways (constriction)
- gastrointestinal (cramps, vomiting, diarrhea)

Central nervous system

- loss of consciousness
- convulsions
- cessation of breathing

Among the nerve-to-nerve effects are stimulation of autonomic ganglia to produce adrenergic effects such as hypertension (high blood pressure) and tachycardia (rapid heart rate). The exact mechanisms in the central nervous system (CNS) are less well defined, but the result is loss of consciousness, seizures, cessation of breathing (apnea) because of depression of the respiratory center, and finally death. Early effects also include stimulation of the vagus nerve, which causes slowing of the heart (bradycardia).

The effects that occur depend on the route of exposure and the amount of exposure. The initial effects from a small amount of vapor are not the same as those from a small droplet on the skin, and the initial effects from a small amount of vapor are not the same as those from a large amount of vapor. Exposure to nerve agent vapor produces effects within seconds of contact. These effects will continue to worsen as long as the patient is in the vapor atmosphere, but will not worsen significantly after the patient is removed from the atmosphere.

Exposure to a small concentration of vapor will cause effects in the sensitive organs of the face that come into direct contact with the vapor, the eyes, the nose, and the mouth and lower airways. Miosis (small pupils) is the most common sign of exposure to nerve agent vapor. Reddened, watery eyes may accompany the small pupils, and the patient may complain of blurred and/or dim vision, a headache, and nausea and vomiting (from reflex mechanisms). Rhinorrhea (runny nose) is also common, and after a severe exposure the secretions might be quite copious. Increased salivation may be present. Agent contact with the airways will cause constriction of the airways and secretions from the glands in the airways. The patient will complain of shortness of breath (dyspnea), which depending on the amount of agent inhaled may be mild and tolerable, or may be very severe. These effects will begin within seconds after contact with the agent. They will increase in severity while the patient is in the vapor, but will maximize within minutes after the patient leaves the vapor.

Sudden exposure to a large concentration of vapor, or continuing exposure to a small amount, will cause loss of consciousness, seizures, cessation of seizures with cessation of breathing and flaccid paralysis, and death. After exposure to a large concentration, loss of consciousness occurs within seconds, and effects progress rapidly to cessation of breathing within 5 to 10 minutes.

FIGURE 6-2   Rescuers operating in Level A suits will find it difficult to remove patients from the hot zone due to physical limitations and stress from suits. *Source:* Courtesy P. Maniscalco

---

Vapor Exposure

Small concentration

- miosis (red eyes, pain, blurring, nausea)
- runny nose
- shortness of breath
- Effects start within seconds of contact.

Large concentration

- loss of consciousness
- convulsions
- cessation of breathing
- flaccid paralysis
- Effects start within seconds of contact.

A very small sublethal droplet of agent on the skin causes sweating and muscular fasciculations in the area of the droplet. These may begin as long as 18 hours after agent contact with the skin and generally will not be noticed by either the patient or medical personnel. A slightly larger, but still sublethal droplet will cause those effects and later cause gastrointestinal effects, such as nausea, vomiting, diarrhea, and cramps. The onset of these are also delayed and may start as long as 18 hours after exposure. A lethal-sized droplet causes effects much sooner, usually within 30 minutes of contact. Without any preliminary signs there will be a sudden loss of consciousness and seizures followed within minutes by cessation of breathing, flaccid paralysis, and death.

Effects from skin contact with a liquid droplet will occur even though the droplet was removed or decontaminated within minutes after contact. Rapid decontamination will decrease the illness but will not prevent it. Nerve agent liquid on skin will cause effects that begin many minutes to many hours after initial contact. After the effects begin they may worsen because of continued absorption of agent through the skin.

Liquid on Skin

Very small droplet

- sweating, fasciculations
- can start as long as 18 hours after contact

Small droplet

- vomiting, diarrhea
- can start as long as 18 hours after contact

Lethal-sized droplet or larger

- loss of consciousness
- convulsions
- cessation of breathing
- flaccid paralysis
- usually starts without warning within 30 minutes

Management of a nerve agent casualty consists of removing the agent from the patient (decontamination) or the patient from the agent, administration of antidotes, and ventilation if needed. EMS providers must have proper personal protective equipment (PPE) during these operations.

For the antidotes to be effective, the patient must be removed from the contaminated area and/or the agent must be removed from the patient's skin. Although the antidotes are quite effective they cannot overcome the effects of the agent while the patient is continuing to breath the agent or while the agent is still being absorbed through the skin. Skin decontamination will not remove agent that has been absorbed into the skin but not yet through the skin, and effects may start as long as several hours after skin decontamination.

Removing the patient from the area of contamination or the vapor area should be rather simple in a normal haz mat incident, but the complexities of a mass casualty terrorism event present some unique challenges. If the agent was released inside a building or other enclosed space moving the patients outside should suffice. If the agent was released outside, patients should be removed and triaged far upwind.

Removal of the agent from the skin must be done as early as possible. It is unlikely that you will see a living patient with visible amounts of nerve agent on his skin. However, if this occurs, remove it (the substance) as quickly as possible. Flushing with large amounts of water or wiping it off with dirt or any other convenient substance will help. If clothing is wet suggesting agent exposure, remove the clothing as quickly as possible. This should be done in the hot zone before the patient reaches the decontamination site in the warm zone. Although agent is removed from the surface of the skin, the agent that has already penetrated into the skin cannot be removed, and absorption will continue; the patient may worsen despite antidotes.

Critical Factor: Treatment for nerve agent exposure involves decontamination, administration of antidotes, and ventilation.

The antidotes for nerve agent poisoning are atropine and an oxime, 2-pyridoxime chloride or 2-PAMCl (Protopam). They act by different mechanisms. Atropine blocks the excess neurotransmitter and protects the site on the organ that the neurotransmitter stimulates. As a result the glands dry and the smooth muscles stop contracting (such as those in the airways and gastrointestinal tract). However, atropine has little effect on the skeletal muscles, and these muscles may continue to twitch despite an adequate dose of atropine.

The initial dose of atropine is 2 mg to 6 mg (page 123). This dose might seem high to those accustomed to administering it for cardiac or other purposes, but it is the amount necessary to overcome a total-body excess of the neurotransmitter. After the initial dose, a dose of 2 mg should be administered every 5 to 10 minutes until (1) the secretions have diminished considerably, and (2) breathing has improved or airway resistance has decreased (if the patient is being ventilated). Atropine can be administered intramuscularly (IM), intravenously (IV), or endotracheally (ET). Atropine administered IV to animals hypoxic from nerve agent poisoning has caused ventricular fibrillation, so good advice is not to administer it by this route until ventilation has begun.

The oxime, 2-PAMCl, attacks the complex of the agent bound to the enzyme and removes the agent from the enzyme. As a result, the enzyme can resume its normal function of destroying the neurotransmitter. Despite the fact that this drug sounds like a very effective antidote, it does not reverse the effects seen clinically in the glands and smooth muscle. It does reduce the skeletal muscle twitching and weakness. It is almost totally ineffective when used against poisoning from one nerve agent, soman (GD), but it is unlikely that identification of the agent will be made before the initial therapy, and use of the oxime in the usual doses will do no harm.

The initial dose of 2-PAMCl is 1 gram given slowly (20 minutes or longer) in an IV drip. More rapid administration will cause hypertension (which can transiently be reversed by phentolamine). 2-PAMCl should not be titrated with the patient's condition as atropine is, but should be administered at hourly intervals for a total of three doses.

A third drug, diazepam (Valium) or a similar anticonvulsant, might be used for prolonged convulsions.

The military has a device called the MARK I with two spring-powered injectors, one containing atropine (2 mg) and the other 2-PAMCl (600 mg).

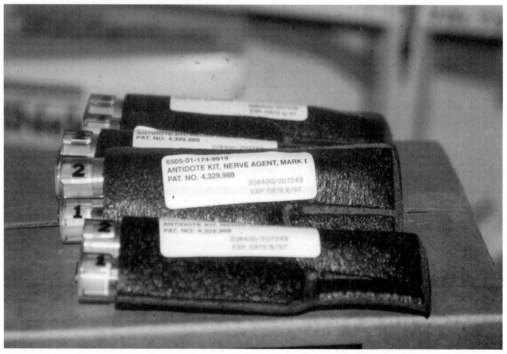

FIGURE 6-3   MARK 1 Auto-Injector Kit. *Source:* Courtesy P. Maniscalco

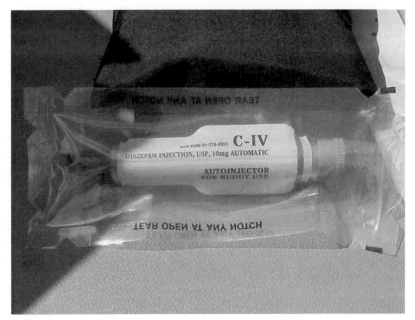

FIGURE 6-4   Valium Autoinjector. *Source:* Courtesy P. Maniscalco

This is a very effective and fast way to administer the antidotes, and use of this causes the drugs to be absorbed faster.

When treating an unconscious patient severely affected by nerve agent poisoning, gasping for air or not breathing, seizing or postical, take care of the airway, breathing, and circulation first. When an airway is inserted and ventilation is attempted in a severe nerve agent patient, the airway resistance will be so great that most devices used for ventilation will not be effective, making ventilation impossible. It might be best to administer (IM) the antidotes first. This ensures that some air will be moved when ventilation is attempted.

The following describes patients that might be seen and the appropriate therapy for each.

A patient exposed to vapor only, who is out of the vapor environment when first seen and is walking and talking, might be relatively asymptomatic or may be very uncomfortable from shortness of breath, but generally is in no danger of loss of life. There may be miosis with red, watery eyes, rhinorrhea, a headache or eye pain, nausea and vomiting, and shortness of breath with auscultatory sounds of airway constriction and secretions. Atropine (2 mg, IM or IV) will reduce or eliminate the shortness of breath and most of the rhinorrhea, but not the eye effects (miosis, pain), or the nausea and vomiting. 2-PAMCl (1 gram, slow IV drip) should be started.

Initial Antidote Use

Vapor exposure

- Miosis and/or runny nose—no antidotes unless eye pain is severe (eye drops)
- Shortness of breath—2 or 4 mg of atropine depending on severity; 2-PAMCl by slow drip
- Unconscious, convulsions, severe breathing difficulty; moderate to severe effects in two or more systems—6 mg of atropine IM; 2-PAMCl by slow drip; ventilation

Liquid on skin

- Local sweating, fasciculations—2 mg of atropine; 2-PAMCl by slow drip
- Vomiting, diarrhea—2 mg of atropine; 2-PAMCl by slow drip
- Unconscious, convulsing, severe breathing difficulty; moderate to severe effects in two or more systems—6 mg of atropine IM; 2-PAMCl by slow drip; ventilation

In all cases, follow with 2 mg of atropine every 5 to 10 minutes until improvement occurs.

Systemic atropine (IM, IV, and ET) has almost no effect on the eyes unless large amounts are administered. If eye pain/headache or nausea and vomiting are severe, these are relieved by topical application of atropine or homatropine eye ointment. These medications will cause severe blurring of vision for about 24 hours, and it is best not to administer them unless the pain or vomiting is severe. The slight reduction in vision (dimness, slight blurring) caused by the agent is less than that caused by the medications. Miosis by itself without pain or nausea, should not be treated.

If dyspnea is more than moderate and if the patient is still capable of walking and talking, the initial dose of atropine should be 4 mg. Whether the initial amount is 2 mg or 4 mg, an additional 2 mg should be administered in 5 to 10 minutes if there is no improvement in the patient's condition. More should be given at similar intervals if necessary, but in most instances the initial 2 mg will reduce the symptoms.

A more severely affected patient will be unable to walk or talk. He will be unconsciousness with severe breathing difficulties or not breathing, perhaps convulsing or postictal with copious secretions and muscular twitching. A severely affected patient may also be one who has moderate or severe signs in two or more organ systems (respiratory, gastrointestinal, muscular, and central nervous systems). Eyes and nose are not considered in this evaluation. This patient should initially be given 6 mg of atropine (IM, not IV), and an IV drip containing 1 gram of the oxime should be started. Ventilation begins after the antidotes are administered. Diazepam or a similar anticonvulsant should be administered. Atropine should be continued at 5 to 10 minute intervals until there is improvement.

A small liquid droplet on the skin will cause sweating and fasciculation at the site, and if this is noted the patient should receive atropine (2 mg, IM) and 2-PAMCl (1 gram in a slow IV drip). A slightly larger droplet initially will cause gastrointestinal effects (nausea, vomiting, diarrhea, cramps), and a patient with these symptoms should receive the same drugs in the same amounts. Either of these patients might worsen, and atropine should be continued at intervals. The onset of these effects may be as long as 18 hours after contact with the agent. Any patient suspected of contacting liquid agent should be kept under observation for 18 hours.

A large, lethal-sized droplet of agent will cause sudden loss of consciousness followed by seizures, cessation of breathing, flaccid paralysis, and death. These effects begin within 30 minutes of contact with the agent, and there are usually no preliminary effects before the loss of consciousness. Management is the same as for severe vapor exposure, with early decontamination if therapy is to be successful.

You can save a patient with a heartbeat by timely and adequate therapy. Occasionally an arrested patient can also be saved. One patient from the Tokyo subway incident had no heartbeat when he was taken into the hospital, but he was adequately treated and he walked out of the hospital several days later.

Cyanide

Cyanide, like the nerve agents, can cause serious illness and death within minutes. Cyanide was not successful as a warfare agent in World War I for several reasons: (1) it is very volatile and tended to evaporate and be blown away by a breeze, (2) it is lighter than air and will not stay close to the ground where it can do damage, and (3) the dose to cause effects is relatively large and, unlike other agents, it causes few effects at lower doses.

Some forms of cyanide are gases under temperate conditions (hydrogen cyanide, cyanogen chloride), and other forms are solids (sodium, potassium, or calcium cyanide). Hundreds of thousands of tons of cyanide are manufactured, shipped, and used in this country annually. It is used in the manufacture of certain synthetic products, paper, and textiles, in tanning, in ore extraction, in printing, and in photography. It is in the seeds of some foods and is in the cassava plant, a staple in certain parts of the world. It is produced when synthetic materials (e.g., plastics) burn. Cyanide has been associated with killing. For centuries it has been used for assassinations and is used in the gas chamber for executions. People sometimes ingest cyanide with suicidal intent. It was taken by the followers of the Reverend Jim Jones for suicide and was illicitly placed in Tylenol bottles in the Chicago area years ago.

The human body has a means of detoxifying or neutralizing small amounts of cyanide, and this is very effective until the system is overwhelmed. The body combines cyanide with a form of sulfur, and the nontoxic product is excreted. When the body runs out of sulfur, effects appear.

Cyanide causes biological effects by combining with an enzyme that is in cells, and stopping or inhibiting its activity. This enzyme normally metabolizes oxygen in the cell so that the cell can function. When cyanide stops the activity of this enzyme, the cell cannot function and dies. There is plenty of oxygen available in the blood, but the cell cannot use it so it does not take it from the blood.

The effects of exposure to a small concentration of vapor or the initial effects from drinking cyanide are relatively nonspecific. They include a brief period of rapid and deep breathing, feelings of anxiety or apprehension, agitation, dizziness, a feeling of weakness, nausea with or without vomiting, and muscular trembling. As more cyanide is absorbed, consciousness is lost, respiration decreases in rate and frequency, and seizures, cessation of breathing, and disturbances in heart rate and rhythm follow. After inhalation of a high concentration of vapor, seizures can occur within 30 seconds, and cessation of breathing and disturbances of cardiac rhythm follow. Death occurs in 6 to 10 minutes after exposure.

Cyanide—Large Amount by Inhalation

Hyperventilation: 15 seconds
Convulsions: 30 seconds
Cessation of breathing: 3–5 minutes
Cessation of heartbeat: 6–10 minutes

Cyanide—Management

Amyl nitrite perle
Sodium nitrite IV (10 ml; 300 mg)
Sodium thiosulfate IV (50 ml; 12.5 gm)

Ventilation with oxygen
Correction of acidosis

Cyanogen chloride causes the effects of cyanide as listed above. However, it is very irritating (similar to the riot control agents), and will produce burning of the eyes, the nose, and airways. It has a pungent odor.

There are few findings on physical examination. The skin is said to be "cherry red" (because of the red, oxygenated venous blood), but this is not always present. The pupils are normal in size or slightly large, secretions are relatively normal, and there are no muscular fasciculations, all of which serve to distinguish cyanide poisoning from nerve agent poisoning.

In the laboratory, cyanide can be measured in blood. Also, there will be more than the normal amount of oxygen in venous blood, and there may be a metabolic acidosis. Management consists of removing the patient from the contaminated atmosphere (or by removing the poison from the patient), and administering antidotes and oxygen.

There are two antidotes that should be administered sequentially. The first, a nitrite, causes the normal hemoglobin in the red blood cells to change to methemoglobin. The methemoglobin thus formed attracts the cyanide in the cells and serves to pull it out of the cells so that the intracellular enzyme is restored to normal. Forming too much methemoglobin will reduce the blood's ability to carry oxygen; the normal dose of the nitrite will cause the formation of about 25 to 30 percent methemoglobin.

There are two forms of nitrite available. The first is amyl nitrite, which is available in a "perle." This should be broken and placed in a breathing bag for the patient to inhale. (Instructions state that this should be held under the patient's nose for him to breath, but if the patient can breath he does not need the antidote.) The second is sodium nitrite, which is packaged in an ampule containing 300 mg in 10 ml for IV administration. Amyl nitrite should be used only until the sodium nitrite can be administered IV.

Critical Factor: Cyanide antidotes are amyl nitrites, sodium nitrite, and thiosulfate.

The second antidote is a sulfur compound, sodium thiosulfate. When this is administered the body can resume the normal process of tying up cyanide with sulfur to form a nontoxic substance. An ampule contains 12.5 grams in 50 ml for IV administration. All of these antidotes, both nitrite and thiosulfate, are in the Pasadena Cyanide Antidote Kit.

These should be given sequentially to a patient who is unconscious and/or not breathing. Oxygen should be administered, even though the oxygen content of blood is normal. The acidosis should be corrected.

Vesicants

Vesicants are things that cause vesicles or blisters. They may be of animal, vegetable, or mineral origin, such as some types of sea creatures, poison ivy, and certain chemicals. Other things, such as sunlight, can produce blisters. Vesicants have been used as chemical warfare agents. Several have been de-

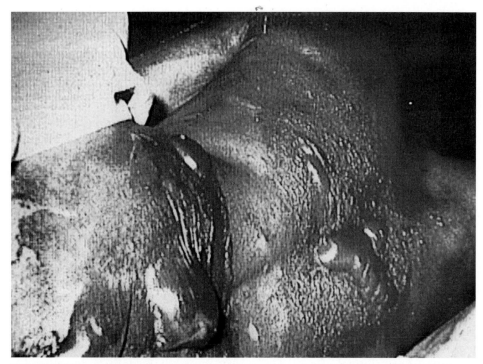

FIGURE 6-5 Victim of vesicant exposure; note the prominent blisters. *Source:* Courtesy U.S. Army, Office of Surgeon General.

veloped for this purpose, but only one, sulfur mustard, has been used. The other major chemical warfare vesicant is lewisite.

Sulfur mustard was first synthesized in the early 1800s and was first used on the battlefield in World War I. During that war it caused more chemical casualties than any other agent; however, only about three percent of these casualties died. Iraq used it extensively during its war with Iran, and pictures of some casualties were in the media during that period. Its use has been alleged in some other conflicts over the past 80 years.

In the early 1940s nitrogen mustard (developed for military use), a close relative of sulfur mustard, was used in the treatment of cancer, the first chemical to be used for that purpose.

Critical Factor: Vesicants cause vesicles or blisters.

Sulfur mustard (mustard) is a light yellowish to brown oily liquid that smells like garlic, onions, or mustard (the reason for its name). Its boiling point is over 200°F, and it freezes at 58°F. The low freezing point hinders its battlefield use in cool weather, and it is often mixed with another chemical to lower the freezing point. It does not evaporate very quickly, but large amounts of mustard, particularly in warm weather, produce a vapor hazard.

Mustard causes cellular damage and death with subsequent tissue damage. The mechanism by which it does this has not been entirely elucidated, but the best evidence suggests that it damages DNA, which then prevents further cellular functioning and leads to cellular death. Although its best-known effects are those on the tissues, the agent directly contacts the skin, the eyes, and the airways. When it is absorbed into the body in adequate amounts, mustard damages many tissues such as bone marrow, lymphoid tissue, and the gastrointestinal tract. Its effects are similar to those caused by radiation, and it is a radiomimetic agent.

Once liquid or vapor mustard is in contact with an epithelial surface, the skin, the eye, or the mucosa of the airways, it penetrates that surface quite rapidly, and enough is absorbed within a minute to cause cellular damage. Decontamination after a minute will not prevent tissue damage, but it will reduce the amount of ensuing damage. Once into tissue, the chemical reactions within the cell that eventually result in clinical effects begin. Once mustard touches a body surface, irreversible damage is done in cells within minutes.

Upon contact with the skin, the eyes, or the airways mustard causes no immediate clinical effects. There is no immediate pain, redness, or blister formation. The patient usually does not know he has been exposed. The itching and pain of erythema, the irritation of the conjunctiva, or the irritation and discomfort in the upper airways do not appear until many hours later. The period without signs or symptoms is called the latent period, and can range from two to 24 hours after contact. Commonly these effects begin in four to eight hours after contact.

At the site of an incident or spill involving mustard, there will be no patients with signs and symptoms of mustard exposure. Hours later the pain, irritation, and discomfort will start, and the patient will seek medical care.

The initial effects in the eyes after exposure to mustard vapor are irritation or burning, and the patient will complain of grittiness in his eyes. The eyes will be red, similar to the appearance of eyes with sand or dust in them. This may progress to a severe conjunctivitis, swelling of the lids, and even corneal edema (seen as an irregular light pattern on the cornea). The patient will complain of pain, irritation, and sensitivity to light. He may also complain of inability to see. This is usually because the lids are shut, either because of swelling or because of involuntary contracture of the muscles around the eye. Rarely, a droplet of mustard will get into the eyes, and this may cause more severe damage to the cornea, including ulceration and perforation.

Mustard contact with skin will initially cause redness, or erythema, which is similar to sunburn with burning and itching. If the contact was to

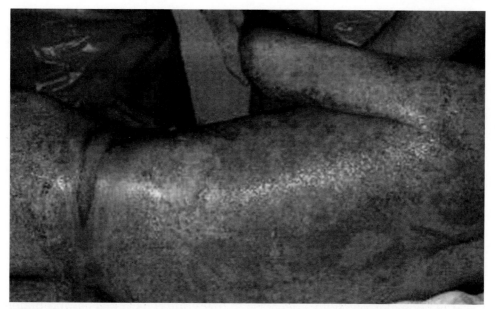

FIGURE 6-6 Victim of mustard exposure. *Source:* Courtesy U.S. Army, Office of Surgeon General.

a low concentration of vapor this may be the extent of the injury, but more commonly small blisters develop around the edges of the redness. These gradually coalesce to form larger blisters, which are generally no worse than second degree burns. Third degree burns are very uncommon and require exposure to a large amount of liquid agent.

Mustard vapor, when inhaled, damages the mucosa or inner layer of the airways. The damage begins at the upper part of the airways (the nose), and descends to the lowermost portion, the terminal bronchioles. The amount of damage depends on the amount of mustard inhaled, which in turn depends on the concentration of the vapor and exposure time. The initial effects are in the nose and sinuses with burning, irritation, and perhaps some nasal bleeding. Pharyngitis, with a sore throat and a nonproductive cough, may appear, followed by laryngitis with hoarseness or complete lack of voice. Mustard damage in the lower airways causes shortness of breath and a cough productive of inflammatory and necrotic material as the agent destroys the inner lining of these small airways. Severe damage provides an ideal setting for infection four or five days later.

Mustard—Initial Effects
No immediate effects; effects start hours after contact.

Skin

- redness (erythema) with burning and itching
- blisters

Eye

- redness with burning and itching

Airways

- nasal and sinus pain
- sore throat, nonproductive cough

Large amounts of absorbed mustard severely damage the precursor cells in the bone marrow, with a decrease in the white cells, red cells, and platelets in the blood. This usually happens four or more days after exposure in a severely exposed patient. The lining of the gastrointestinal tract is also severely damaged after absorption of a large amount of mustard with subsequent loss of fluid and electrolytes starting days after exposure. Again this effect is similar to that seen after radiation exposure.

Critical Factor: Immediate decontamination is very important with mustard exposure.

Immediate decontamination—within a minute—should be performed to minimize the damage, but responders will not be on the scene that quickly. Skin damage will be reduced by decontamination of the contact site on the skin if done within 30 minutes, but not beyond that time.

A patient returning hours after the incident with red skin (erythema) needs no immediate care, although soothing lotions (e.g., calamine) can be applied to reduce the burning and itching. Later, areas of blistering or denuded skin must be irrigated frequently, with the application of topical antibiotics three or four times a day to these areas. Fluids do not need to be replaced in large amounts as they do after thermal burns, because mustard burns do not cause the amount of fluid loss seen in thermal burns. Care must be taken not to overhydrate these patients. They will not need significant fluid replacement unless they are dehydrated from other causes.

A patient with red eyes (conjunctivitis) who is complaining of burning or irritation in the eyes should have the eyes washed out and a soothing ophthalmic ointment or drops applied. Since the lesion appears hours after contact with the agent, the agent is no longer in the eye because of absorption and evaporation, and the purpose of eye irrigation is to wash out inflammatory debris. Later eye care consists of regular application of a topical antibiotic and a mydriatic (to prevent future adhesions between lens and iris). Vaseline should be applied regularly to the edges of the lids (to prevent adhesions). Some believe that topical steroids used within the first 24 hours only will reduce inflammation, but this should be done by an ophthalmologist.

A suggestion of airway involvement by the agent, such as nasal or sinus irritation or a sore throat with a dry hacking cough, calls for the use of steam inhalation and cough suppressants. Laryngeal damage with voice changes or hoarseness accompanied by signs of beginning lower airway damage is an indication for the immediate insertion of an endotracheal tube. Later, more severe damage will necessitate assisted ventilation including positive end-expiratory pressure (PEEP) and frequent sputum examinations for infecting organisms.

Bone marrow depression and severe gastrointestinal damage occur days after the initial exposure in an already severely ill patient.

All patients must be decontaminated before they enter a medical facility. When signs and symptoms appear hours after the initial agent contact, the agent will be gone from exposed surfaces by evaporation or by absorption. Later decontamination will not prevent further injury to the patient. However, liquid may be in clothing or the agent (liquid or condensed vapor) may be in hair.

Lewisite

Lewisite was developed late in World War I but was not used in that war. Japan possibly used it against China in the late 1930s; otherwise it has not been used on the battlefield. Some countries are known to have military stockpiles of lewisite.

Lewisite is an oily liquid with the odor of geraniums. Its freezing point is below 0°F, it boils at 190°F, it contains arsenic, a heavy metal, and it is more volatile than mustard.

Lewisite participates in many biological reactions, but the mechanism of cellular injury is unknown. It damages cells causing cellular death, and its biological effects are similar to those of mustard with topical damage to eyes, skin, and airways. It does not damage marrow, the gastrointestinal tract, or lymphoid tissue, but it does damage systemic capillaries allowing leakage of intravascular volume. This can culminate in hypovolemic shock in severe cases.

Critical Factor: Lewisite causes eye and upper airway irritation and pain on contact.

An important initial clinical distinction between lewisite and mustard is that lewisite vapor causes immediate irritation of eyes, skin, and upper airways. Lewisite liquid causes pain or burning on whatever surface it contacts within seconds. The patient is alerted to its presence and will leave the area or remove the liquid. Mustard causes no clinical effects until the lesions develop, hours after contact.

Lewisite causes topical damage to eyes (conjunctivitis and more severe damage), skin (erythema and blisters), and airways (damage to the lining or mucosa) similar to that of mustard. Severe lewisite exposure may cause pulmonary edema, which is very uncommon after mustard exposure. Generally, the lesions from lewisite are deeper with more tissue damage than those from mustard.

Management of a patient with lewisite exposure is similar to the management of a patient with mustard exposure. The patient will usually self-decontaminate quickly because of the pain or irritation. In addition to the measures recommended for mustard lesions, there is a specific antidote for the systematic (non-topical) effects of lewisite. This is British-Anti-Lewisite (BAL), a drug used for several other types of heavy metal poisoning, and is for hospital use only.

Pulmonary Agents

Pulmonary agents are chemicals that produce pulmonary edema (fluid in the lung), with little damage to the airways or other tissues. The best known and most studied of these is phosgene (carbonyl chloride), although other chemicals (e.g., chlorine) behave in this manner.

Phosgene and chlorine were major agents in World War I until the use of mustard. Their usefulness as warfare agents has diminished since then, and now they are not considered important militarily. However, both are important in industry, and large amounts of both are manufactured and shipped annually.

After inhalation of phosgene, the carbonyl part of the molecule causes damage in the thin wall between the blood vessels (capillaries) and the air sac (alveolus). As a result of this damage the watery part of the blood leaks into the alveoli. When these become filled with fluid, air cannot enter to deliver oxygen to the blood, oxygen cannot be delivered to other tissues, and the patient suffocates in a sense. This fluid in the lungs is similar to that seen in drowning. Damage by these agents is sometimes called "dry-land drowning." Another name for this is noncardiac pulmonary edema, which is pulmonary edema (fluid in the lung tissue) caused by something other than heart failure.

A high concentration of phosgene causes an immediate irritation in the eyes, nose, and upper airways. This is usually transient and is followed later by pulmonary edema. An extremely high concentration will cause laryngeal edema and death within a short period of time, but this is very uncommon. The usual circumstance is that the patient inhales phosgene without immediate effects. Anywhere from two to 24 hours later, the patient begins to become short of breath. Initially, he notices the shortness of breath only with walking or other exertion, but as time passes it is present at rest. His cough brings up clear, frothy sputum, the fluid that leaked into his lungs. If the symptoms begin late, after six or eight hours, the damage is usually not severe enough to cause death, but if the effects begin early, from an hour or two after exposure to six hours, the lung damage is often severe enough to cause death despite medical care.

Pulmonary Agents—Initial Effects
No immediate effects; effects begin hours after exposure.

- shortness of breath with exertion, later at rest
- cough, later with production of frothy sputum

A responder at the site will see few or no symptomatic patients, except possibly some with irritation of the eyes and upper airways or some exposed to extremely high concentrations who will soon have laryngeal edema. Most casualties will be asymptomatic, and the tendency might be to discharge them from care. This could well be a mistake. Symptoms can start suddenly, and if they begin within the first several hours, death may occur within the next several hours. Anyone who has been exposed to one of these agents must be kept under medical observation for at least six hours.

A patient exposed to a pulmonary agent will have two major problems for hospital management. The first is the fluid in the lungs (pulmonary edema) with resulting lack of oxygen (hypoxemia). The second is loss of fluid from his intravascular space (hypovolemia) which may lead to hypotension, shock, and organ damage.

Initial management of these patients is twofold. The first and hardest to remember is that anyone possibly exposed to one of these agents should be kept at absolute rest with absolutely no exertion. He must be carried, not walked, to the ambulance. It is hard to tell a healthy person at the site of a spill or incident who has no symptoms that he cannot walk, but it must be done. World War I experience with these casualties shows that a patient breathing comfortably in bed might collapse and die if allowed to walk down the hall to the bathroom.

The second and more obvious is to provide oxygen to anyone who is short of breath. This usually will not happen while the responder is on the scene initially, but may happen when the responder provides transport later.

Critical Factor: Treatment for exposure to pulmonary agents is oxygen and immediate rest (no exertion).

Riot Control Agents

Most people are familiar with riot control agents otherwise known as tear gas or irritants. Three are in common use in this country. (CS) is used by law enforcement agencies and the military; (CN) (Mace) was used in World War I and is now in small spray devices carried for self-protection; and pepper spray, which is replacing the others for both law enforcement and military use and for self-protection.

Unlike other agents that are liquids, these are solids. The powdery particles are suspended in liquids when they are in spray devices. These agents have much in common. Their effects begin within seconds of contact, the effects last only a few minutes after the person is in fresh air, they are effective in small concentrations, and the lethal concentration is thousands of times higher than the effective concentration, which means that accidentally producing an overdose is very unlikely.

These agents cause irritation, pain, or burning on surfaces they contact, including the eyes, the nose, the mouth and airways, and the skin. Eye effects include burning, tearing, redness, and an initial involuntary temporary closing of the eyes (blepharospasm). While the eyes are closed the patient cannot see and might be considered incapacitated. The interior of the nose burns, and there are secretions from the nose. There are secretions from the mouth, and the interior of the mouth burns. If the agent is inhaled there will be coughing and perhaps a feeling of shortness of breath. There is an initial burning or tingling on the skin accompanied by a mild redness.

Sometimes a high concentration will cause retching or gagging. The effects will gradually recede in 15 to 30 minutes after exposure has ceased.

There are potential complications that seem to be rare. If the face is close to the agent when is dispersed with force (e.g., a spray device), the force may drive the particles into the eye. This necessitates flushing with copious amounts of water or manual removal of the particle by an ophthalmologist. The agent might precipitate a severe reaction in a person with chronic lung disease including hyperactive airways. The use of oxygen, assisted ventilation, and bronchodilators might be indicated. A person exposed to a high concentration in a hot and humid environment might develop a delayed dermatitis beginning about six hours after contact, with erythema developing into blisters.

People can develop tolerance to these agents. With continued exposure, the effects lessen and the exposed people can open their eyes and function relatively normally.

Triage

Triage is an ongoing process that begins with the first person to see the patient and continues through hospital management. The responder will triage at several places, including in the hot zone, in the cold zone after the patient has been decontaminated, and possibly in between. In the hot zone the responder is encumbered with protective clothing and patient examination is not ideal. There are four triage categories generally used: immediate, delayed, minimal, and expectant.

Critical Factor: Effective triage of chemically exposed patients is crucial in mass patient events.

An immediate patient is one who is in danger of loss of life unless there is intervention within a short period of time. Intervention generally has to do with airway, breathing, and circulation (the ABCs), and to this the administration of antidotes should be added. A delayed patient is one who can wait for intervention, and this wait will not affect the outcome of his care. The patient is stable but will require further care. A minimal patient is one who requires care for a relatively minor injury. The care can be done quickly, the injury is not life threatening, and the patient is unlikely to require long-term care, (i.e., hospitalization). An expectant patient is one who cannot be saved with the resources available or resources cannot be made available within the time he needs them.

Nerve Agents

An immediate patient is one (a) who is unconscious, is apneic or struggling to breath, is convulsing or has convulsed, and has muscular twitching or is flaccid, or (b) who has moderate or severe signs in two or more organ systems (respiratory, gastrointestinal, muscular, and central nervous system). This patient should be given three MARK's Is or 6 mg of atropine—IM not IV—and a 20 to 30 minute drip of one gram of 2-PAMCl.

A delayed patient is one who is recovering from moderate or severe effects or from the effects of several doses of antidote.

A minimal patient is one who is walking and talking. He may be severely short of breath or vomiting, but still has muscle strength and control and can understand the spoken word enough to respond. Generally, this patient should be given one MARK I (or 2 mg of atropine with a drip of 2-PAMCl). If he is extremely short of breath, 4 mg of atropine should be administered. Despite the shortness of breath, this patient is not immediate.

An expectant patient is one who is not breathing and is without a heartbeat. However, if he has been without cardiac activity for a very brief period of time, every attempt should be made to resuscitate him.

Cyanide

Cyanide patients can die within minutes after inhaling a large concentration of the agent. Those who are unconscious and not breathing, but who still have a heartbeat should be classified as immediate, and the antidotes should be given as soon as possible. If a patient is conscious he will be minimal and will not need antidotes. An expectant patient is one who has been without a heartbeat for many minutes.

Vesicants

Almost all vesicant patients will be delayed. They will need no immediate care, but they will need further care for their eye, skin, or airway injuries. An exception is a patient with moderate to severe airway effects including shortness of breath. They are immediate and need intensive pulmonary care.

Pulmonary Agents

Although shortness of breath, the major symptom from these agents, can be faked by a malingerer, anyone complaining of shortness of breath within six hours of exposure should be classified as immediate for intensive pulmonary care. A patient with shortness of breath beginning later than six hours post exposure will also need care but not as urgently. He will be delayed. A patient with severe shortness of breath and copious frothy sputum within an hour after exposure is expectant, although an attempt should be made to provide maximum care.

Riot Control Agents

With the exception of a patient who has a severe airway reaction to these agents and needs immediate care, these patients will be classified as minimal.

Early Recognition

When first responders in protective gear first enter the hot zone, they usually will not know what the toxic agent is and usually will not have a detector to tell them. They must quickly evaluate the patients based on what is seen and heard and take appropriate action. Early therapeutic intervention is needed for only two types of agents—nerve agents and cyanide. A patient exposed to a large concentration of a pulmonary agent may be in severe respiratory distress, but there is nothing that can be done in the hot zone; if the effects started before the responder arrived, probably nothing can be done elsewhere.

In most chemical mass casualty situations, the patients will exhibit a spectrum of effects. Some patients will be quite severe, and others will have minor effects, and the responder must quickly evaluate this spectrum. For example, if some patients are convulsing or are unconscious and appear to be postictal, the responder should look at other patients. The presence of miosis, runny noses, and shortness of breath or any one or two of these strongly suggests that nerve agents were the offending substance. If the conscious patients are relatively normal with a few nonspecific complaints, cyanide should be considered. If all patients are conscious with no complaints, the responder should consider that (1) no chemical agent was present, or it was it was present in concentrations too low to produce effects, or (2) the agent was one that produces delayed effects only, such as mustard or the pulmonary agents.

If many patients are complaining of irritation or burning in the eyes and nose, on the mucous membranes of the mouth, and on the skin, one might consider

1. riot control agents (in which case the patients will improve with fresh air),
2. phosgene (this effect will improve, but there will be later, more severe ones),
3. cyanogen chloride (the irritation will gradually decrease and if the patient is conscious when help arrives it is unlikely that a lethal concentration was present), or
4. lewisite (the effects will worsen).

Reader's note—refer to Table 6-1 as a summary of signs, symptoms, and decontamination procedures for chemical agents.

Table 6-1   Chemical Agents: Symptoms and Treatment

Nerve Agents (GA, GB, GD, GF, VX)	Mustard (HD, H)
Signs and Symptoms: *Vapor: Small exposure*—Miosis, rhinorrhea, and mild dyspnea. *Large exposures*—Sudden loss of consciousness, convulsions, apnea, flaccid paralysis, copious secretions, and miosis. *Liquid on skin: Small to moderate exposure*—Localized sweating, nausea, vomiting, and feeling of weakness. *Large exposure*—Sudden loss of consciousness, convulsions, apnea, flaccid paralysis, and copious secretions.	**Signs and Symptoms:** Asymptomatic latent period (hours). Erythema and blisters on the skin, irritation, conjunctivitis, corneal opacity, and damage in the eyes; mild upper respiratory signs, marked airway damage; also gastrointestinal effects and bone marrow stem cell suppression.
Decontamination: Large amounts of water with a hypochlorite solution.	**Decontamination:** Large amounts of water with a hypochlorite solution.
Immediate Treatment/Management: Administration of atropine and pralidoxime chloride (2PAM); diazepam in addition if casualty is severe; ventilation and suction of airway for respiratory distress.	**Immediate Treatment/Management:** Decontamination immediately after exposure is the only way to prevent/limit injury/damage. Symptomatic management of lesions.
Lewisite (L)	**Phosgene Oxime (CX)**
Signs and Symptoms: Lewisite causes immediate pain or irritation of skin and mucous membranes. Erythema and blisters on the skin and eyes and airway damage similar to those seen after mustard exposure develop later.	**Signs and Symptoms:** Immediate burning and irritation followed by wheal-like skin lesions and eye and airway damage.
Decontamination: Large amounts of water with a hypochlorite solution.	**Decontamination:** Large amounts of water.
Immediate Treatment/Management: Immediate decontamination; symptomatic management of lesions the same as for mustard lesions; a specific antidote British Anti-Lewisite (BAL) will decrease systemic effects.	**Immediate Treatment/Management:** Immediate decontamination; symptomatic management of lesions.

Table 6-1 Continued

Cyanide (AC, CK)	Pulmonary Agents (CG)
Signs and Symptoms:	**Signs and Symptoms:**
Few. After exposure to high estimated dose (Ct): seizures, respiratory and cardiac arrest.	Eye and airway irritation, dyspnea, chest tightness, and delayed pulmonary edema.
Decontamination:	**Decontamination:**
Skin decontamination is usually not necessary because agents are highly volatile. Wet, contaminated clothing shoud be removed and the underlying skin decontaminated with water or other standard decontaminates.	Vapor: fresh air. Liquid: copious water irrigation.
Treatment:	**Treatment:**
Antidote intravenous sodium nitrite and sodium thiosulfate. Supportive care: oxygen and correct acidosis.	Termination of exposure, ABCs of resuscitation, enforced rest and observation, oxygen with or without positive airway pressure for signs of respiratory distress, other supportive therapy as needed.

Riot Control Agents (CS, CN)	
Signs and Symptoms:	
Burning and pain on exposed mucous membranes and skin, eye pain and tearing, burning nostrils, respiratory discomfort, and tingling of the exposed skin.	
Decontamination:	
Eyes: thoroughly flush with water, saline or similar substance. *Skin:* flush with copious amounts of water, alkaline soap and water, or a mildly alkaline solution (sodium bicarbonate or sodium carbonate). Generally decontamination is not required if wind is brisk.	
Treatment:	
Usually none is necessary; effects are self-limiting.	

Source: Medical Management of Chemical Casualties Handbook, 2nd edition, Chemical Casualty Care Office, United States Army Medical Research Institute of Chemical Defense, Aberdeen Proving Ground, Maryland, 1995.

CHAPTER SUMMARY

Chemical warfare agents are not new, and terrorist organizations have access to these weapons as demonstrated by the use of sarin in a Tokyo subway attack in 1995. Many industrial chemicals such as chlorine, cyanide, phosgene, and pesticides are readily available in large quantities.

It is essential that response agencies be prepared for a chemical attack. Effective planning must include protective equipment, decontamination procedures, and antidotes. A deliberate incident is a crime scene, usually with mass numbers of patients.

Nerve agents are toxic materials that produce injury or death in seconds to minutes. Nerve agents are similar to insecticides, but are more toxic.

Very good antidotes are available for nerve agents, but they must be administered quickly. Common nerve agents include tabun (GA), sarin (GB), soman (GD), and GF and VX. Effects from nerve agents show very quickly. Management of nerve agent exposure consists of decontamination, administration of antidotes, and ventilation.

The antidotes for nerve agent poisoning are atropine and an oxime, 2-pyridoxime chloride or PAMC1 (Protopam). Atropine blocks the excess neurotransmitters. Protopam removes the nerve agent from enzymes, allowing the enzymes to block the neurotransmitters. Valium can be used as an anticonvulsant.

Cyanide can cause serious illness and death within minutes. Cyanide affects the ability of the cells to metabolize oxygen. Treatment of cyanide poisoning includes amyl nitrite perle, or sodium nitrite IV and sodium thiosulfate IV. Patients should be ventilated with oxygen and acidosis should be corrected.

Vesicants are agents that cause vesicles or blisters. The most common vesicant agents are sulfur mustard and lewisite. Mustard does not cause an immediate effect; the common latent period is four to eight hours after contact. Inhaled vapor causes damage to the airway and bronchioles. Patients must be decontaminated immediately and the eyes irrigated.

Lewisite produces instant effect on contact. The symptoms include immediate pain, eye damage, and airway injury.

Pulmonary agents produce pulmonary edema. The best-known agents are phosgene and chlorine. The effects of these agents are not immediate, but shortness of breath followed by pulmonary edema follows hours after exposure. Exposed patients with no symptoms must be kept under medical observation for at least six hours. Initial patient treatment includes keeping the patient at rest and administering oxygen.

Riot control agents are known as tear gas and irritants. They include CS (tear gas), CN (Mace), and pepper spray. Effects begin in seconds, but last only a few minutes after the patient is removed to fresh air. These agents cause pain, burning, and irritation to the contact body surfaces. The use of oxygen is indicated.

Triage is an ongoing process in a chemical exposure incident. The triage categories are immediate (critical), delayed, minimal (walking wounded), and expectant. A critical patient is one that is unconscious, apneic, or convulsing. Almost all vesicant patients will be delayed. Most riot control patients will be in the minimal or walking wounded category.

Early responders must quickly evaluate the scene. Rapid therapeutic intervention is only needed for nerve agents and cyanide. If a patient shows severe pulmonary distress from a pulmonary agent, nothing can be done in the pre-hospital setting. In mass casualty incidents patients will exhibit a spectrum of effects. Responders must quickly don protective equipment and triage all patients to determine treatment categories.

CHAPTER QUESTIONS

1. Discuss several reasons why your community should be prepared for a terrorist chemical attack.
2. What are the methods of disseminating a chemical agent?
3. Define nerve agents. How do they act on the body?
4. List five common nerve agents.
5. What are the symptoms of nerve agent exposure?
6. What is the treatment for nerve agent exposure? Name three antidotes.
7. Discuss the symptoms of cyanide poisoning. What are the antidotes for cyanide exposure?
8. What are vesicants? What are the signs of vesicant exposure? What is the treatment?
9. Define lewisite. How does lewisite exposure differ from mustard exposure?
10. What are pulmonary agents? What is the treatment for severe exposure?
11. List three riot control agents. What are the symptoms and treatment for exposure?
12. What are the triage categories in a chemical attack with mass patients? Describe typical symptoms of a patient in each category.

Simulation

You are an EMS training officer in an organization that has no chemical attack training program. Your goal is to develop guidelines for first response fire units and EMS units. Develop a written chemical response guideline that includes the following key elements:

- Categories of chemical agents
- Symptoms at onset and long-term symptoms
- Advanced life support (ALS) care for each type of agent including antidotes
- Basic life support (BLS) treatment procedures
- Safety guidelines
- Triage categories and related symptoms

7
Weapons of Mass Effect: Biological Terrorism

Dr. Charles Stewart
Paul M. Maniscalco

Source: Courtesy American Cancer Society

Chapter Objectives

After reading this chapter, you will be able to accomplish the following objectives:

1. Define the concept of biological warfare.
2. Have awareness of the history of biological warfare.
3. Understand the concepts of biological threat assessment.
4. Recognize the importance of biological protective equipment.
5. Outline the types of biological agents including the chemical effects, detection, and prophylaxis/treatment of botulinum toxins, clostridium toxins, ricin, saxitoxin, staphylococcal enterotoxin, tetrodotoxin, and trochothecene mycotoxins.
6. List and define live bacteriologic agents including anthrax, brucellosis, cholera, Ebola virus, plague, psittacosis, Q fever, smallpox, and tularemia.

Introduction

The uses of biological substances as weapons pose a unique problem for the emergency response and public health communities. Unlike the consequences of a chemical attack or an explosion, which are essentially "in your face" events, biological terrorism creates a slow-motion riot that builds with each hour after the event. The inability to readily identify what has occurred immediately allows the maturation process to continue while increasing the risk to a vulnerable population.

Biological terrorism is the use of etiological agents (disease) to cause harm or kill a population, food, and/or livestock. Biological terrorism includes the use of organisms such as bacteria, viruses, and rickettsia and the use of products of organisms—toxins.

Biological terrorism has recently become more threatening to the world. One only needs to consider the current state of technology, the future possibilities of biotechnologies, and what appears to be a readiness on the part of some individuals/countries to utilize this technology as a weapon.

Successful genetic engineering has arrived and advances are being achieved almost daily. It requires a relatively easy process and only crude technology to manufacture a lethal organism/toxin in vast quantities, some specifically designed to be resistant to antibiotics, for use as a horrible weapon. It has been said that "if you can make beer, you can make bugs (biological weapons)." This is an oversimplification, but it provides a vivid picture of what a motivated person with modern technology is capable of achieving. While wreaking considerable havoc and death among humans resulting in a medical disaster, should bio-weapons be employed against livestock or vegetation, the results would be an economic disaster.

Biological weapons (BW) are more deadly and financially efficient, pound for pound and dollar for dollar, than chemical agents or even nuclear weapons. It has been estimated that 10 grams of anthrax could kill as many people as a metric ton of the nerve agent sarin. Biological weapons are relatively inexpensive, easy to manufacture, and dispersal devices can be disguised as agricultural or pest-control sprayers. Unfortunately for the law-abiding world, it is very difficult, if not impossible, for an intelligence service to detect research, production, or transportation of these agents for rogue intentions. It is equally hard to defend against these agents once they have been employed due to the inability to recognize a delivery.

History of Biological Agents as Weapons

The use of biological agents as a warfare weapon has a long and deadly past. In fact, history has shown us that use of bio-weapons can be found more than 2,000 years ago. Some examples of its use are:

1. In the sixth century B.C., Assyrians poisoned enemy wells with rye ergot;[1] Solon's use of the purgative hellebore during the siege of Krissa.[2]
2. Persian, Greek, and Roman authors quote the use of animal cadavers to contaminate water supplies. In 1155, Barbarossa used the bodies of dead soldiers to poison the wells at the battle of Tortona.
3. The Scythian archers would dip their arrows in blood mixed with manure or in decomposing cadavers.
4. The Mongols in the 1300s catapulted plague victim corpses into the city of Kaffa to infect the defenders.[3] The besieged town was rapidly devastated by disease.
5. British and early American settlers gave American Indians blankets used for victims of smallpox. The resultant infection decimated the defenseless American Indian tribes.
6. In 1941 the Allies tested anthrax on Gruinard Island off the shore of Scotland. The organism persisted in a virulent state until 1988, when the island was finally declared safe for unprotected humans.[4]
7. During World War II the Allies administered 235,000 doses of antitoxin to Allied troops and deliberately leaked this to the Nazis. Simultaneously they told the Nazis that the Allies were prepared to use biological weapons if they were employed in the war.[5]
8. During World War II on the Pacific front, the Japanese tested biological weapons on prisoners of war in China, killing more than 1,000 people. In fact, it has been reported that the Japanese had

"stockpiled 400 kilograms of anthrax to be used in specially designed fragmentation bombs."[6]

9. In the 1991 Persian Gulf War, the threat that Iraq would use biological warfare was a major concern for all coalition forces and extensive preparations were made for this threat. Iraq was believed to have an active development program in biological weapons (particularly anthrax and botulinum toxins). Unclassified information from CIA and Defense Intelligence Agency documents indicates that numerous rogue states such as Iran, Iraq, Libya, and North Korea have or are pursuing BW programs.[7]

10. Presently, Iraq has become the center of attention once again for their involvement with WMD. Although the Iraqi government reports that this is bad intel, United Nations Special Commission (UNSCOM) inspectors have gathered what they feel is sufficient evidence that agents such as anthrax, sarin, VX, and BZ class agents (Agent 15) and others have been manufactured and are being stockpiled for possible use.

In recent times, the military has examined the possibility of biological actions against the United States. In the 1950s Serratia and Bacillus species were released from ships in the San Francisco Bay area and caused at least one death.[8] In the 1960s, military researchers introduced *Bacillus subtilis* into New York City subway ventilator shafts. Both passengers and security guards were oblivious to the danger.[9] The bacteria were rapidly spread to the ends of the subway system.

The United States Office of Technology Assessment has estimated that a small private plane with 220 pounds of anthrax spores, flying over Washington, D.C., on a windless night, could kill between 1 and 3 million people and render the city uninhabitable for years. In April 1996 a Department of Defense (DoD) report noted that the Iraqis claimed to have manufactured over 8,300 liters of anthrax.[10]

Other countries are certainly continuing to develop bio-warfare capabilities. The Soviets and their allies employed a trichothecene mycotoxin dubbed "yellow rain" in Laos, Cambodia, and Afghanistan, and the former USSR and Iraq have independently developed anthrax species. In 1979 an outbreak of inhalational anthrax occurred in Sverdlovsk. This outbreak resulted from an accident at a Soviet bio-warfare research facility.

Treaties

The Geneva Convention (1925) prohibited the use of biological and chemical warfare and in 1972, more than 118 nations ratified an agreement to stop research on offensive biological weapons. Among the 118 signatory nations are Russia, the United States, and Iraq. Research for defensive purposes is still allowed and continues across the globe. Treaties and multilateral agreements cannot rid the world of chemical and biological weapons, which are simple, inexpensive, and produced by widely available technology. Here are some recent examples:

Paris Police, in 1984, raided a suspected safe house for the German Red Army Faction. During the search they found documentation and a bathtub filled with flasks containing *Clostridium Botulinum*.[11]

Russia's biological warfare technology may be vulnerable to leakage to third parties through either theft or outright sale (like nuclear materials), as a result of the financial crises that exist. Open source intelligence reports that army personnel and scientists have been known to sell off military equipment to get money to feed their families. In some cases, reports have been received that these individuals, in "critical and sensitive" positions, have not been paid in months, making them vulnerable to recruitment by rogue organizations or nations.

The Aum Shinrikyo cult members (famous for the sarin gas attack in Tokyo subways) were found to have anthrax and botulinum cultures when the Japanese National Police conducted their raid of the Aum base camp at the foot of Mount Fuji. They had constructed dedicated laboratories and had purchased a helicopter equipped with a spraying apparatus. The Aum had also visited Zaire during the Ebola outbreak to collect specimens of Ebola virus.[12]

In the town of The Dalles, Oregon, in 1984, 750 people became sick after eating in four different restaurants. The illness was traced to the Bagwan Sri Rajneesh Sect, which had spread salmonella on salad bars in the four restaurants.[13] The intention of this group was to sicken many of the community to prevent them from going to the polls, thus interfering with the political process and manipulating a local election.

A United States microbiologist named Larry Wayne Harris fraudulently ordered bubonic plague cultures by mail in 1995.[14] Harris allegedly has ties to militarist right-wing groups in the United States. The ease with which he obtained these cultures prompted new legislation to ensure that biologic materials are destined only for legitimate medical and scientific purposes. Even scarier is the fact that these products are often shipped via commercial delivery companies such as UPS and Fed Ex, which is perfectly legal.

Threat Assessment

Although the conclusion that the United States is very vulnerable to a bio-warfare attack or terrorism is indisputable, the BW programs of the 1950s and 1960s were appropriately criticized for the unethical exposure of unwitting test subjects. Despite the escalating BW threats, our level of preparedness has not changed significantly since the 1960s. It is only recently that we are witnessing a renewed interest in preparing for the response that will be required should an attack occur.

With limited capacity to anticipate a biological attack, little or no ability to detect one if it occurs (unless the perpetrators decide to announce the release and take credit for their demonic acts), and a diminished ability to effectively manage the consequences if attacked, this problem poses a series of complex issues that need immediate review. There are a number of reasons for this unpreparedness:

Intelligence

When a BW manufacturing facility can be constructed in the area of a large garage, law enforcement/intelligence services are confronted with great difficulty in locating them. As has been noted, most of these agents are not controlled and can be found endemically throughout the world. Accessing cultures is not nearly as expensive or tracked as well as nuclear material. BW culture processing requires equipment that would be considered suitable for a well-equipped hospital laboratory or academic research facility, and is thus easily ordered and diverted. If this does not sound credible, please note that Saddam Hussein bought his original anthrax cultures from a mail order house in the United States and had it shipped by a commercial overnight carrier.[15]

Detection

Detection of biological agents occurs most often after a release. Quite simply, there is no way to detect the deployment of a bacterial agent in the civilian community under normal operating procedures, or with the technology presently available for civilian use. The only way to detect the agent is through the clinical presentation of patients, and that will be retrospective for most of the casualties. Some limited battlefield detection devices exist, but these are unusable in the majority of United States cities. These devices are effective for special events such as the Olympics, inauguration, or where crowds are moderately constrained, but due to cost and availability have limited benefit to local emergency response organizations. When threat assessments are quite high and advance notice of the threat exists, use of these items through the WMD Civil Support Teams or through the Department of Defense (DoD) is highly recommended.

Biological warfare agents are almost undetectable during transit. Likewise, there is no mechanism using routine customs, immigration, drug scan, or bomb search procedures to identify the agent. The only way to find it would be a physical search by a very well trained and very lucky searcher.[16] Indeed, the agent could be simply sent using Fed Ex or similar overnight carrier from one point to another. Even in an event where a package is broken and product is leaked, you may have a high index of suspicion, but identification of the agent will usually take place at a laboratory not in the field.

Again, the bio-terrorism threat may not be directed toward humans. Livestock and crops are strategic targets and vulnerable to attack. As an example, it is not inconceivable that a rogue individual or group could attempt to destroy all pork and pork products in the United States. Although not a fatal blow for the United States on the whole, it would certainly not help the U.S. economy to have a porcine plague. Detection of this plague would be very difficult indeed prior to symptoms in a substantial number of the affected animals. (Few communities include veterinarians in their biological surveillance plans.)

Control of Supplies

A military commander maintains the luxury of knowing that his troops are under threat of attack. The civilian emergency response chief does not usually have this warning and the targets for introduction of a biological agent are almost unlimited. To a large extent, the battlefield commander controls the food and water supply of his troops. To institute such control in the civilian sector would mean martial law and this is unacceptable in a free society. This level of freedom is not without cost and it creates vulnerability for rogue groups to exploit.

Personal Protective Equipment

Biological terrorism is most likely to be executed covertly and sick individuals may be the initial "detector" that an attack has occurred. If delivered effectively, a large number of casualties can be generated in a short period of time. In the midst of treating the casualties, the emergency responder and organization must not only provide effective care, but also protect themselves and their members.

This will be difficult most of the time due to the unknowing responders believing that they are operating at another "sick job." Remember after the release of a bio-agent, there is an incubation period in the new host prior to its clinical manifestation. In some cases this period may be more than 72 hours, and in some agents two weeks or more.

One of the limiting factors of personal protective equipment (PPE) is that military issued gear is not OSHA certified. OSHA certification is a standard requirement for civilian use of any PPE. It is only recently that some of this equipment is being considered for civilian use or "technology transfer," hence some testing for OSHA standard compliance is taking place.

Much of the civilian PPE that is available for bio-agents is designed for use in the static environment of the laboratory and not the street. This, too, is another issue that will require a cooperative public/private working arrangement. Clearly, with the threat to the civilian responder escalating, continuing research for better PPE should be expected with the expertise of military, academia, and private industry pooling talent and resources to create a successful resolution of this conundrum.

Even if the military or CDC provide gear, it must be prepositioned and issued after either a significant threat or after the first casualties have been identified. In either case, there is a significant risk that many of the emergency services (including emergency physicians, nurses, paramedics, and EMTs) will be exposed and become unknowing casualties prior to the arrival of protective gear.

Prophylaxis

The United States General Accounting Office (GAO) found that at the beginning of the Gulf War in 1991, the U.S. Army's stockpiles of vaccine for anthrax and botulism had fallen far short of what was needed to protect U.S. troops. Indeed, the GAO felt that at least 20 percent and perhaps 40 percent of the military's bio-war budget was *not* directed at diseases or toxins that were identified as threats by the military's own intelligence agencies.[17] Recently, (December 1997) heated discussions have again taken place at the Pentagon and in the media about the need to provide vaccinations to all uniformed service members as a provision of force protection in the face of these BW threats.

There are other issues that this poses including the limited production facilities and stock on hand for vaccines. For instance, the Michigan Biological Products Institute (a part of the Michigan Department of Health) under contract to the Defense Department exclusively produces the anthrax vaccine. Virtually all vaccine produced is under Defense Department contract for primarily military use and a small number of other official uses. So the question that must be asked is whether the emergency response communities' needs would be met on short notice should a decision to vaccinate all emergency responders from a specific region be made.

Protecting the armed forces against a biologic attack is less challenging and complex than protecting the civilian population. Military personnel are a captive audience who have little choice regarding whether they will receive an immunization.

At present we do not have sufficient emergency providers with enough immunizations to care for the population of a U.S. city (if attacked), without additional harm coming to the rescuers. These providers will be at the highest risk if active agents are employed.

Training

Emergency responders will be called upon to know treatments for exotic diseases that they are unlikely to have ever encountered. Emergency service members must be aware of symptoms and epidemiological patterns that may indicate a biologic attack and many have never been taught these techniques of pattern identification. This creates a heavy reliance upon the expertise of the EMS medical director (MD) and the public health director for guidance as well as assembling and analyzing epidemiological data of the community. These resources may not be available on off hours such as nights, weekends, or holidays.

Emergency medical service, fire, police, and even hospitals must purchase PPE and train employees for work in protective gear that has only been found in the military and specialized hazardous materials teams. Emergency service organizations must realize that they are significant targets for primary/secondary attacks and conduct their routine operations appropriately while ensuring that the proper security measures are implemented.

Interagency Coordination and Public Perception

It is always difficult to balance the perceived needs of multiple population groups. Drawing the fine line between antiwar protesters who feel any research into bio-war techniques should be forbidden, and hawks who look for a threat around any corner is always difficult. When given the choice of where to spend defense money, it is easier to put it into real and "visible" tools such as guns, battleships, and troops playing the CNN factor to the max. Bluntly speaking, in the past, cruise missiles on CNN are easier to sell to a congressional committee than protective garments for use in Bloomfield, New Jersey. Hopefully the momentum that has been attained regarding community bio-terrorism preparedness will continue to grow and change this view.

Likewise, control of a program that will spend millions of dollars brings a smile to many bureaucrats' faces. Will infighting between bureaucratic agencies dissipate any real effort to protect the United States? These issues require policy decisions with requisite directives to be issued in an effort to set the tone, from the top, ensuring that the process does not become mired in rivalry and competition.

When a bio-terrorism incident occurs, who will be in charge? Will it be local EMS, fire, police, or emergency management personnel? Will it be medical authorities from state, county, or local departments of health? Will it be medical authorities from the CDC or even military specialists from one of the bio-war development centers at Dugway or Fort Detrick? Will the FBI attempt to assume control of the scene to preserve evidence? Will FEMA attempt to usurp control of the incident? Will martial law be declared with the military in control of a city? Even with the issuance of Presidential Decision Directive 39 (PDD 39), these questions have not been fully and adequately answered. Control issues are always best answered in advance of the incident, rather than during the emergency. Having a comprehensive emergency action annex to the existing community emergency management plan for bio-terrorism incidents is strongly encouraged. Determination of issues such as these should be accomplished before "game day," not on the field of play. The latter will contribute to the chaos one can expect at this type of incident.

Deniability

The existence of naturally occurring or endemic agricultural pests or diseases and outbreaks will permit an adversary to use bio-terrorism with completely plausible denial. Such biological warfare attacks could be against the food supply or crops. The effects of such biological and economic warfare could bring devastation to the affected nation. The Russian wheat aphid in the United States cost American agriculture over $600 million in damage and lost revenues. What would a focused and deliberate release of a BW agent mean to the economics of the United States?

Response Time

Even if an astute emergency physician notes that an unusual number of patients have certain symptoms and contacts the CDC for help, the crisis is immediately recognized as a bio-terrorism event, and help is dispatched immediately, the lag time may be unwieldy. Remember that with some of the agents that have been identified, there is an incubation period that exceeds three days from time of agent distribution until the first cases occur and some agents carry a 20 day period. The patient may be quite infective during latter parts of this incubation period with emergency personnel and hospital staff unaware of the jeopardy they are in. The mortality rate approaches 100 percent when symptoms present with some of these agents.

Given an absolute best case scenario from notification, it will take at least two hours for a qualified team of predesignated physicians and pre-hospital providers (paramedics and EMTs) to assemble (if a community has had the foresight to convene a team prior to the event), ready gear, and respond to the deployment assembly point. It will take another few hours to assess the situation, draw appropriate clinical samples, and formulate an idea of what illness or toxin was employed. During this time, others will be exposed and potential carriers may be leaving the city, bound for other destinations.

When casualties exceed the available medical resources, additional resources must be identified, summoned, and either the patient transported to them or the providers and equipment transported to the patients. This scenario will warrant the deployment of federal assets in the form of National Disaster Medical System (NDMS) Disaster Medical Action Teams (DMATs) and military units such as the deployable medical teams from Ft. Sam Houston or the USMC CBIRF. This could take many hours or days.

If news services broadcast any warning, one can expect a panic-stricken response that may cause gridlock on the roads and further complicate any response team's travel to the area (not to mention what this will do to the standard 911 call volume). Essentially, we are looking at a regional, if not national or international health, crisis emerging. This statement is very real. Consider the fact that in a conventional weapon attack such as the World Trade Center bombing, patients were tracked as far west as Pennsylvania and as far north as New Haven, Connecticut.

Realities and Costs

It is unlikely that a rational foreign government would hazard potential military reprisals and the political/economic sanctions that overt use of biologic warfare against the United States would bring.[18] Covert action or deniable independent rogue factions and transnational terrorist organizations do not have the same constraints. Terrorist organizations operating in a civilian environment have freedom of movement, and the ability to use commercially available equipment for development and discharge of their weapon. They are not constrained by a need for precise targeting or predictable results. A determined transnational organization or rogue individual may not be deterred, may escape detection and intelligence gathering activities, and may succeed in releasing a biologic agent in a susceptible target area. Lastly, transnational terrorist groups generally are not affiliated with one country or organization, and as such have no "return address" making a military retaliation very complicated if not impossible.

The effects of a bio-terrorism incident are catastrophic. In a paper by researchers at the CDC, the projected economic impact alone ranges from $477 million per 100,000 people exposed to brucellosis to $26.2 billion in the case of anthrax.[19] Over 30,000 deaths were predicted if anthrax was used as the biologic agent. The paper consistently used the lowest possible expense for all factors that affected costs, including the virulence of the disease. Costs of both preparedness and intervention were significant. It would be clear that this would not be the case in a "real" disaster of this magnitude. Even so, the researcher concluded that reducing preventable losses has a significantly greater impact than reducing the probability of an attack through intelligence gathering and related activities.

The authors also noted that the best possible measures to decrease both costs and deaths were those that would enhance rapid response to an attack. "These measures would include developing and maintaining laboratory capabilities for both clinical diagnostic testing and environmental sampling, developing and maintaining drug stockpiles, and developing and practicing response plans at the local level."

Possible Biotoxins

Until recently, toxins were of interest only to the toxicologist, the rare patient who ingested or was exposed to these toxins, and the even rarer writer who discussed toxicological environmental emergencies. Unfortunately, several simultaneous political and scientific events have moved these toxins to a more prominent medical and social position.

Discovery that some of these toxins have been used as agents in warfare or have been stockpiled to use in warfare has given the medical community an impetus to learn more about the effects and production of toxins for biological warfare. New uses for old toxins include botulinum therapy for spastic muscles and dystonia. For an overview of biological symptoms and treatments, refer to Table 7-1 (pg. 177).

Botulinum Toxins

Botulinum neurotoxin is among the most potent toxins known. The mouse lethal dose is less than 0.1 nanograms per 100 grams. It is over 275 times more toxic than cyanide.

Mueller (1735 to 1793) and Kerner (1786 to 1862) in Germany first described botulism. They associated the disease with ingestion of insufficiently cooked "blood sausages" and described death by muscle paralysis and suffocation. In the early 1900s, botulism occurred commonly in the United States and nearly destroyed the canned food industry.[20]

The major source of botulinum toxin is the organism *Clostridium botulinum.* There are seven serotypes produced by clostridia species. These serotypes are similar, but do not cross-react to immune reactions. They are released as a single polypeptide chain of about 150,000 Daltons, which is cleaved to generate two disulfide linked fragments. The heavy fragment (H1 100,000 Daltons) is involved in cell binding and penetration, while the light chain is responsible for the toxic intracellular effects.

CLINICAL EFFECTS

Two natural types of poisoning occur. In the first type, food tainted with clostridia species is stored or processed in a way that allows the anaerobic organisms to grow and multiply. As they grow, they produce and release toxin. If the food is not subsequently heated to destroy the toxin, clinically significant amounts can be consumed. The toxin passes through the gut into the general circulation and is distributed throughout the body. In the second type, usually found in infants, the organisms colonize and produce their toxin in the gut. The clinical effects of the two types of botulism are the same.

After intoxication with botulinum toxin, the victim will develop diplopia and ptosis, difficulty speaking and swallowing, decreased bowel function, and muscle weakness that can progress to a flaccid paralysis. The patient will generally be awake, oriented, and afebrile. Development of respiratory failure may be quite rapid after initial symptoms develop.

It is sometimes difficult to distinguish organophosphate nerve agent poisoning from botulism. The copious secretions of the nerve agent will be the significant clue to the differential. Isolated cases have a wider differential diagnosis including Guillain-Barre syndrome, myasthenia gravis, and tick paralysis.

Botulinum toxin penetrates into the cell and blocks release of acetylcholine, preventing neuromuscular transmission and leading to muscle weakness.[21] Botulinum toxin is thought to preferentially affect active neuromuscular fibers and has been shown in rats to have a greater affect when nerve activity is greater.[22] It may also affect the central nervous system.[23] The local injection of botulinum toxin has been used clinically to treat involuntary focal muscle spasms and dystonic movements.

Botulinum toxin was used to assassinate Reinhard Heydrich, a Nazi leader and probable successor to Hitler. The Czechoslovakian underground used a grenade impregnated with botulinum toxin made by English researchers in Porton Down. Although Heydrich's wounds were relatively minor, he died unexpectedly several days after the attack.[24]

DETECTION

Current laboratory tests are unhelpful in the clinical course. Detection of botulinum may be done by mouse bioassay or by liquid chromatography. Uses of radioimmunoassay and radioreceptor assays have also been reported. A DNA probe has been designed for detection of botulinum toxin, which would markedly expedite diagnosis. An immunoassay has been developed by Environmental Technologies Group, Incorporated in Baltimore, Maryland, to detect this toxin, ricin, and staphylococcal enterotoxin.

Survivors will probably not develop an antibody response due to the small amount of toxin required for lethality.

PROPHYLAXIS AND TREATMENT

Treatment is supportive. Respiratory failure will require prolonged (weeks to months) ventilatory support. If ventilatory support is available, then fatalities are likely to occur in less than five percent of the exposed population.

An equine antitoxin is available and may be of some help in both food-borne and aerosol botulism. This is available from the CDC and protects against A, B, and E toxins. It has been used for treating ingestion botulism and should be given as soon as the diagnosis is made. It does not reverse paralysis, but does prevent progression of the disease. There is no human-based antitoxin currently available, but human-based antitoxin testing is now in progress. Obviously, it will not help in types C and D intoxication.

Penicillin has been recommended, but is controversial, because release of toxin in the gut may worsen neurologic symptoms through lysis of bacterial cells in the gut or wound.[25] It would be ineffective if the problem is a direct toxin release.

Botulinum toxoid vaccine is available.[26] [27] The CDC provides a pentavalent vaccine that gives protection from toxin types A, B, C, D, and E but provides no protection against the F and G type toxins. The military believes that F and G type toxins are unlikely to be used in warfare because the strains of C. botulinum that produce toxins F and G are difficult to grow in large quantities. If new techniques allow production of toxins F and G in large quantities, the pentavalent vaccine will be useless. A heptavalent antitoxin against types A through G is available in limited supply at the U.S. Army Medical Research Institute of Infectious Diseases (USAMRIID) in Fort Detrick, Frederick, MD.

Clostridium Toxins

Tetanus neurotoxin is secreted by Clostridium species in similar fashion to botulinum. The toxin is a single 150,000 Dalton polypeptide that is cleaved into two peptides held together by disulfide and noncovalent bonds. The intoxication occurs at extremely low concentrations of toxin, is irreversible, and like botulism, requires activity of the nerve to cause toxicity to that nerve.

Clostridium perfringens also secretes at least 12 toxins and can produce gas gangrene (clostridial myonecrosis), enteritis necroticans, and clostridium food poisoning. One or more of these toxins could be produced as a weapon. The alpha toxin is a highly toxic phospholipase that could be lethal when delivered as an aerosol.

CLINICAL EFFECTS

Where botulinum toxin causes a flaccid paralysis, tetanus causes spastic paralysis. The tetanus neurotoxin migrates retroaxonally (up the nerve fiber) and by transcytosis, reaches the spinal inhibitory neurons, where it blocks neurotransmitter release and thus causes a spastic paralysis. Despite the seemingly different actions of tetanus and botulism, the toxins act in a similar way at the appropriate cellular level. The clinical effect in humans is well documented and includes twitches, spasms, rictus sardonicus, and convulsions.

Clostridium perfringens alpha toxin would cause vascular leaks, pulmonary damage, thrombocytopenia, and hepatic damage. Inhaled clostridium perfringens would cause a serious respiratory distress.

Detection
Acute serum and tissue samples should be collected for further testing. Specific immunoassays are available for both perfringens and tetani species. Bacteria may be readily cultured. As with most of these toxins and diseases, specific laboratory findings may be too late to be of clinical use.

Prophylaxis and Treatment
Clostridium perfringens and tetanus are generally sensitive to penicillin and this is the current drug of choice. There is some data that treatment with either clindamycin or rifampin may decrease clostridium perfringens toxin production and give better results.

Every medical provider is aware of the schedule for tetanus immunizations. It is unlikely that there will be any use of this toxin in the United States due to widespread tetanus immunization.[28] This may not be true in other countries, and in the United States there has been no published program about clinical syndromes of overwhelming amounts of tetanus toxin. Although the U.S. military apparently discounts this toxin, it is so easy to make and spread and so lethal that it would make a useful biologic toxin.

There is no specific prophylaxis against most of the clostridium perfringens toxins. Some toxoids for enteritis necroticans are available for humans. Veterinary toxoids are in wide use.

Ricin

Ricin is a type II ribosome inactivating protein produced by the castor bean plant and secreted in the castor seeds. The toxin is a 576 amino acid protein precursor weighing 65,000 Daltons. Once inside the cell, ricin depurinates an adenine from rRNA and thereby inactivates the ribosome, killing the cell.

Ricin is available worldwide by simple chemical process of the castor bean. Although ricin is only a natural product of the castor bean plant, ricin has been produced from transgenic tobacco using gene transfer principles. Large amounts of toxin would be able to be produced easily by this transgenic method.[29]

Clinical Effects
The clinical picture depends on the route of exposure. Castor bean ingestion causes rapid onset of nausea, vomiting, abdominal cramps, and severe diarrhea followed by vascular collapse. Death occurs on the third day. Inhalation of ricin will cause nonspecific weakness, cough, fever, hypothermia,

and hypotension, followed by cardiovascular collapse about 24 to 36 hours after inhalation. Death will occur about 36 to 48 hours after inhalation. High doses by inhalation appear to produce severe enough pulmonary damage to cause death.

At least one fatality has been documented as a direct result of ricin employed in bio-warfare. In 1978, ricin-impregnated pellets were fired from an umbrella at Georgi Markov and Vladimar Kostov. The pellets were coated with wax designed to melt at body temperature and release the ricin. At least six other assassinations have used the same technique according to intelligence sources.

DETECTION

ELISA for blood or histochemical analysis may be useful in confirming ricin intoxication. Ricin causes marked immune response and sera should be obtained from survivors for measurement of antibody response. An immunoassay technique has been developed by Environmental Technologies Group for ricin.

Standard laboratory tests are of little help in diagnosis of ricin intoxication. The patient may have some leukocytosis, with neutrophil predominance. The pleomorphic picture of ricin intoxication would suggest many respiratory pathogens and may be of little help in diagnosis.

PROPHYLAXIS AND TREATMENT

There is no approved immunologic or chemoprophylaxis at this time. Respiratory protection will prevent inhalation exposure and is the best prophylaxis currently available. Ricin has no dermal activity and is not transported through the skin.

There is ongoing effort to produce both active immunization and passive antibody prophylaxis suitable for humans. These techniques have been used in animals.

Treatment is supportive and includes both respiratory support and cardiovascular support as needed. If oral ingestion is suspected, then lavage followed by charcoal is appropriate.

Saxitoxin

Saxitoxin is a dinoflagellate toxin responsible for paralytic shellfish poisoning. It is also found in several species of puffers and other marine animals and was originally discovered in 1927.[30] The toxin is very soluble in water, heat stable, and is not destroyed by cooking. The lethal dose is 1 to 2 mg. There are multiple related toxins with substitutions at key positions.

Clinical Effects

It is similar in effects and treatment to tetrodotoxin. Onset of symptoms is within minutes of exposure. Death may occur within 2 to 24 hours. If the patient survives, then normal functions are regained within a few days.

Detection

A mouse unit is the minimum amount of toxin that will kill a 20 gram mouse within 15 minutes. There is a standardized mouse assay for routine surveillance and immunoassays are available.

Prophylaxis and Treatment

There is no antidote, so symptomatic treatment is appropriate. Antibodies for tetrodotoxin will frequently protect against saxitoxin.[31]

Staphylococcal Enterotoxin

Staphylococcal food poisoning is familiar to emergency physicians. The disease is changed when the enterotoxin is delivered by aerosol. The organism that produces this agent is readily available and could be tailored to produce large quantities of the toxin.

Clinical Effects

The disease begins 1 to 6 hours after exposure with the sudden onset of fever, chills, headache, myalgias, and a nonproductive cough. The cough may progress to dyspnea and substernal chest pain. In severe cases, pulmonary edema may be found. Nausea, vomiting, and diarrhea are common (as in the poisoning familiar to emergency physicians). The only physical finding of note is conjunctival injection.

In food-borne staphylococcal enterotoxin B (SEB), fever and respiratory involvement are not found and the gastrointestinal symptoms predominate.

Sickness may last as long as 2 weeks and severe exposures may cause fatalities.

Detection

The lab is not helpful. Erythrocyte Sedimentation rate (ESR) may be elevated, but this is a nonspecific finding. A chest x-ray is usually normal, but may have increased intrastitial markings and possibly pulmonary edema. An immunoassay has been developed by Environmental Technologies Group from Baltimore, Maryland,[32] that is cost efficient and usable in the field environment.

PROPHYLAXIS AND TREATMENT

There is no significant treatment regimen available. Therapy is entirely supportive. There is no current prophylaxis available, although experimental immunization has been reported.

Tetrodotoxin

Tetrodotoxin is a potent neurotoxin produced by fish, salamanders, frogs, octopus, starfish, and mollusks, notably the puffer (also called the globefish or blowfish).[33] The dangers of tetrodotoxin poisoning were known by the ancient Egyptians (2400 to 2700 B.C.). All organs of the fresh water puffer are toxic with the skin having the highest toxicity followed by gonad, muscle, liver, and intestine. In saltwater puffers, the liver is the most toxic organ. The lethal dose of tetrodotoxin is only 5 micrograms per kilogram in the guinea pig.

Puffer intoxication is a serious public health problem in Japan and over 50 people each year are intoxicated. Raw puffer fish, commonly called fugu, is a delicacy in several Southeast Asian countries including Japan. Consumption of fugu causes a mild tetrodotoxin intoxication with a pleasant peripheral and perioral tingling sensation. Improperly prepared fugu may contain a lethal quantity of tetrodotoxin. Fatalities have gradually decreased because of the increased understanding of the toxin and careful preparation of the puffer for food.[34] Cooking the food will not dissipate the toxin. Tetrodotoxin is heat stable.

There are several microbial sources of tetrodotoxin including *Pseudomonas, Vibrio, Listonella,* and *Alteromonas* species. Although there is only one known bacteria that has produced tetrodotoxin toxicity in humans, there is a significant potential for genetic alteration of common species of bacteria to produce tetrodotoxin.[35]

Tetrodotoxin is well known for its ability to inhibit neuromuscular function by blocking the axonal sodium channels.[36] Cranial diabetes insipidus has been reported in critically ill patients. Mortality from tetrodotoxin is thought to be due to hypoxic brain damage from prolonged respiratory paralysis.

CLINICAL EFFECTS

The clinical symptoms and signs of tetrodotoxin poisoning are similar to those of the acetylcholinesterase poisons.[37] Clinical symptoms include nausea, vomiting, vertigo, perioral numbness, unsteady gait, and extremity numbness. Clinical symptoms begin within 30 minutes of ingestion. The speed of onset depends on the quantity of the toxin ingested. The symptoms progress to muscle weakness, chest tightness, diaphoresis, dyspnea, chest pain, and finally paralysis. Hypotension and respiratory failure are seen in severe poisonings.

Patients will frequently complain of a sensation of cold or chilliness. Paresthesias spreads to the extremities with symptoms often more pronounced distally. Death can occur within 17 minutes after ingestion of tetrodotoxin.

DETECTION

Detection of tetrodotoxin is by mouse bioassay [Association of Official Analytical Chemists (AOAC) Official Methods of Analysis 18.086–18.092, 1984, pp. 344–345] or by liquid chromatography. Use of radioimmunoassay and radioreceptor assays have also been reported. An *in vitro* colorimetric cell assay against a rabbit antiserum has been developed and may be more rapid than older methods, but is not yet publicly available.[38]

PROPHYLAXIS AND TREATMENT

At present, there is no known antidote for tetrodotoxin intoxication. There are numerous anecdotal treatments of survivors with supportive therapy alone. Certainly respiratory support and airway management will be lifesaving for a majority of these patients. Gastric lavage will remove unabsorbed toxin from the gut and is used in puffer fish intoxication. Activated charcoal has been reported to effectively bind the toxin and may be employed in ingestions.

4-Aminopyridine has been used to treat tetrodotoxin intoxication in laboratory animals.[39] 4-Aminopyridine is a potent potassium channel blocker and enhances impulse evoked acetylcholine release from presynaptic motor terminals. There have been no human studies of its use as an antidote. 4-Aminopyridine can cause muscle fasciculation and seizures in a dose-dependent phenomenon.

Naloxone has been proposed as a possible antidote, since the opiates and tetrodotoxin have similar molecular configurations.[40] There are no reports of this in either laboratory or clinical use.

Active and passive immunization has been demonstrated in laboratory animals, although there is no known available human immunization for tetrodotoxin.[41] Tolerance does not develop on repeated puffer fish exposure. Monoclonal antibodies have been produced and protected laboratory animals against lethal doses of tetrodotoxin.[42 43]

Trichothecene Myocotoxins (T2)

The trichothecene mycotoxins are produced by fungi and achieved fame in the 1970s as the best candidates for the infamous "yellow rain" found in Laos, Cambodia, and Afghanistan. Naturally occurring trichothecenes have caused moldy corn toxicosis in animals.

They are potent inhibitors of protein synthesis, inhibit mitochrondrial respiration, impair DNA synthesis, and destroy cell membranes.

Clinical Effects

Consumption of trichothecenes causes weight loss, vomiting, bloody diarrhea, and diffuse hemorrhage. The onset of the illness occurs within hours, and death within 12 hours. Inhalation adds respiratory distress and failure to the picture. Survivors have reported a radiation-sickness-like disease. This has included fever, nausea, vomiting, leukopenia, diarrhea, bleeding, and finally sepsis. Painful skin lesions also occur in survivors.

Detection

There is no readily available diagnostic test, although reference laboratories may be able to help with gas-liquid chromatography. There are some polyclonal and monoclonal antibodies for detection in liquid or solid samples. Urine samples are most useful for this purpose because the metabolites can be detected as long as 28 days after exposure to the agent.

Prophylaxis and Treatment

Ascorbic acid has been proposed to decrease the lethality. This has been studied in animals only, but since ascorbic acid has few side effects and is cheap, it should be used in all suspected cases.

Dexamethasone (1 to 10 mg IV) has also been shown to decrease lethality as late as three hours after exposure to these toxins.

In ingestions, charcoal or superactivated charcoal will absorb remaining toxin and decrease lethality.

Possible Live Bacteriologic Warfare Agents

This group covers only a few diseases that have been researched. Much information was obtained from *The United States Army Field Manual 8–9; Handbook on the Medical Aspects of NBC Defensive Operations* (FM 8–9), Part II, Biological, United States Government Printing Office, 1996 (also available on the Internet at http://www.nbc-med.org/FMs). Other diseases have been proposed and researched as a result of multiple sessions with interested colleagues or my travels to the city of Sverdlovsk in the USSR.

Although these diseases have been proposed by the U.S. military and others as possible biologic warfare agents, there is no question that the list is neither exhaustive nor all inclusive. Other diseases that have been considered include typhoid fever, Ebola virus, melioidosis, Rift Valley fever, epidemic typhus, Rocky Mountain spotted fever, scrub typhus, coccidiomycosis, histoplasmosis, Chikun-Gunya fever, Congo Crimean fever, Lassa fever, dengue fever, Eastern equine encephalitis, Western equine encephalitis, Venezuelan encephalitis, Omsk hemorrhagic fever, Korean hemorrhagic fever, and many others (at least 60). The astute reader can recognize the potential for bio-warfare in almost any disease that can possibly afflict humans. Numerous other diseases could be used as bio-warfare agents against selected crops or livestock.

With the current level of gene manipulation, it is easy to foresee a chimera tailored bacteria or rickettsia that has characteristics of one disease, with tailored resistance to all usual antibiotics, yet responsive to an unusual antibiotic that the designer has stockpiled. It is equally easy to think of a tailored virus that has unusual mortality for white Anglo-Saxon males, but has little mortality for Asian or African American stock. One does not have to imagine an increase in lethality in order to find substantial bio-warfare applications. A rapidly spreading upper respiratory illness—the common cold—that merely causes 3 days of cough, fever, rhinorrhea, and malaise could be incapacitating if an entire army caught it simultaneously. A city's police force would be unable to deal with terrorists effectively if over three-fourths of the entire city's population had uncontrollable diarrhea for a two- or three-day course.

The problems cited in dispersal, control, mutability, and side effects that have been previously discussed are entirely too applicable for live bio-warfare agents.

Anthrax

Anthrax is caused by *Bacillus anthracis.* Under usual (non-wartime) conditions, humans become infected by contact with an infected animal or contaminated animal by-products. Anthrax is also known as "wool-sorter's disease." There are three forms of anthrax: cutaneous, inhalation, and gastrointestinal. Almost all naturally occurring cases of anthrax are cutaneous or gastrointestinal.

Anthrax was proposed and investigated as a bio-weapon by both the Allies in World War II and by the communists in the former USSR. Indeed, an epidemic that caused 96 cases of human anthrax in the city of Ekatrinburg (formerly Sverdlosvk) in spring of 1979 has been traced to an escaped Russian bio-weapon strain of anthrax. In these patients the pathogen was airborne. Although medical records were confiscated by the KGB, investigators have pieced together the epidemiology and the source of the epidemic.[44] Following the epidemic, thousands of citizens were immunized against anthrax, the exteriors of the buildings and trees were washed by local fire brigades, and several unpaved streets were asphalted. Notably absent in the public health response was a military component. In 1992 Russian President Boris Yeltzin admitted that the military was the source of the outbreak. Peristroika and the downfall of the former communist empire has led to greater release of information, but the staff of city hospital number 40 where the victims were cared for remains quite sensitive in discussions about this event.[45]

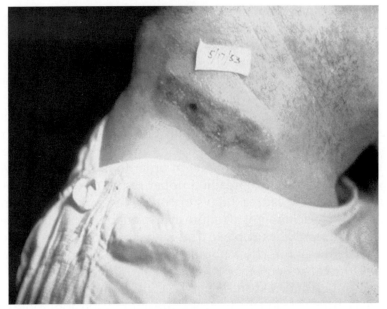

FIGURE 7-1 Anthrax Skin Lesion on Neck of Man. *Source:* Courtesy C.D.C./P.H.I.L.

Anthrax is likely to be disseminated as an aerosol of the very persistent spores. The incubation time is from 1 to 6 days, but as the Ekatrinburg incident showed, anthrax may have a prolonged incubation period of up to 2 months. The longer incubation periods are seen most frequently when partial treatment has been given. The spores can be quite stable, even in the alveolus. The duration of the disease is between 2 and 5 days.

PRESENTATION

The inhalation form of anthrax is particularly uncommon and particularly lethal. In its early presentation, inhalation anthrax could be confused with a plethora of viral or bacterial respiratory illnesses. The patient progresses over 2 to 3 days and then suddenly develops respiratory distress, shock, and death within 24 to 36 hours. Widening of the mediastinum on chest radiograph is common. Evidence of infiltrates on the chest x-ray are uncommon. Other suggestive findings include chest wall edema, hemorrhagic pleural effusions, and hemorrhagic meningitis.

DIAGNOSIS

Diagnosis can be made by culture of blood, pleural fluid, or cerebrospinal fluid. The blood culture is most often positive. In fatal cases, impressions of mediastinal lymph nodes or spleen will be positive. Anthrax toxin may be detected in blood by immunoassay.

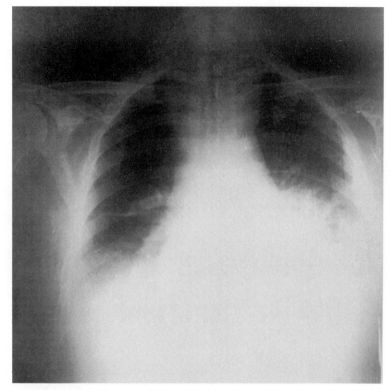

FIGURE 7-2   Chest radiograph showing widened mediastinum due to inhalation anthrax. Radiograph taken 22 hours before death. *Source:* Courtesy C.D.C./P.H.I.L.

The cases in Ekatrinburg were diagnosed on autopsy by a pathologist who noted a peculiar "cardinal's cap" meningeal inflammation typical in anthrax. Inhalational anthrax may be diagnosed at autopsy by the mediastinal inflammation.

Environmental Technologies Group, Incorporated, has developed an immunoassay for anthrax.

THERAPY

Penicillin is considered the drug of choice for treatment of naturally occurring anthrax. However, penicillin-resistant strains do exist, and one could expect that anthrax used for a biologic weapon would be developed as penicillin resistant. Tetracycline and erythromycin have been used for patients who are allergic to penicillin. Induction of resistance to these antibiotics is an easy exercise for genetic manipulation and warfare strains should be presumed to be resistant to these antibiotics until proven otherwise. Chloramphenicol, gentamycin, and ciprofloxacin would be appropriate choices for initial therapy.

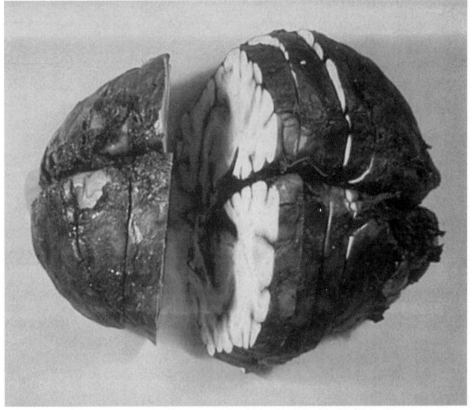

FIGURE 7-3 Gross pathology of fixed cut brain showing hemorrhagic meningitis due to inhalation anthrax. *Source:* Courtesy C.D.C./P.H.I.L.

The U.S. military recommends oral ciprofloxacin or intravenous doxycycline for initial therapy.[46] This therapy is not appropriate for those under 18 years of age or for pregnant females. Supportive therapy for airway, shock, and fluid volume deficits are appropriate.

PROPHYLAXIS

Two types of anthrax vaccine for human use are available in the United States and United Kingdom, albeit in totally insufficient quantities for a civilian biological warfare challenge. Both are based on the partially purified protective antigen of the *B. anthracis* adsorbed to an aluminum adjuvant. The usual immunization series is six 0.5 ml doses over a span of 18 months. The military feels that a primary series of three 0.5 ml doses (0, 2, and 4 weeks) will be protective against both cutaneous and inhalation anthrax for about 6 months after the primary series. These immunizations were given to many coalition troops during the Gulf War in anticipation of Saddam Hussein's employment of this agent. Large quantities of antigen are presumed to be stockpiled for military use since this agent has been a recurring threat. Unless civilian immunizations start about 1 month prior to a terrorist attack, EMS and medical providers will be essentially unprotected.

Although "minor" reactions to the vaccine are common (6 percent of immunized population), major reactions are uncommon. Obviously, the vaccine is contraindicated for those who are known to be sensitive to it and those who have already had clinical anthrax. The choice between immunization and some allergic reaction and no immunization in the face of a serious bio-warfare threat presents a difficult clinical dilemma.

A live anthrax vaccine is used in Russia to immunize both livestock and human beings. It is a spore vaccine with both STI-1 and strain 3 mixtures. The Russians feel that this vaccine is superior at stimulating cell-mediated immunity.[47] There would be considerable resistance to use of the Russian vaccine in Western countries because of concerns over purity and residual virulence of a live vaccine.

There is no available evidence that these vaccines will adequately protect against an aerosol challenge. New vaccines with a highly purified protective antigen or designer attenuated strains have both been used in laboratories, but are not commercially available.[48] [49]

Antibiotic prophylaxis with ciprofloxacin (500 mg PO bid), or doxycycline (100 mg PO bid) is also recommended by the U.S. military for imminent attack by a biological weapon. Should the attack be confirmed as anthrax, then antibiotics should be continued for four weeks for all who are exposed. Those exposed should also be started on anti-anthrax vaccine with the standard schedule (if it is available), if they have not been previously immunized. Those who have received fewer than three doses of vaccine prior to exposure should receive a single booster injection. If vaccine is not available, then antibiotics should be continued until the patient can be safely and closely observed when the antibiotics are discontinued. Inhaled spores are not destroyed by antibiotics and may persist beyond the course of antibiotics recommended.

Brucellosis

Brucellosis is a zoonotic disease caused by a small nonmotile coccobacilli. The natural reservoir is domestic herbivores such as goats, sheep, cattle, and pigs. There are four species: *Brucella melitensis, B. abortus* (cattle), *B. suis* (pigs), and *B. canis* (dogs). Humans become infected when they ingest raw infected meat or milk, inhale contaminated aerosols, or through skin contact. Human infection is also called undulant fever. Human to human transmission is rare if it occurs at all.

Brucella species have been long considered as biological warfare agents because of the stability, persistence, and ease of infection without human-to-human transfer. Brucellosis can be spread by aerosol spray or by contamination of food supply (sabotage). There is a long persistence in wet ground or food.

PRESENTATION

The incubation period is about 8 to 14 days, but may be considerably longer. Clinical disease is a nonspecific febrile illness with headache, fatigue, myalgias, anorexia, chills, sweats, and cough. The fever is often up to 105°F. The disease may progress and include arthritis, lymphadenopathy, arthralgias, osteomyelitis, epididymitis, orchitis, and endocarditis. Disability is pronounced, but lethality is about 5 percent or less in usual cases. The disease may be followed by recovery and relapse. The duration of the disease is usually a few weeks, but brucellosis may last for years.

DIAGNOSIS

Diagnosis of this disease is by blood culture, bone marrow culture, or by serology. There are no other laboratory findings that contribute to a diagnosis of brucellosis.

THERAPY

The U.S. military recommends doxycycline (100 mg BID) plus rifampin (900 mg/day) for six weeks. These antibiotics are generally available in sufficient quantity in the United States. Alternative therapy proposed has been doxycycline (100 mg BID) for six weeks and streptomycin (1 gm/day) for three weeks. TMP-SMX has been given for 4 to 6 weeks, but is thought to be less effective. Relapse and treatment failure is common.

PROPHYLAXIS

There is no information available about chemoprophylaxis for this disease. Human vaccines are not routinely available in the United States, but have been developed by other countries. A variant of Brucella abortus, S19-BA, has been used in the former USSR to protect occupationally exposed groups. Efficacy is limited and annual revaccination is needed. A similar vaccine is available in China. Neither of these two vaccines would meet Western requirements for safety and effectiveness.[50]

Cholera

Cholera is a well-known diarrheal disease caused by *Vibrio cholera* acquired in humans through ingestion of contaminated water. The organism causes a profound secretory "rice water" diarrhea by elaborating an enterotoxin.

Although cholera can be spread by aerosols, more likely terrorist or military employment would be contamination of food or water supplies. There is negligible direct human-to-human transmissibility. The bacterium does not have long persistence in food or pure water and is not persistent when applied by aerosols.

PRESENTATION

Cholera can cause a profuse watery diarrhea that causes hypovolemia and hypotension. Without treatment, cholera can rapidly kill adults and children alike from severe dehydration and resultant shock. The incubation period is 1 to 5 days and the course of the illness is about 1 week.

The patient may have vomiting early in the illness. There is little abdominal pain associated with the disease.

DIAGNOSIS

Gram staining of the stool sample will show few or no red or white cells. Renal failure may complicate severe dehydration. Electrolyte abnormalities are common with the profound fluid loss; generally hypokalemia predominates.

Rotavirus, E. coli, and toxic ingestions such as staphylococcal food poisoning, Bacillus cereus, or even clostridia species can all cause similar watery diarrhea. Bacteriologic diagnosis of cholera diarrhea has been well studied for decades. Vibrio species can be seen and identified readily with dark field or phase contrast microscopes. Culture will prove the diagnosis, but is not necessary for the treatment.

THERAPY

Treatment of cholera is mostly supportive. Although most U.S. emergency physicians are used to treating significant hypovolemia with intravenous fluid replacement, it is unlikely to be readily available if an epidemic of cholera is caused by terrorist or enemy action. The World Health Organization (WHO) oral rehydration formula is appropriate, but generally not stocked in sufficient quantities in most cities. Pedialyte and sport drinks such as Gatorade will provide interim oral hydration. If a cholera epidemic is treated, then intravenous fluids should be reserved for those patients who are vomiting and cannot tolerate oral rehydration, patients who have more than 7 liters per day of stool, and patients who have such hypovolemia that they are in shock.

Tetracycline and doxycycline have both been found to shorten the course of the diarrhea. Other effective drugs include ampicillin (250 mg every 6 hours for 5 days) and TMP-SMX (1 tablet every 12 hours). Appropriate scale should be used for pediatric doses.

Prophylaxis

The currently available vaccine is a killed suspension of *V. Cholera*. It provides incomplete protection and lasts for no longer than 6 months. It requires two injections with a booster dose every 6 months. Improved vaccines are being tested but are not yet available.

Ebola Virus

The Ebola virus is a member of a family of RNA viruses known as filoviruses. When magnified several thousand times by an electron microscope, these viruses have the appearance of long filaments or threads. Ebola virus was discovered in 1976 and was named for a river in Zaire, Africa, where it was first detected.[51]

Ebola virus has been covered significantly in the popular literature and in several books and movies (*Outbreak*). The Aum Shirinkyo cult visited Zaire to collect Ebola. This virus is well spread by body fluids, particularly blood. It is quite dangerous for the health care provider because human-to-human contact will rapidly spread the disease. It is capable of aerosol spread.

Use of this virus (with greater than 90 percent lethality) would be considered a "doomsday" operation by the military. There is no guarantee that this virus would be able to be contained if spread to a modern city. The persistence is low, but the transmissibility is so high that this is immaterial.

Presentation

Ebola virus is a viral hemorrhagic fever. It can be spread by blood and blood products, secretions, and by aerosol transmission.[52] It is highly lethal (more than 90 percent) with a rapid course.

Diagnosis

A diagnosis is made by detection of Ebola antigens, antibody, or genetic material, or by culture of the virus from these sources. Diagnostic tests are usually performed on clinical specimens that have been treated to inactivate (kill) the virus. Research on Ebola virus must be done in a special, high-containment laboratory to protect scientists working with infected tissues.

Therapy

Therapy is supportive only. There is no known therapy for this disease.

Prophylaxis

There is no known prophylaxis for Ebola virus. Sera from survivors have been obtained and it is possible that passive protection could be developed.

Plague

Plague is a zoonotic disease caused by *Yersinia pestis*. It is naturally found on rodents and prairie dogs and their fleas. Under normal conditions, three syndromes are recognized: inhalational (pneumonic), septicemic, and bubonic. The usual first infection is the bubonic form.

In 1994 defectors revealed that the Russians had conducted research on *Yersinia pestis*, the plague bacterium, to make it more virulent and stable in the environment. The plague can retain viability in water for 2 to 30 days, moist areas for up to 2 years, and in near freezing temperatures for several months to a year.

Plague could be spread by either infected vectors such as fleas, or by an aerosol spray. Person-to-person transmissibility is high and the bacterium is highly infective. The persistence is low, but the transmissibility is so high that this is immaterial.

PRESENTATION

In bubonic plague, the incubation period is from 2 to 10 days. The onset is acute with malaise, fever (often quite high), and purulent lymphadenitis. The lymphadenitis is most often inguinal, but cervical and axillary nodes are also involved. As the disease progresses, the nodes become tender, fluctuant, and finally necrotic. The bubonic form may progress to the septicemic form with seeding of the CNS and lungs. If the organisms are seeded to the lungs, then the pneumonic form follows and the patient becomes contagious through coughing and droplet spread. The course of the disease is 2 to 3 days and the disease is quite lethal.

In primary pneumonic plague, the incubation period is 2 to 3 days. The onset is acute and fulminant with malaise, fever, chills, cough with bloody sputum, and toxemia. The pneumonia progresses rapidly to respiratory failure with dyspnea, stridor, and cyanosis.

In untreated patients, the mortality is over 50 percent for the bubonic and septicemic forms. In the pneumonic form, the mortality approaches 100 percent. The terminal events are circulatory collapse, hemorrhage, and peripheral thrombosis in septicemic plague. In pneumonic plague, the terminal event is often respiratory failure as well as circulatory collapse.

DIAGNOSIS

A presumptive diagnosis can be made by finding the typical safety pin bipolar staining organisms in Giemsa-stained specimens. Appropriate specimens are lymph node aspirate, sputum, or cerebral spinal fluid (CSF). Immunofluorescent staining is available and helpful if readily accessible. *Y. pestis* can be readily cultured from any of these sources.

Environmental Technologies Group, Incorporated, has developed an immunoassay for plague.

THERAPY

This disease is readily contagious and strict isolation of patients is essential. Streptomycin, tetracycline, and chloramphenicol are all useful if given within the first 24 hours after symptoms of pneumonic plague begin. Supportive therapy of complications is essential.

PROPHYLAXIS

Plague vaccine is available, but probably does not protect against an aerosol exposure and subsequent pneumonic plague. The plague vaccine is a whole cell formalin-killed product. The usual dose is 0.5 ml given at 0,1, and 2 weeks.

Plague vaccines providing protection against aerosol exposure are not yet available, but are under development.[53] Current whole-cell plague vaccines stimulate immunity against the bubonic form, but are probably not effective for the pneumonic form.[54][55]

Q Fever

PRESENTATION

Q fever is a rickettsial zoonotic disease caused by *Coxiella burnetti.* The usual animals affected are sheep, cattle, and goats. Human disease is usually caused by inhalation of particles contaminated with *Coxiella.*

Q fever is a self-limiting febrile illness of 2 days to 2 weeks. The incubation period is about 10 to 20 days. The patient is usually ill, but uneventful recovery is the rule. Q fever pneumonia is a frequent complication and may be noted only on radiographs in most cases. Some patients will have nonproductive cough and pleuritic chest pain. Other complications are not common and may include chronic hepatitis, endocarditis, meningitis, encephalitis, and osteomyelitis.

DIAGNOSIS

Q fever's presentation as a febrile illness with an atypical pneumonia is similar to a host of other atypical pneumonias, including mycoplasma, legionnaire's disease, chlamydia pneumonia, psittacosis, or hantavirus.

The diagnosis can be confirmed serologically and other laboratory findings are unlikely to be helpful. Most patients with Q fever will have slightly elevated liver enzymes. It is difficult to isolate rickettsia, and Q fever is no exception.

Therapy

As with other rickettsial diseases such as Rocky Mountain Spotted Fever, the treatment of choice is tetracycline, doxycycline, or erythromycin. Although not tested, azithromycin and Biaxin would be expected to be effective.

Prophylaxis

A formalin-inactivated whole cell vaccine is available as an investigational drug in the United States and has been used for those who are at risk of occupational infection with Q fever.[56] One dose will provide immunity for an aerosol challenge within 3 weeks.

Skin testing is required to prevent a severe local reaction in previously immune individuals. A live attenuated strain (M44) has been used in the former USSR.[57]

Smallpox

Smallpox was used as a biologic weapon in the United States during the French and Indian War. Smallpox is an orthopox virus that affects primates, particularly man. The disease was declared eradicated in the world in 1977, and the last reported human case occurred in a laboratory in 1978. Theoretically, the virus exists in only two laboratories in the world, one in the United States and one in Russia. The virus can be transmitted by face-to-face contact, secretions, and aerosols. It is a durable virus and can exist for long periods outside the host. It is remotely possible that it is still living outside of the repository labs. A very closely related disease, monkeypox, cannot be easily distinguished from smallpox.

Presentation

Smallpox has a long incubation period of about 10 to 17 days. The illness has a prodrome of 2 to 3 days with malaise, fever, headache, and backache. Over the next 7 to 10 days, all of the the characteristic lesions erupt, progress from macules to papules to vesicles to pustules and then crust and scarify. The lesions are more numerous on the extremities and face than on the trunk. The disease is fatal in about 35 percent of cases. Some patients will develop disseminated intravascular coagulopathy. Other complications include smallpox pneumonia, arthritis (may have permanent joint deformities), and keratitis (may cause blindness).

Diagnosis

Like many viral diseases, the diagnosis is best made by clinical impression. Routine labs are not helpful, although leukopenia is frequent. Clotting factors may be depressed and thrombocytopenia may be found. Diagnosis may be made with immunofluorescence, electron microscopy, or culture.

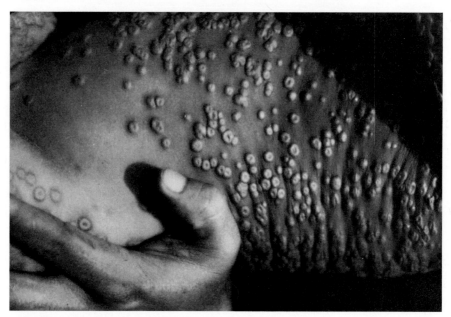

FIGURE 7-4   Smallpox Lesions on Skin of Trunk. *Source:* Courtesy C.D.C./ P.H.I.L.

THERAPY

Therapy is entirely supportive.

PROPHYLAXIS

Prophylaxis against smallpox has been available since the time of Jenner and is well documented. Since smallpox is presumed to have been eradicated worldwide, there is no recommendation or requirement for routine vaccination. Adequate stocks of smallpox vaccine are probably not available for exposure of large portions of the population.

Objects in contact with a contaminated patient need to be cleansed with live steam or sodium hypochlorite solution.

Tularemia

Tularemia or "rabbit fever" is caused by *Francisella tularensis,* a gram negative bacillus. Humans can contract this disease by handling an infected animal or by the bites of ticks, mosquitoes, or deerflies. The natural disease has a mortality rate of 5 to 10 percent. As few as 50 organisms can cause disease if inhaled.

PRESENTATION

Like plague, tularemia has an ulceroglandular form, a pneumonic form, and a septicemic form. Two additional forms also occur with tularemia. Oculoglandular tularemia occurs when the innoculum is in the eye. Gastro-

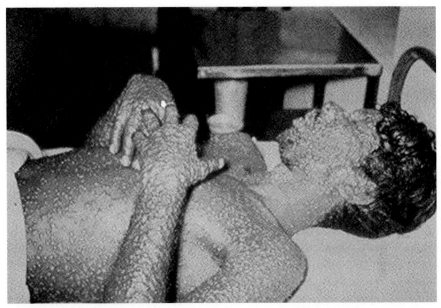

FIGURE 7-5 Adult Male with Smallpox Lesions. *Source:* Courtesy C.D.C./ P.H.I.L.

intestinal tularemia occurs when tularemia bacilli are ingested. It may also infect the oropharynx.

The septicemic form can occur in 5 to 15 percent of natural cases. The clinical features include fever, prostration, and weight loss.

The pneumonic form may occur by inhaling contaminated dusts or by a deliberate aerosol. The resulting pneumonia is atypical and may be fulminant. Fever, headache, malaise, substernal discomfort, and cough are prominent. The cough is often nonproductive. A chest x-ray may or may not show a pneumonia.

DIAGNOSIS
As noted, the diagnosis of pneumonic tularemia will be difficult clinically, with several types of atypical pneumonia as differential diagnoses. The laboratory is unhelpful early in this disease.

THERAPY
Human-to-human spread is unusual and isolation is not required.

Treatment is streptomycin or gentamycin for 10 to 14 days. Tetracycline and chloramphenicol are also useful, but the military reports that there has been a significant relapse rate.

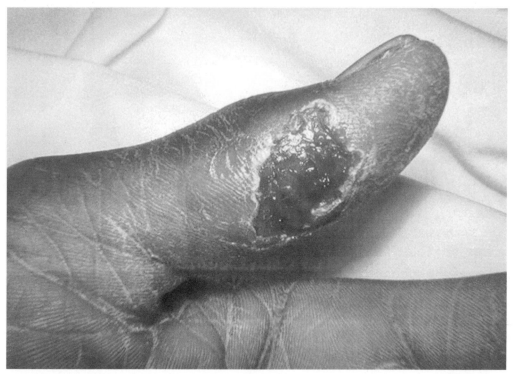

FIGURE 7-6 Thumb with Skin Ulcer of Tularemia. *Source:* Courtesy C.D.C./P.H.I.L.

Prophylaxis

A live vaccine strain is available to United States military personnel. This vaccine is delivered intradermally and provides protection to an aerosol challenge by the third week post immunization. Protection is dependent on the inhaled dose of tularemia, and inhalation of massive quantities of bacteria may overwhelm the protective effects of the vaccine.[58] Protection falls after 14 months, suggesting that a booster dose is appropriate.

This vaccine is not available for civilian use.

Chapter Summary

As the emergency service community continues to be subjected to reports of escalating bio-terrorism threats, the increase of media attention on the bio-threat (such as the five-day ABC Nightline Special broadcast in 1999) results in heightened fears and public questions of readiness. Events such as the 1999 West Nile Virus outbreak in the New York Metropolitan area immediately result in the press and public fearing that the event was a deliberate act rather than a naturally occurring event.

Table 7-1 Biological Agents: Symptoms and Treatment

Anthrax	Cholera
Signs and Symptoms:	**Signs and Symptoms:**
Incubation period is 1–6 days. Fever, malaise, fatigue, cough, and mild chest discomfort are followed by severe respiratory distress with dyspnea, diaphoresis, stridor, and cyanosis. Shock and death occur within 24–36 hours of severe symptoms.	Incubation period is 1–5 days. Asymptomatic to severe with sudden onset. Vomiting, abdominal distention, and pain with little or no fever followed rapidly by diarrhea. Fluid losses may exceed 5–10 liters per day. Without treatment, death may result from severe dehydration, hypovolemia, and shock.
Diagnosis:	**Diagnosis:**
Physical findings are nonspecific. Possible widened mediastinum. Detectable Gram stain of the blood and by blood culture in the course of illness.	Clinical diagnosis; Watery diarrhea and dehydration. Microscopic exam of stool samples reveals few or no red or white cells. Can be identified in stool by dark field or phase contrast microscopy and can be grown on a variety of culture media.
Treatment:	**Treatment:**
Although usually not effective after symptoms are present, high dose antibiotic treatment with penicillin, ciprofloxacin, or doxycycline should be undertaken. Supportive therapy may be necessary.	Fluid and electrolyte replacement. Antibiotics such as tetracycline, ampicillin, or trimethoprim-sulfamethoxazole will shorten the duration of diarrhea.
Prophylaxis:	**Prophylaxis:**
A licensed vaccine for use in those considered at risk for exposure. Vaccine schedule is 0, 2, and 4 weeks for initial series, followed by boosts at 6, 12, and 18 months, and then a yearly booster.	A licensed, killed vaccine is available, but provides only about 50% protection that lasts no more than 6 months. Vaccination schedule is at 0 and 4 weeks with booster doses every 6 months.
Decontamination:	**Decontamination:**
Secretion and lesion precautions should be practiced. After an invasive procedure or autopsy is performed, the instruments and area used should be thoroughly decontaminated with a sporicidal agent such as iodine or chlorine.	Personal contact rarely causes infection; however, enteric precautions and careful hand washing should be frequently employed. Bactericidal solutions such as hypochlorite would provide adequate decontamination.

Table 7-1 Continued

Plague	Tularemia
Signs and Symptoms:	**Signs and Symptoms:**
Pneumonic plague: Incubation period is 2–3 days. High fever, chills, hemoptysis, toxemia, progressing rapidly to dyspnea, stridor, and cyanosis. Death results from respiratory failure, circulatory collapse, and bleeding diathesis. Bubonic plague: Incubation period is 2–10 days. Malaise, high fever, and tender lymph nodes (buboes); may progress spontaneously to the septicemic form, with spread to the central nervous system, lungs, and elsewhere	Ulceroglandular tularemia presents with a local ulcer and regional lymphadenopathy, fever, chills, headache, and malaise. Typhoidal or septicemic tularemia presents with fever, headache, malaise, substernal discomfort, prostration, weight loss, and a nonproductive cough.
Diagnosis:	**Diagnosis:**
Clinical diagnosis. A presumptive diagnosis can be made by Gram or Wayson stain of lymph node aspirates, sputum, or cerebral spinal fluid. Plague can also be cultured.	Clinical diagnosis; Physical findings are usually nonspecific. Chest x-ray may reveal pneumonic process, mediastinal lymphadenopathy or pleural effusion. Routine culture is possible but difficult. The diagnosis can be established by serology.
Treatment:	**Treatment:**
Early administration of antibiotics is very effective. Supportive therapy for pneumonic and septicemic forms is required.	Administration of antibiotics with early treatment is very effective.
Prophylaxis:	**Prophylaxis:**
A licensed, killed vaccine is available. Initial dose followed by a second smaller dose 1–3 months later, and a third 3–6 months later. A booster dose is given at 6, 12, and 18 months, and then every 1–2 years. This vaccine may not protect against aerosol exposure.	A live attenuated vaccine is available as an investigational new drug. It is administered once by scarification. A two-week course of tetracycline is effective as prophylaxis when given after exposure.
Decontamination:	**Decontamination:**
Secretion and lesion precautions with bubonic plague should be practiced. Strict isolation of patients with pneumonic plague. Heat, disinfectants, and exposure to sunlight render bacteria harmless.	Secretion and lesion precautions should be practiced. Strict isolation of patients is not required. Organisms are relatively easy to render harmless by heat and disinfectants.

Table 7-1 Continued

Q Fever	Smallpox
Signs and Symptoms:	**Signs and Symptoms:**
Fever, cough, and pleuritic chest pain may occur as early as 10 days after exposure. Patients are not generally critically ill and the illness lasts from 2 days to 2 weeks.	Clinical manifestation begins acutely with malaise, fever, rigors, vomiting, headache, and backache. 2–3 days later lesions appear, which quickly progress from macules to papules and eventually pustular vesicles. They are more abundant on the extremities and face and develop synchronously.
Diagnosis:	**Diagnosis:**
Q fever is not a clinically distinctive illness and may resemble a viral illness or other types of atypical pneumonia. The diagnosis is confirmed serologically.	Tests of electron and light microscopy are not capable of discriminating variola from vaccinia, monkeypox, or cowpox. The new PCR diagnostics techniques may be more accurate in discriminating between variola and other orthopox viruses.
Treatment:	**Treatment:**
Q fever is generally a self-limiting illness even without treatment. Tetracycline or doxycycline are the treatments of choice and are orally administered for 5–7 days. Q fever endocarditis (rare) is much more difficult to treat.	At present there is no effective chemotherapy and treatment of a clinical case remains supportive.
Prophylaxis:	**Prophylaxis:**
Treatment with tetracycline during the incubation period may delay but not prevent the onset of symptoms. An activated whole-cell vaccine is effective in eliciting protection against exposure, but severe local reactions to this vaccine may be seen in those who already possess immunity.	Immediate vaccination or revaccination should be undertaken for all personnel exposed. Vaccinia-immune globulin (VIG) is of value in postexposure prophylaxis of smallpox when given within the first week following exposure, and with vaccination.
Decontamination:	**Decontamination:**
Patients who are exposed to Q fever by aerosol do not present a risk for secondary contamination or re-aerosolization of the organism. Decontamination is accomplished with soap and water or by the use of weak (0.5%) hypochlorite solutions.	Strict quarantine with respiratory isolation for a minimum of 16–17 days following exposure for all contacts. Patients should be considered infectious until all scabs separate.

Table 7-1   Continued

Venezuelan Equine Encephalitis	Viral Hemorrhagic Fevers
Signs and Symptoms:	**Signs and Symptoms:**
Sudden onset of illness with general malaise, spiking fevers, rigors, severe headache, photophobia, and myalgias. Nausea, vomiting, cough, sore throat, and diarrhea may follow. Full recovery takes 1–2 weeks.	Viral hemorrhagic fevers (VHFs) are febrile illnesses that can be complicated by easy bleeding, petechiae, hypotension, and even shock, flushing of the face and chest, and edema. Constitutional symptoms such as malaise, myalgias, headache, vomiting, and diarrhea may occur in any hemorrhagic fevers.
Diagnosis:	**Diagnosis:**
Clinical diagnosis; Physical findings are usually nonspecific. The white blood cell count often shows a striking leukopenia and lymphopenia. Virus isolation may be made from serum, and in some cases throat swab specimens.	Clinical diagnosis; Watery diarrhea and dehydration. Microscopic exam of stool samples reveals few or no red or white cells. Can be identified in stool by dark field or phase contrast microscopy and can be grown on a variety of culture media.
Treatment:	**Treatment:**
Supportive therapy only.	Intensive supportive care may be required. Antiviral therapy with ribavirin may be useful in several of these infections. Convalescent plasma may be effective in Argentine hemorrhagic fever.
Prophylaxis:	**Prophylaxis:**
A live, attenuated vaccine is available as an investigational new drug. A second, formalin-inactivated killed vaccine is available for boosting antibody titers in those initially receiving the live vaccine.	The only licensed VHF vaccine is yellow fever vaccine. Prophylactic ribavirin may be effective for Lassa fever, Rift Valley fever, Congo-Crimean hemorrhagic fever (CCHF) and possibly hemorrhagic fever with renal syndrome (HFRS).
Decontamination:	**Decontamination:**
Blood and body fluid precautions (body substance isolation {BSI}) should be employed. Human cases are infectious for mosquitoes for at least 72 hours. The virus can be destroyed by heat (80°C (176°F) for 30 minutes) and ordinary disinfectants.	Decontamination with hypochlorite or phenolic disinfectant. Isolation measures and barrier nursing procedures are indicated.

Table 7-1 Continued

Botulinum Toxins	Staphylococcal Enterotoxin B
Signs and Symptoms:	**Signs and Symptoms:**
Ptosis, generalized weakness, dizziness, dry mouth and throat, blurred vision and diplopia, dysarthia, dysphonia, and dysphagia followed by symmetrical descending flaccid paralysis and development of respiratory failure. Symptoms begin as early as 24–36 hours, but may take several days after inhalation of toxin.	From 3–12 hours after aerosol exposure, sudden onset of fever, chills, headache, myalgia, and nonproductive cough. Some patients may develop shortness of breath and retrosternal chest pain. Fever may last 2–5 days, and cough may persist up to 4 weeks. Patients may also present with nausea, vomiting, and diarrhea if they swallow the toxin. Higher exposure levels can lead to septic shock and death.
Diagnosis:	**Diagnosis:**
Clinical diagnosis; No routine laboratory findings. Bio-terrorism/warfare should be suspected if numerous collocated casualties have progressive descending bulbar, muscular, and respiratory weakness.	Clinical diagnosis; Patient presents with a febrile respiratory syndrome without chest x-ray (CXR) abnormalities. Large numbers of patients presenting with typical symptoms and signs of SEB pulmonary exposure would suggest an intentional attack with this toxin.
Treatment:	**Treatment:**
Intubation and ventilatory assistance for respiratory failure. Tracheostomy may be required. Administration of botulinum antitoxin (IND product) may prevent or decrease progression to respiratory failure and hasten recovery.	Treatment is limited to supportive care. Artificial ventilation might be needed for very severe cases and attention to fluid management is essential.
Prophylaxis:	**Prophylaxis:**
Pentavalent toxoid (types A, B, C, D, and E) is available as an IND product for those at high risk of exposure.	Use of protective mask. There is currently no vaccine available to prevent SEB intoxication.
Decontamination:	**Decontamination:**
Hypochlorite (0.5% for 10–15 minutes) and/or soap and water. Toxin is not dermally active and secondary aerosols are not a hazard from patients.	Hypochlorite (0.5% for 10–15 minutes) and/or soap and water. Destroy any food that may have been contaminated.

Table 7-1   Continued

Ricin	Trichothecene Mycotoxins (T2)
Signs and Symptoms:	**Signs and Symptoms:**
Weakness, fever, cough, and hypothermia about 36 hours after aerosol exposure, followed in the next 12 hours by hypotension and cardiovascular collapse.	Exposure causes skin pain, pruritus, redness, vesicles, necrosis, and sloughing of epidermis. Effects on the airway include nose and throat pain, nasal discharge, itching, and sneezing, cough, dyspnea, wheezing, chest pain, and hemoptysis. Toxin also produces effects after ingestion or eye contact. Severe poisoning results in prostration, weakness, ataxia, collapse, shock, and death.
Diagnosis:	**Diagnosis:**
Signs and symptoms noted above in large numbers of geographically clustered patients could suggest an exposure to aerosolized ricin. The rapid time course to severe symptoms and death would be unusual for infectious agents. Laboratory findings are nonspecific except for specific serum ELISA. Acute and convalescent sera should be collected.	Should be suspected if an aerosol attack occurs in the form of "yellow rain" with droplets of yellow fluid contaminating clothes and the environment. Confirmation requires testing of blood, tissue, and environmental samples.
Treatment:	**Treatment:**
Patient management is supportive. Presently there is no available antitoxin. Gastric decontamination measures should be employed if the toxin is ingested.	There is no specific antidote. Superactive charcoal should be given orally if swallowed.
Prophylaxis:	**Prophylaxis:**
Presently there is no vaccine or prophylactic antitoxin available for human use. Use of a protective mask (respirator) is currently the best protection against inhalation if an attack/exposure is anticipated.	The only defense is to wear personal protective equipment during an attack. No specific immunotherapy or chemotherapy is available for use in the field.
Decontamination:	**Decontamination:**
Weak hypochlorite solutions and/or soap and water can decontaminate skin surfaces. Ricin is not volatile, so secondary aerosols are generally not a danger to health care providers.	Outer garments should be removed and exposed skin should be decontaminated with soap and water. Eye exposure should be treated by copious saline irrigation. Once decontamination is complete, isolation is not required.

Source: Medical Management of Biological Casualties Handbook, 2nd edition, United States Army Medical Research Institute of Infectious Diseases, Ft. Detrick, Fredrick, Maryland, 1996.

While the probability of a high-impact or widespread attack using biological weapons is low, the yield from such an event could be devastating. It is for this reason that being prepared to respond to the aftermath of a biological attack is critical. As highlighted in this chapter, there are many complex issues that nonmilitary responders are not familiar with that accompany the planning and response phases of confronting a bio-terrorism event. We strongly recommend that this matter be given the attention it deserves so that your members can be protected and a level of response effectiveness to your community can be maintained. Remember that this type of incident can rapidly swell to a level that will overwhelm your EMS system as well as local health care resources. Incorporating the talents of your public health and hospital officials in the planning process will provide you with the ability to develop comprehensive and cohesive contingency plans.

CHAPTER QUESTIONS

1. Define biological warfare. How is a biological attack different from other more traditional forms of terrorism?
2. Briefly discuss the history of biological terrorism.
3. What are the major steps in a biological threat assessment?
4. What types of personal protective equipment are necessary for emergency responders in a biological event?
5. Discuss the clinical effects, detection, and prophylaxis/treatment for the following biological toxins:
    a. Ricin
    b. Botulinium toxin
    c. Clostridum toxin
6. Discuss the clinical effects and treatments for:
    a. Anthrax
    b. Ebola virus
    c. Tularemia
    d. Q fever

NOTES

[1]*Medical Management of Biological Casualties Handbook,* 2nd ed., U.S. Army Medical Research Institute of Infectious Diseases, Ft. Detrick, Fredrick, Maryland, March 1996.

[2]Ibid.

[3]Ibid.

[4]Bernstein, B. J. "The birth of the U.S. Biological-Warfare program." *Scientific American,* 1987, 94–99.

[5]Mobley, J. A. "Biological warfare in the twentieth century: Lessons from the past, challenges for the future." *Mil. Med.,* 1995, 160, 547–553.

[6]*Medical Management of Biological Casualties Handbook,* 2nd ed., U.S. Army Medical Research Institute of Infectious Diseases, Ft. Detrick, Fredrick, Maryland, March 1996.

[7]Horrock, N. "The new terror fear: biological weapons." *US News and World Report* (online version), issue 970512/12biow.htm

[8]Cole, L. A. "Cloud cover: The army's secret germ warfare test over San Francisco." *Common Cause Magazine,* 1988, 14, 16–37.

[9]Cole, L. A. "Operation bacterium: Testing germs on the A train." *Washington Monthly,* 1985, 17.

[10]Stephenson, J. "Confronting a biological Armageddon: Experts tackle prospect of bioterrorism," *JAMA,* 1996, 276, 349–351.

[11]Douglas, J. D. *America the Vulnerable: The Threat of Chemical/Biological Warfare, the New Shape of Terrorism and Conflict.* Lexington Books; Lexington, MA, 1987, 29.

[12]Flanagin, A. and Lederberg, J. "The threat of biological weapons—prophylaxis and mitigation." *JAMA,* 1996, 276, 410–411.

[13]Cole, L. A. "The specter of biological weapons." *Scientific American,* 1996, http://www.sciam.com/1296issue/1296cole.html.

[14]Horrock, N. "The new terror fear: Biological weapons." *US News and World Report* (online version), issue 970512/12biow.htm, OP CIT.

[15]Harris, R. and Paxman J. *A Higher Form of Killing.* Wang and Hill, New York, 1982, 75–81.

[16]Mayer, T. N. "The biological weapon: A poor nation's weapon of mass destruction," in *The Battlefield of the Future.* www.airpower.maxwell.af.mil/airchronicles/battle/chp8.html

[17]"Biowarfare wars: Critics ask whether the army can manage the program." *Scientific American,* 1994, January, 22.

[18]Lebeda, F. J. "Deterrence of biological and chemical warfare: A review of policy options." *Mil. Med.* 1997, 162, 156–161.

[19]Kaufmann, A. F., Meltzer, M. I., and Schmid, G. P. "The economic impact of a bioterrorist attack: Are prevention and post-attack intervention programs justifiable?" *Emerg. Infect. Dis.,* 1997, 3, 83–94.

[20]Meyer, K. F. "The status of botulism as a world health problem." *Bull. WHO.* 1956, 15, 281–298.

[21]Jankovic, J. and Brin, M. "Therapeutic uses of botulinum toxin." *N. Engl. J. Med.,* 1991, 324, 1186–1194.

[22]Hughes, R. and Whaler, B. C. "Influence of nerve-ending activity and of drugs on the rate of paralysis of rat diaphragm preparations by *Cl. botulinum* type A toxin." *J. Physiol.* (London), 1962, 160, 221–233.

[23]Hallett, M., Glocker, F. X., and Deuschl, G. "Mechanism of action of botulinum toxin." *Ann. Neurol.* 1994, 36, 449.

[24]Mobley, J. A. "Biological warfare in the twentieth century: Lessons from the past, challenges for the future." *Mil. Med.,* 1995, 160, 547–553.

[25]Hatheway, C. L. "Botulism: The present status of the disease." *Current Topics in Microbiology and Immunology,* 1995, 195, 55.

[26]*Biological Defense: Vaccine Information Summaries.* Publication of USAMRIID. Frederick Maryland, Fort Detrick, 1994.

[27]Wiener, S. L. "Strategies for prevention of a successful biological warfare aerosol attack." *Mil. Med.,* 1996, 161, 251–256.

[28]Lebeda, F. J. "Deterrence of biological and chemical warfare: A review of policy options." *Mil. Med.,* 1997, 162, 156–161.

[29]Sehnke, P. C., Pedrosa, L., Paul, A. L., et al. "Expression of active processed ricin in transgenic tobacco." *J. Biologic Chem.,* 1994, 269, 22473–22476.

[30]Sato, S., Kodama, M., Ogata, T., et al. "Saxitoxin as a toxic principle of a freshwater puffer Tetradon fangi, in Thailand." *Toxicon* 1997, 35, 137–140.

[31]Kaufman, B., Wright, D. C., Ballou, W. R., and Monheit, D. "Protection against tetrodotoxin and saxitoxin intoxication by a cross-protective rabbit anti-tetrodotoxin antiserum." *Toxicon,* 1991, 29, 581–587.

[32]Environmental Technologies Group, Inc. 1400 Taylor Avenue., P.O. Box 9840, Baltimore, Maryland 21284-9840.

[33]Lange, W. R. "Puffer fish poisoning." *Amer. Fam. Phys.* 1990, 42, 1029–1033.

[34]Laobhripatr, S., Limpakarnjanarat, K., Sanwanloy, O., et al. "Food poisoning due to consumption of the freshwater puffer *Tetradon fangi* in Thailand." *Toxicon,* 1990, 28, 1372–1375.

[35]Nozue, H., Hayashi, T., Hasimoto, Y., et al. "Isolation and characterization of She-wanell alga from human clinical specimens and emendation of the description of S. Alga Simidu et al." *Int. J. Syst. Bacteriol.* 1990, 42, 628–634.

[36]Tambyah, P. A., Hui, K. P., Gopalakrishnakone, N. K., and Chin T. B. "Central nervous system effects of tetrodotoxin poisoning." *Lancet* 1994, 343, 538–539.

[37]Mackenzie, C. F., Smalley, A. J., Barnas, G. M., and Park, S. G. "Tetrodotoxin infusion: Nonventilatory effects and role in toxicity models." *Academ. Emerg. Med.,* 1996, 3, 1106–1112.

[38]Kaufman, B., Wright, D. C., Ballou, W. R., and Monheit, D. "Protection against tetrodotoxin and saxitoxin intoxication by a cross-protective rabbit anti-tetrodotoxin antiserum." *Toxicon,* 1991, 29, 581–587.

[39]Chang, F. T., Bauer, R. M., Benton, B. J., et al. "4-Aminopyridine antagonizes saxitoxin and tetrodotoxin induced cardiorespiratory depression." *Toxicon,* 1996, 34, 671–690.

[40]Sims, J. K., Ostman, D. C. "Pufferfish poisoning: Emergency diagnosis and management of mild tetrodotoxication." *Ann. Emerg. Med.,* 1986, 15, 1094–1098.

[41]Fukiya, S., and Matsumura, K. "Active and passive immunization for tetrodotoxin in mice." *Toxicon,* 1992, 30, 1631–1634.

[42]Matsumura, K. "A monoclonal antibody against tetrodotoxin that reacts to the active group for the toxicity." *Eur. J. Pharm.* 1995, 293, 41–45.

[43]Rivera, V. R., Poli, M. A., and Bignami, G. S. "Prophylaxis and treatment with a monoclonal antibody of tetrodotoxin poisoning in mice." *Toxicon,* 1995, 33, 1231–1237.

[44]Meselson, M., Guillemin, J., Hugh-Jones, M., et al. "The Sverdlovsk anthrax outbreak of 1979." *Science,* 1994, 266, 1202–1208.

[45]Personal interviews during visit to University of Urals, Ekatrinburg, 1996.

[46]*Handbook on the Medical Aspects of NBC Defensive Operations* (FM 8-9), Part II—Biological (Annex B), United States Government Printing Office, 1996.

[47]Shlyakhov, E. N., and Rubinstein, E. "Human live anthrax vaccine in the former USSR." *Vaccine,* 1994, 12, 727–730.

[48]Coulson, N. M., Fulop, M., and Titball, R. W. "Bacillus anthracis protective antigen expressed in Salmonella typhimurium SL 3261, afford protection against spore challenge." *Vaccine,* 1994, 12, 1395–1401.

[49]Ivins, B., Fellows, P., Pitt, L., et al. "Experimental anthrax vaccines: Efficacy of adjuvants combined with protective antigen against an aerosol Bacillus anthracis spore challenge in guinea pigs." *Vaccine,* 1995, 13, 1779–1794.

[50]"Vaccines against bacterial zoonoses." *J. Med. Microbiol.,* 1997, 46, 267–269.

[51]Ebola Virus Hemorrhagic Fever: General Information, http://www.cdc.gov/ncidod/publications/brochures/ebolainf.htm

[52]Jaax, N., Jahrling, P., Geisbert, T., et al. "Transmission of Ebola virus (Zaire strain) to uninfected control monkeys in a biocontainment laboratory." *Lancet,* 1995, 356, 1669–1671.

[53]Oyston, P. C. F., Williamson, E. D., Leary, S. E., et al. "Immunization with live recombinant Salmonella typhimurium aroA producing F1 antigen protects against plague." *Infect. Immun.* 1995, 63, 563–568.

[54]Meyer, K. F. "Effectiveness of live or killed plague vaccines in man." *Bull. World Health Organ.,* 1970, 42, 653–666.

[55]Russel, P., Eley, S. M., Hibbs, S. E., et al. "A comparison of Plague Vaccine, USP and EV76 vaccine induced protection against Yersinia pestis in a murine model." *Vaccine,* 1995, 13, 1551–1556.

[56]Ackland, J. R., Worswick, D. A., and Marmion, B. P. "Vaccine prophylaxis of Q fever. A follow-up study of the efficacy of Qvac (CSL) 1985–1990." *Med. J. Aust.,* 1994, 160, 704–708.

[57]Genig, V. A. "Experience on mass immunization of human beings with the M-44 live vaccine against Q fever. Report 2. Skin and oral routes of immunization." *Vopr. Virosol.* 1965, 6, 703–707.

[58]Hornick, R. B., and Eigelsbach, H. T. "Aerogenic immunization of man with live tularemia vaccine." *Bact. Rev.,* 1966, 30, 532–538.

8

Weapons of Mass Effect: Cyber-Terrorism

Hank T. Christen
James P. Denney
Paul M. Maniscalco

Source: Courtesy Index Stock Imagery Inc.

Chapter Objectives

After reading this chapter, you will be able to accomplish the following objectives:

1. Recognize and understand the concept of information operations becoming a new "battlespace."
2. Define and discuss the critical infrastructure and the public safety infrastructure.
3. List the critical elements in the public safety infrastructure.
4. Understand the vulnerability of the information infrastructure.
5. Understand the importance of questioning computer data when inaccuracy is suspected.
6. Recognize that public records laws can require the release of information that increases a community's vulnerability.

Introduction

In December 1998 the Department of Defense (DoD) announced that information warfare was being institutionalized as the new operational battlespace (Information Operations Department, School of Information Warfare and Strategy, National Defense University). The three traditional battlegrounds were land, sea, and air. These elements are now called battlespaces to incorporate the fourth element of information warfare. In essence, cyberspace is formally recognized as permeating all DoD battlespaces.

Battlespaces do not exist in the emergency response world. However, we organize response units in an operations section, with response branches being EMS, fire/rescue, law enforcement, public health, and public works. These critical branches are similar to battlespaces, are dependent on information operations, and are vulnerable to information warfare. Like the DoD, civilian emergency agencies are subject to cyber attack, and must protect this essential battlespace element.

Sample Scenario

At 4:30 hrs you are dispatched to a chemical release as the result of a train derailment at a railroad crossing. A tank car has ruptured and is leaking an unknown substance at approximately 10 gallons per minute down the street and into a drain. The red placard reads 1092 Inhibited Acrolein ($CH_2 CHCHO$ or C_2H_3CHO). The wind is out of the north at 10 to 15 mph. You give your size-up to the dispatcher and request a haz mat response team.

The dispatcher notifies you that a nursing home 1,000 ft to the east has reported several residents having difficulty breathing. You see two of your responders go down. Suddenly, the railroad crossing is silenced and the flashing lights go out. The dispatcher is cut off in the middle of a dispatch. You notice that the streetlights have gone out and traffic signals do not appear to be working. The dispatcher fails to respond to your repeated calls. You have responded to a binary act of terrorism involving chemical release and power disruption as a result of cyber terrorism and electronic warfare.

Adapting to Change

As the millennium approached we were constantly reminded that things were changing in the world, that things are different in this century, that new technologies will revolutionize the future. Of course they will. We witnessed miraculous changes over the last 100 years and there are no indications that this century will be any different. All of the factors influencing future global social, economic, technologic, information, and political evolution are currently in place or in development. For public safety the concern is not the changes that are occurring, but in which forms they will manifest and whether any potential negatives are associated with them.

For example, we know that information technology will have a profound positive effect on emergency services, yet we also faced the first crisis of the new millennium: the possibility of failure of the information technology platform. The Year 2000 (Y2K) computer problem, also known as the millennium bug, was potentially catastrophic to the national infrastructure of industrial economies. On January 1, 2000, we realized that "apocalypse now" was really "apocalypse not." However, the Y2K nonevent was not an imagined threat. Cyber awareness, preparation, and an intense focus on the information infrastructure saved the day. The transformation to a technology-based society brought not only solutions, but also unanticipated problems. This paradox illustrates that we will always encounter complex future management issues involving critical infrastructure components.

What Is Critical Infrastructure?

The critical infrastructure consists of, in part, information and financial management systems, telecommunications, dispatch centers, cable television, power production, water service, and natural gas and its storage facilities, transportation, and distribution mechanisms. Protecting this infrastructure against physical and electronic attack and ensuring the availability of the infrastructure is a complex issue. In 2000, as a result of a survey

FIGURE 8-1   The preparations taken in response to the "millennium bug" need to be sustained. The rapidly evolving world of information manifests itself with increasing vulnerability to our digital infrastructure. *Source:* Courtesy Jean-Francois Podevin/Index Stock Imagery, Inc.

conducted by the Computer Security Institute to determine an estimated impact of these attacks, alarming findings were revealed. The highlights of the "2000 Computer Crime and Security Survey" include the following:

- Ninety percent of respondents (primarily large corporations and government agencies) detected computer security breaches within the last twelve months.
- Seventy percent reported a variety of serious computer security breaches other than the most common ones of computer viruses, laptop theft, or employee "net abuse"; for example, theft of proprietary information, financial fraud, system penetration from outsiders, denial of service attacks, and sabotage of data or networks.
- Seventy-four percent acknowledged financial losses due to computer breaches.
- Forty-two percent were willing and/or able to quantify their financial losses. The losses from the 273 respondents totaled $265,589,940 (the average annual total over the last three years was $120,240,180).

The critical infrastructure is currently vulnerable to attack. While this in itself poses a national security threat, the linkage between information systems and traditional critical infrastructures has increased the scope and potential for the use of cyber-terrorism. For economic reasons, increasing deregulation and competition created an increased reliance on information systems to operate, maintain, and monitor critical infrastructures. This, in turn, creates a tunnel of vulnerability previously unrealized in history.[1]

The critical infrastructure of our communities is basically transparent; we cannot function without it, yet we do not see it or realize that it is there. Almost every aspect of twenty-first-century life revolves around the efficiency of zeros and ones flowing at near light speed through microchips. Transportation is an example. Air, ground, and water transportation systems depend on computer traffic control. Software, data systems, and communications guide the entire air traffic control system.

The nation's power grid is another example. The system is a complex matrix of generating systems, switching systems, and distribution systems, all computer controlled (sometimes called a system of systems). Major blackouts and brownouts have occurred because of a software failure or the bisection of a fiber-optic cable.

A third example is the world financial system. As reported in a Learning Channel™ documentary, 90 percent of the world's wealth is digital. Individual and business financial holdings are essentially an account number associated with a fiscal amount in a financial database.

In the previous examples, accidental electronic failures have disrupted the systems. Intentional cyber attacks have likewise caused system disruption. An information attack can halt air traffic, gridlock a ground transportation system, cause a regional power failure, or cripple a financial system. These events cause multiple deaths and injuries.

The Public Safety Infrastructure

The elements of the critical infrastructure that are most important to emergency responders are the essential systems in the public safety infrastructure. The general areas of our infrastructure are communications, computer-aided dispatch (CAD) systems, geo-based information systems (GIS), e-mail, and informational databases.

The most critical system is communications. A modern handheld radio has more computer memory than a Commodore 64 computer (64K ram) from the 1970s. The communications dispatch system includes repeaters, consoles, enhanced trunking systems, transmitters, and receivers. All of these elements are electronic, and susceptible to data corruption. The public also has a communications system, namely, 911. All aspects of an enhanced

911 system are computer driven including electronic switching, automatic location identification, and automatic call routing. Most systems also have a database of caller medical information, hazardous materials data, location descriptions, and premise histories.

There are many cases of hackers corrupting the telephone switching system for the purpose of making free toll calls. At a more serious level, 911 systems have been hacked for malicious purposes. The result has been missed calls, system outages, and confusion coupled with a diminished or absent capacity for responding to emergencies.

Computer-aided dispatch systems include electronic mapping, system status software, automatic vehicle location software, and databases of call information. A failure of any of these systems results in downgrading dispatching to a manual mode. Calls are missed, unit locations erased, and call data lost during critical peak periods. Information operations sabotage during a terrorist incident could greatly inhibit the ability of the public safety delivery system to effectively respond.

Information databases and decision systems have progressed from an oddity to a necessity in less than a decade. Some of the information now routinely used includes:

- Medical protocols
- Logistics data
- Disaster plans
- Personnel information
- Hazardous materials response guidelines
- Building pre-plans
- Criminal histories
- Financial reports and spreadsheets

A loss of information in any of these systems results in significant reduction of efficiency. More importantly, an intentional manipulation of data may go unnoticed for a significant duration, and result in poor decisions being made from inaccurate data.

In summary, the critical infrastructure in your community is hardly visible. It is as simple as a chip or a computer disk. We must remember that these systems are essential and must be protected.

Cyber Warfare: Incident and Response

Terrorism, as a tool of the disenfranchised, the disenchanted, and the just plain destructive will undergo fundamental changes during the next decade. Data packets may very likely replace explosives as the favored implement of destruction. TCP/IP (computer protocol) will be preferred over the Kalishnikov (AK- 47) and modems rather than suicide bombers will deliver chaos to the world's governments, local communities, and institutional infrastructures. The advent of the cyber warrior is at hand. As just described above, it is predicted that acts of terrorism will be binary events that couple information and electronic warfare with other activities such as explosives or chemical releases.

The most disconcerting aspect of this threat shift in the nature of terrorism is the magnitude of the destruction that can be inflicted by a single individual with a simple keystroke rather than a detonator. Disruption of the world's financial markets, chaos in the public safety system, reduction in commercial productivity, depletion of health services, and the downing of telecommunication networks will be only the beginning.

Lt. General Kenneth A. Minihan, while director of the U.S. National Security Agency, stated "the threat that is posed by potential cyber attacks against the U.S. Military and computer system networks is now growing beyond the computer hacker stage."[2] He said that groups potentially hostile to the United States are developing, or attempting to develop offensive information warfare capabilities. The Minihan warning is viewed as validating what the prestigious U.S. Defense Science Board has called "a recipe for national disaster." The following illustration is a depiction of local, national, and global information interdependence:

The Rand Corporation was a little more forthcoming. Their report, "Strategic Information Warfare" bluntly states, "Many U.S. allies and coalition partners will be vulnerable to information warfare attacks on their core infrastructures."[3] Lieutenant General Patrick M. Hughes, while he was Director of the Defense Intelligence Agency, clarified the threat further in his statement before the Senate Select Committee on Intelligence when he stated, in part:

> Transnational Infrastructure Warfare involves attacking a nation's or sub-national entity's key industries and utilities; to include telecommunications, banking and finance, transportation, water, government operations, emergency services, energy and power, and manufacturing. These industries normally have key linkages and interdependencies, which could significantly increase the impact of an attack on a single component. Threats to critical infrastructure include those from nation-states, state-sponsored sub-national groups, international and domestic terrorists, criminal elements, computer hackers, and insiders.[4]

Governments, at all levels, have an obligation to secure their information systems and prepare for continuous infrastructure threats; having the will to do so is another matter entirely. Systems security is not a "do it once and you're done" proposition. An effective security plan is similar to an

FIGURE 8-2   Kevin Mitnick, who has been called a "computer terror-ist" by the Department of Justice is perhaps the most high-profile com-puter hacker in the world today. *The United States v. Kevin David Mitnick* case became a rally for action cause in the hacker world, resulting in many unauthorized computer incursions, Web site defacements and dis-tributed denial of service attacks on the Internet in response to his arrest, conviction and incarceration. *Source:* Courtesy Bob Jordan/AP/Wide World Photos.

effective response plan. It must evolve and develop as the threat to systems evolves and develops. A firewall that kept the world out yesterday can be Swiss cheese tomorrow.

The lack of geographical, spatial, and political boundaries precludes conventional preventive measures. Attribution is second to information stability and therefore the majority of effort is placed on denying unauthorized access and system manipulation. Information warfare is attractive because it is relatively cheap to wage and offers an asymmetrical return on investment for resource-poor adversaries.

When information systems are under attack, the demand for information will increase while the capacity of the information infrastructure will concurrently decrease. The law, particularly international law, is ambiguous regarding criminality in, and acts of war on, information infrastructures. This ambiguity, coupled with a lack of clearly designated responsibilities for electronic defense, hinders the development of remedies and limits response options.

The vulnerability of our information systems was painfully demonstrated by the "denial of service" attacks that occurred during a three-day period in February, 2000. A new breed of hacker called a "cracker" prompted this attack. Crackers are sophisticated computer terrorists that attempt to disrupt or totally shut down networks or systems, whereas hackers are satisfied with just breaking into a system. According to Knight Ridder, in the 2000 attacks, major providers such as Yahoo!, eBay, Amazon.com and CNN were shut down.

The real-time use of information assumes the availability of information and information technology. The operational implications of a failure of information and information technology must be addressed in an organized, sequential manner. Redundant capability in command and control capacity must be built into the system. Emergency communications plans must consider the extended system and its processes, and prepare for the eventuality of widespread system failure.

Globalization

Globalization is changing the context in which terrorists operate.[5] A transnational cast of characters that cannot be controlled by governments, either individually or collectively, increasingly affects even so-called domestic terrorism. Information technology has effectively removed the ability of countries to isolate themselves. Information and communication control is difficult, if not impossible to achieve because the information revolution has resulted in democratic access to technology. An important result is that free speech and civil liberty have been given an inexpensive international medium with which to voice discontent with existing government.

The notional concept of a centrally controlled international terrorist network, previously investigated during the 1960s and 1970s, was deemed to be unlikely due to conflicting ideologies, motivating factors, funding, arming, and training among global practitioners. However, networks are now quite possible with the advent of public access to the Internet, the ability to transfer funds and conduct banking electronically, the international arms market, encrypted digital communication technology, and the emergence of stateless terrorism. An important result is that instant global communication between offensive action cells and their controllers is now possible. Controllers now have global reach and can run multiple independent cells from a single location with no interaction between the cells. They can also contract terrorism services utilizing the local indigenous practitioners in a given target community.

The complexity of weapons acquisition, production, transportation, lethality, and delivery platform has been diminished. Information management technology has also resulted in a reduced requirement for infrastructure, security, and detection avoidance and has resulted in an asymmetry between cause and effect.

The Questioning of Computer Data

Data on a computer screen has a high degree of credibility. Anyone born after 1950 was raised in front of a television screen. Anyone born after 1970 was raised in front of a computer screen along with the television. As a result, data on a screen has a very high degree of believability.

The habit of accepting electronic data without question must change, especially during tactical operations. When data does not agree with reasonable expectations, the data must be questioned, and data corruption suspected. In other words, the data must be "in the ballpark." For example, a chemical database that indicates procedures that appear inaccurate or unsafe should be compared with a printed source or another data system. In another case, if the CAD system suddenly indicates grossly inaccurate unit status, the information should be checked and corruption suspected. Any uses of electronic data by tactical decision makers should observe the following guidelines:

1. Do not blindly trust data screens.
2. Evaluate tactical data as a reality check.
3. Check other sources when data corruption is suspected.

FIGURE 8-3   In the present as well as the future the "digital battlespace" will increasingly become the theater for terrorists, criminals, and malicious acts of disruption or destruction. *Source:* Courtesy Jonnie Miles/PhotoDisc, Inc.

In a terrorism/tactical violence event, consider the possibility of a coordinated information attack. Suspect intentional data manipulation when there is a mysterious communications failure. Maintain low tech information sources (books and paper) as an alternative to vulnerable electronic information.

Data Theft

Recent literature is inundated with articles about sophisticated intrusion methods. Sophisticated hackers (local and international) are able to crack passwords or find a back door route through a security firewall. However, simple theft is still an easy way to use low technology for high technology data corruption.

How easy is it to walk into your agency, remove disks stacked on a desk, and walk out? If the data is removed, altered, and discretely returned, great damage may result. If backup disks are removed, followed by a system attack, provisions for storing the system may be lost. There are several steps that should be taken in any public or private agency to protect vital data from simple theft:

1. *Office design*—no one should be able to freely enter an office area; place a door between the entry lobby and data storage areas.
2. *Entry control*—anyone entering an office or data storage area should encounter a receptionist/secretary or security officer before proceeding.
3. *ID badges*—visitors and guests should sign in and be issued a security badge; employees should wear identification badges; procedures should require that any person without a badge be questioned.
4. *Information storage*—stored data should be locked and never stored on a desktop or in an open area; critical backup data should be stored in a safe at an off-premise location. This is good advice for fire and severe weather protection as well as theft.
5. *Electronic entry*—sensitive areas should be controlled by electronic entry; systems should provide printouts of all names, dates, and times of entries.

Data Protection Standards

Emergency agencies are very familiar with standards. In the fire service there are National Fire Protection Association (NFPA) standards; in EMS there are the NHTSA EMT and paramedic training standards. Law enforcement is governed by Department of Justice standards. The Occupational Safety and Health Administration (OSHA) standards govern most of us. However, there are no present standards for data and information storage/security.

During the Twentieth National Information System Security conference (Baltimore, 1997), Robert T. Marsh (keynote address) stated, "For example, we recommend the National Institute of Standards and Technology (NIST) and the National Security Agency (NSA) jointly set standards and publish best practices for information security, and then share these with federal, state, and local governments, as well as private industry."

Most local government response agencies are merely end users of electronic data. They lack the sophistication of federal government agencies and private organizations regarding protection of critical data. Standards and protocols are needed in the following areas:

- Data storage procedures
- Detection of running system attacks (real time)
- System restoration (disaster recovery)
- Physical security protocols
- Training standards for information technology security specialists

In the future we may have standards and protocols on information operations that rival tactical procedures. Presently, no such standards exist. At best, there is an inadequate mix of guidelines borrowed from private industry and federal agencies.

Public Information vs. Data Protection

A major information services issue in state and local government agencies is public disclosure of information. Most state laws are very liberal in their definition of "public information." State legislation usually defines exact types of documents that are confidential; all other documents not specified as confidential are public documents.

In many states, public documents, including electronic data, must be released within the same business day they are requested. Only reasonable charges for copying or duplication are allowed. You may be shocked to discover that the following information is public in your state:

- Names and addresses of all employees, including elected officials, managers, and emergency response personnel.
- Standard operating procedures for response agencies and special teams.
- Locations and descriptions of emergency response units, including equipment inventories.
- Radio frequencies, codes, and dispatch procedures.
- Driver's license lists, including social security numbers and pictures.
- Mutual aid contracts and mutual aid procedures.
- Incident reports and after-action reports.
- Building pre-plans and building layout graphics.

Hazardous materials information is a classic example. Federal legislation passed in the 1980s referred to as "The Community Right-to-Know

Act" requires that information relating to the storage of hazardous materials be available to any member of the public who seeks the information. In many locales, this information is available at the public library. The data includes storage locations of reportable quantities, layout drawings of the storage sites, transportation routes, materials safety data sheets (MSDS), and descriptions of storage containers.

Right-to-know legislation directly conflicts with information security. The intentional release of industrial hazardous materials provides an effective terrorism/tactical violence weapon. Because of public records laws and right-to-know legislation, domestic and foreign enemies can use our information to attack us.

Changing the law is admittedly a long and often painful process, but reducing the tactical information available as public record is a worthy goal. As times change and the terrorism/tactical violence threat increases, the naïve twentieth-century public records laws must be altered.

Many emergency response agencies are knowingly releasing sensitive information through the Internet. Agency homepages include links to computer-aided dispatch data screens, tactical response information, and links to communications centers that include real-time audio radio traffic. The motive is usually an attempt to generate positive public relations. Unfortunately, this type of information is very helpful to an adversary. Take another look at your organization's Internet links and homepages, and remove information that may make your system or your people vulnerable.

Infrastructure Protection—A Public/Private Partnership

Most of the critical infrastructure is owned and operated by private business entities or utilities. Private industry shares government's concern about infrastructure protection. The private sector has the advantage of funding, the ability to spend millions on information security problems.

One of the key recommendations of the President's Commission on Critical Infrastructure Protection is a program of public/private partnering (http://www.pccip.gov). The commission's most serious challenge was achieving private sector partnering. For many years, the private sector has conducted research and implemented procedures to protect it from local threats. The federal government has a more national objective; the government must protect the citizens of the United States and all of the country's systems from cyber intrusion or dysfunction. A sharing of information is in the interest of both parties.

In January 2000, President Bill Clinton announced a $2 billion proposal to combat cyber-terrorism. The proposal establishes the Institute for Information Infrastructure Protection. The objective of the institute is to establish a public/private partnership for infrastructure protection research.

Other aspects of the president's proposal include increased funding for research and development, and increased computer security.

Local governments have similar information concerns because of ownership of the public safety infrastructure. However, local governments do not have the funds or expertise to conduct research in the information protection arena. Local government must depend on spin-offs from the public/private partnerships at the national level.

Information Security Management

In the future, the director of information security management (ISM) will be a new position in progressive response agencies. Physical security is commonplace; information security will be just as important.

Presently, information security is haphazard at best, and certainly not a prominent unit in public safety organizational charts. In most agencies, security is relegated to someone in the information services (IS) department, who usually has many other duties. In the ideal model, information security should pervade the organization. This means an information security department managed by a professional ISM. This department must be high in the management hierarchy and operate by professional standards and protocols.

An effective ISM department should have the following goals:

1. Develop and maintain systems for real-time detection of running cyber attacks.
2. Conduct ongoing educational awareness programs for all internal agencies.
3. Stay informed regarding national research and development efforts.
4. Maintain the standards and best practices of the information technology industry.
5. Maintain an intelligence system for crisis information about cyber threats.
6. Conduct aggressive investigations on all incidents relating to system attacks or data disruption.

CHAPTER SUMMARY

Information operations, information warfare, and cyber attacks are twenty-first-century concerns. The Department of Defense has added information operations as a fourth battlespace. Information operations is a large part of the nation's critical infrastructure, which consists of our financial systems, transportation systems, utility systems, and communications systems.

The public safety infrastructure includes components that are essential to public safety operations and includes 911, communications, computer-aided dispatch, informational databases, geo-based information systems, and electronic mail. All of these systems are based on software and electronic data systems and must be protected from intrusion and data corruption. The disruption of these systems inhibits emergency response capabilities and causes death and injury.

Electronic data has a high degree of trust. Response agencies must recognize that this trusted data is decision-making material that is vulnerable. Tactical decision makers should be trained to perform reality checks on suspicious data and maintain low technology sources of backup information.

Data theft is a simple method of deleting or corrupting sensitive data. Stored data on disks and tapes should be protected from theft and/or tampering. Office design that prevents unescorted entry, and includes electronic entry control and security badge identification is one method of securing sensitive data.

Presently, there are no national standards for critical data protection and security. There must be national standards, developed by public/private partnerships, that address data storage procedures, real-time detection of running system attacks, system restoration, physical security protocols, and information technology training standards.

Protection of response data and tactical information often conflicts with public records laws. For example, community right-to-know legislation requires that all citizens (including terrorists) have access to information on the storage and transportation of hazardous materials. The public safety community, through the legislative process, must initiate a concentrated effort to protect information that makes the community vulnerable to attack.

CHAPTER QUESTIONS

1. What is the significance of classifying information operations as an operational battlespace?
2. Define the national critical infrastructure. List at least four examples of major systems in the critical infrastructure.
3. What is the public safety critical infrastructure? Discuss three systems in the public safety infrastructure.
4. Why should computer data be questioned? What policies should be implemented to ensure cross-checking of suspicious data?
5. List several policies/procedures that should be implemented to reduce data theft.
6. Discuss the present data protection standards in the United States. What elements should be included in an ideal standard?

7. What are the prevailing issues relating to public records laws versus the need for protection of sensitive data?
8. Discuss the concept of information security management in public safety agencies.

Simulation I

Conduct basic research on the history of information system failures. Restrict this project to the past three years and cite at least four major cases. Answer the following questions:

1. Was the system failure an attack or a natural failure?
2. What were failure outcomes such as cost, loss of life/property, reduced response capabilities, political perceptions, and effects on public confidence?
3. What actions and protocols can be implemented to prevent future attacks or prevent a system failure?

Simulation II

Conduct an information security analysis of a public safety agency. Answer the following in detail:

1. What are the critical information systems in the organization?
2. What is the history of system vulnerability related to system attacks or critical failures?
3. What systems and procedures are in place for information system security?
4. What are your recommendations for protocols, policies, and procedures for improvement?

Notes

[1] INFORMATION WARFARE—DEFENSE, 1996: Defense Science Board Task Force on IW-D, OUSD-A&T.
[2] Lt. General Kenneth Minihan, Information System Security: National Security Agency, 1998.
[3] R. C. Molander, A. S. Riddle, and P. Wilson, Strategic Information Warfare: A New Face Of War: Rand Corporation, DocNo: MR-661-OSD, 1996.
[4] Lt. General Patrick M. Hughes, Global Threats and Challenges: The Decade Ahead, hearing of the Senate Select Committee on Intelligence, January 28, 1998.
[5] Strategic Management and Policy-Making, Globalization: What Challenges and Opportunities for Government, Department of State, 1997.

9
Weapons of Mass Effect: Radiation

Susan McElrath

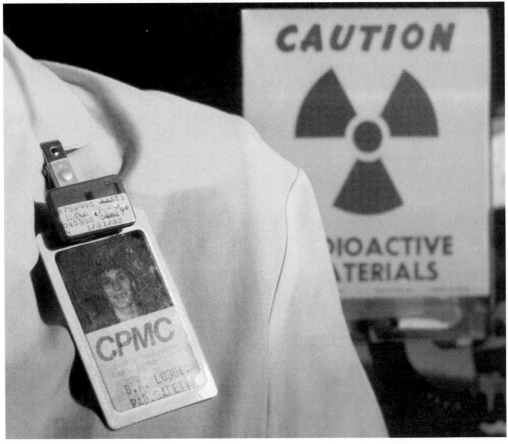

Source: Courtesy Yoav Levy/Phototake N4C

Chapter Objectives

After reading this chapter, you will be able to accomplish the following objectives:

1. Understand the difference between a radiation incident and a traditional hazardous materials incident.
2. Recognize the three types of radiation.
3. Differentiate between the terms dose and exposure.
4. Understand the basic facets of radiation measurement.
5. Recognize the distinction between acute and delayed effects of radiation exposure.
6. Explain the difference between radiation exposure and contamination.
7. Outline the first responder considerations in a radiological terrorism incident.

Introduction

First responders need to know several things to safely and effectively respond to an emergency involving radiation. This chapter provides an explanation of radiation and the types and hazards of radiation exposure. Other topics include assessment of radiation hazards using standard emergency response equipment, uses and limitations of survey instruments and other measuring devices, how radiation exposure is assessed, response to radiation incidents, contamination control techniques, and decontamination procedures.

Radiation incidents are similar to those involving hazardous materials, referred to as *haz mat* or *haz mat incidents* in this chapter. Radiation incidents may be considered a special type of haz mat incident because the two have common factors. These factors are internal exposure pathways, contamination concerns, decontamination techniques, and PPE requirements. If you are familiar with chemical and biological agents, you will note their similarities to haz mat incidents as well.

Characteristics of Radiation

Despite the similarities to haz mat incidents, radiation incidents have a unique characteristic that first responders must understand. The primary difference between a radiation incident and a haz mat (chemical or biological) incident is that radiation exposure may occur without coming in direct contact with the source of radiation. In order to be exposed to haz mat (chemical or biological agents), the material must be inhaled, ingested, injected, absorbed through the skin, deposited on unprotected skin, or be introduced onto or into the body by some means. Exposure to radiation, however, does not require the body to come into direct contact with the radiation source.

Critical Factor: Direct contact is not needed for a radiation exposure to occur.

To understand the mechanism for radiation exposure, an explanation of *radiation* is necessary. Many people incorrectly think radiation is a mysterious chemical substance. Although chemicals may be radioactive, or emit radiation, radiation itself is simply energy in the form of invisible electromagnetic waves or extremely small, energetic particles. Waveforms of radiation are x-ray and gamma ray. Particle forms are alpha, beta, and neutron (see Figure A11-6). X-ray machines and similar equipment commonly found in medical and industrial facilities manufacture radiation, and it is emitted by a wide variety of radioactive materials such as uranium and plutonium.

You should be aware that there are different types of radiation and they have different penetrating abilities and present different hazards. However, it is not necessary to include a detailed discussion of the types of radiation and their characteristics at this point in the chapter. A detailed discussion is provided in Appendix F of this book. Radiation terms are covered in the Glossary.

Machine-generated radiation does not involve the use of radioactive materials. Invisible electromagnetic x-ray waves are generated when the machine is energized and cease to be emitted when the machine is shut down. Objects exposed to common machine-generated radiation do not become radioactive. The most familiar radiation-generating machine is the diagnostic x-ray machine. X-ray energy passes through the body and exposes special x-ray-sensitive film. Dense areas of the body absorb the x-ray energy so they appear lighter on film than less dense areas that allow the energy to pass through. A terrorist or accident scenario involving a radiation-generating machine is difficult to imagine and probably implausible. Therefore, our discussion of possible terrorist scenarios is limited to those involving radioactive materials, not machine-generated radiation.

Radioactive materials may be naturally occurring or manufactured. Radioactive materials emit one or more forms of radiation, typically two forms, one an electromagnetic wave and the other a particle form. Contrary to popular science fiction, radioactive materials do not glow, nor do they have any special characteristics that make them readily distinguishable from nonradioactive materials with the five human senses. They may be liquid, solid, or gas.

Radioactive cobalt looks like and has the same chemical properties as nonradioactive cobalt. Radioactive water, known as tritium, cannot be readily distinguished from regular, nonradioactive water. The difference lies in the atomic structure of the material, the very tiny particles that make up the material. In basic chemistry we learn that all matter is composed of atoms. The structure of these atoms is responsible for the characteristics of the material. A material's stability, density, and several other properties are determined by its atomic structure.

Basic Radiation Physics

An atom is composed of protons and neutrons contained in its nucleus. The only exception to this is the naturally occurring hydrogen atom, which contains no neutrons. Protons and neutrons are virtually the same size. Electrons, which are much, much smaller than protons and neutrons, orbit the nucleus of the atom. The chemical behavior of an atom depends on the number of protons, which are positively charged, and the number of electrons, which are negatively charged. Neutrons, which have no electrical charge, do not play a role in the chemical behavior of the atom.

Elements that have identical atomic structures except for their numbers of neutrons are known as *isotopes* of one another. Not all isotopes are radioactive. Radioactive isotopes are also called *radionuclides.* For example, phosphorous-32 is a manufactured radioactive isotope of phosphorous. It is commonly used in research applications. Iodine-131 is a manufactured radioactive isotope commonly used in nuclear medicine applications. Potassium-40 is a naturally occurring radioactive isotope.

Cobalt-60 is a radioactive isotope used for many purposes. Though it comes in several forms, one typical type of cobalt-60 source is a metal cylinder. The cobalt-60 is contained within a stainless steel cylinder that is welded closed. The cobalt-60 emits gamma rays, a waveform of radiation, and beta particle radiation. The stainless steel cylinder absorbs most of the beta particle energy, but allows the gamma ray radiation to pass through.

In facilities that use radioactive materials, the standard radioactive symbol is used to label the materials so they can be readily identified. Special placards are also required for transporting certain quantities or types of radioactive materials. This information is useful when considering response to an accident involving radioactive materials. However, when considering a terrorist event, you cannot rely on the presence of labels or placards to help identify the hazards involved.

Radiation Effects on the Body

No matter the form or the source of the radiation, radiation energy is deposited in the body when the body is exposed to it. The amount of energy deposited in the body by a radiation source varies widely. It depends largely on the energy of the radiation involved, its penetrating ability, and whether or not the source of radiation is located outside or inside the body. Exposure to radiation from a source outside the body is known as *external exposure.* Exposure to radiation from a source within the body is known as *internal exposure.*

Let us return to the example of the cobalt-60 source discussed earlier. A person located within several feet (the distance depends on the strength of the source) of the cobalt-60 source will be exposed to the gamma radiation emitted from the source without having to directly touch the source. This exposure is an external exposure. If the source somehow becomes damaged, the cobalt-60 could leak from the container. In addition to continuing the external exposure, this would present an internal exposure hazard though it would not necessarily cause an internal exposure. In order to cause an internal exposure, the cobalt-60 would have to be taken into the body via inhalation, ingestion, or some other means.

Another important concept involving radioactive materials can be demonstrated with the cobalt-60 source. *Radioactive contamination* is the presence of radioactive material in a location where it is not desired. Radioactive contamination results from the spillage, leakage, or other dispersal of unsealed radioactive material. The presence of radioactive contamination always presents an internal exposure hazard because of the relative ease with which it can be incorporated into the body. Depending on the radioactive material involved, it may also present an external exposure hazard.

In the cobalt-60 example, damage to the source caused the radioactive material to leak. Anywhere the material is deposited is then contaminated. The contamination is spread by a number of methods including air currents, water run-off, and persons walking through it, touching it, and then cross-contaminating other objects and areas as they touch and walk through them.

So far we have only used the term exposure when discussing radiation and its interaction with the body. *Exposure* is the irradiation of any object, living or inanimate. The term *RAD* (Radiation Absorbed Dose) is the unit of measure for radiation exposure. An exposure of one RAD results in the absorption of 100 ergs of energy per gram of tissue exposed. In the international system of units, the unit of exposure is the Gray. One Gray equals 100 RAD.

The term dose must now be introduced to further discuss radiation's effects on living persons. Radiation *dose* is a calculated measurement of the amount of energy deposited in the body by the radiation to which the person is exposed. The unit of dose is the *REM* (Roentgen Equivalent Man). *Roentgen* is another unit for radiation exposure (but it is not necessary to define the term in order to explain the concept of dose). The REM is derived from this unit of measure by taking into account the type of radiation producing the exposure. In the international system of units, the unit of radiation dose is the *sievert*. One sievert equals 100 REM.

The REM is approximately equivalent to the RAD for exposure to external sources of radiation. When considering the magnitude of radiation exposure encountered in an accident or terrorist activity, it is not really necessary for a first responder to distinguish between exposure and dose. It is important to know how to measure external radiation levels if you are called to the scene of an incident involving radiation. It is equally important to develop an understanding of the dangers associated with different levels of exposure. Your own agency should develop policies regarding acceptable doses for emergency responders. These policies should be consistent with other risks normally encountered in the course of the first responder's duties.

Radiation Measurements

Radiation levels are measured with survey instruments designed for that purpose. Survey instruments usually read out in units of R/hr, in which the R stands for either RAD or REM. The unit R/hr is an exposure (or dose) rate. An instrument reading of 50 R/hr means you will have to stay in that area for one hour to receive a 50 RAD exposure. Dividing the unit will determine the exposure for shorter or longer periods of time (e.g., a six-minute stay in the area will result in a 5 RAD exposure.) An exposure (or dose) rate can be compared to a speedometer. A speed of 80 mph means you have to travel an hour at that rate to go 80 miles. Traveling half an hour at that rate will cover a distance of 40 miles.

Some instruments can measure the dose over a period of time. These can be compared to an odometer, which measures total miles traveled regardless of the speed. Though handheld survey instruments may have this capability, they are more useful in an emergency situation for measuring the exposure rate. Radiation dosimeters, however, are useful for measuring the exposure received over time. Personnel responding to an incident wear dosimeters, which should be checked frequently to determine the exposure received by a first responder at an incident scene.

Survey instruments and dosimeters have certain limitations. Like all electronic equipment, survey instruments may drift over time. They should be recalibrated and inspected at regular intervals. Survey instruments also use batteries that must be checked and replaced when necessary. Dosimeters must be zeroed and checked on a regular basis. Survey instruments and dosimeters are useless if they are not functioning properly.

Internal Radiation Exposure

For internal radiation exposure, the terms RAD and REM are not synonymous. First responders need not be concerned with measuring internal exposure on the scene, however. It is important for first responders to know whether an internal exposure hazard exists and how to protect themselves from it by using PPE, including respirators, as used in protection against haz mat.

Internal exposure assessment is very complicated due to the large number of factors involved. Some of these factors are the chemical form of the material, what type of radiation is emitted, how the material entered the body, and the physical characteristics of the exposed person. Months of study may be required to determine an internal dose. Common methods for assessing internal exposure are to sample blood, urine, feces, sweat, and mucus for the presence of radioactive material. Special radiation detectors may also be used to make a direct measurement of the radiation emitted by radioactive materials deposited within the body. By considering the results of these measurements along with the characteristics of the material and the body's physiology, a fairly precise measurement of radiation dose from internal sources can be made.

Radiation Health Effects

A radiation incident may present both internal and external exposure hazards, or it may present either one. The hazards may be large or possibly insignificant. First responders may be exposed to these hazards while working at the scene, and should be aware of radiation's health effects and risks. How much radiation is too much? That is a tough question. There is a substantial body of scientists and academics who would argue that *any* exposure is dangerous, and that extraordinary precautions are necessary so no one is exposed. At the other end of the spectrum there are credible scientists and academics who argue that some radiation exposure is necessary to life and perhaps even beneficial. Then there is the majority of scientists and academics, this author included, who encourage a common sense approach along with a healthy respect for radiation and its associated dangers.

Critical Factor: Radiation incidents require a common sense, healthy respect approach.

Health effects of radiation exposure have been studied for years. No one will argue that at high levels of exposure, serious health effects occur. These effects are called *prompt* or *acute* effects because they manifest themselves within hours, days, or weeks of the exposure. Acute effects include death, destruction of bone marrow, incapacitation of the digestive and nervous systems, sterility, and birth defects in children exposed *in utero*. A localized high exposure could result in severe localized damage requiring amputation of the affected area. These effects are clearly evident at high exposures such as those produced by an atomic bomb detonation or serious accident involving radioactive materials. These effects are seen at short-term exposures of about 25 RAD and above. The severity and onset of the effect is proportionate to the exposure. Effects of radiation exposure that are not manifest within a short period of time are called *latent* or *delayed* effects. The most important latent effect is a statistically significant increase in the incidence of cancer in populations exposed to high levels of radiation.

The health effects of low exposures are not obvious and are subject to wide debate in scientific and academic circles. Low exposures do not cause obvious bone marrow damage or digestive or nervous system effects. They have not been shown to cause cancer or birth defects. Localized low exposures to the hands, feet, arms, and legs do not cause obvious harm. Information from persons exposed to high levels of radiation has been used to predict possible health effects to persons exposed to low levels. The primary concern is the incidence of cancer. Since high exposures cause a statistically significant increase in the incidence of cancer, low-level exposure is thought to possibly cause an increase in the risk of cancer—perhaps proportionate to the increase seen at high exposures, perhaps disproportionate. To minimize risks, occupational dose limits for persons who work with radiation are set at 5 REM per year. This is not a dividing line between a safe and unsafe dose; it is a conservative limit set to minimize risk.

The basis for the "common sense, healthy respect" approach to radiation exposure held by many scientists, academics, and radiation safety professionals is a solid one. Earlier in the chapter we discussed sources of radiation and mentioned that there are several naturally occurring ones such as uranium, radioactive potassium, and radon gas. Earth is also constantly bombarded by radiation from outer space. These sources contribute to radiation exposure of the general population known as *background exposure*. Additional sources of background exposure are atomic weapons testing fallout, industrial, academic, and military uses of radiation and radioactive material, and radioactive materials and radiation used in medicine. All of these sources combined cause, on average, an annual radiation dose of approximately 0.360 REM per person per year, the majority of which is due to natural radiation.

Exposures similar in magnitude to those caused by background radiation cause no acute effects. Though unproven, low-level exposures are thought to cause a small increase in the risk of cancer. This risk is expressed as an additional 0.0004 cases of cancer per REM. For example, the number of statistical cancer fatalities expected for a population of 10,000 persons is about 20 percent, or 2,000 cancer fatalities. If this population also received a radiation dose of one REM, the expected cancer fatalities would increase by four, causing a total of 2,004 cancer fatalities instead of 2,000 under normal conditions. All these statistics are theoretical, based on information obtained by studying populations exposed to high doses.

Because there are no acute health effects resulting from low-level exposures and because the latent health risks appear to be very small, if they exist at all, the common sense, healthy respect approach to radiation protection allows for the benefits of radiation in society and minimizes the associated risk. Common sense dictates that radiation exposures should be kept to a minimum, but that extraordinary measures to avoid all exposure are unnecessary. Risks associated with low-level radiation exposure are small. They are comparable, if not smaller than risks accepted in normal, everyday life. The risks are much smaller than those encountered in hazardous occupations. Keeping the risks in perspective is important to successfully respond to an incident involving radiation. Radiation should be respected but not feared.

Critical Factor: Minimize exposure. Remember time, distance, and shielding.

Radiation Case Example

The involvement of radiation or radioactive materials at an incident does not justify an inappropriate or inadequate response. Safety of the responder is the first consideration, whether hazards exist from radiation or some other source. Obvious hazards must be dealt with to ensure the best possible outcome. Do not allow an unfounded fear of radiation to cloud your judgment and negatively affect your response. The following anecdote conveys this point very well.[1]

At a large nuclear facility there was a flammable storage area—a structure type known as a "pole barn," which is a covered area of drums containing flammable, toxic, radioactive liquids. The area was surrounded by a dike to contain any spills, but no walls. One day a forklift operator skewered a drum with a tine, and did not notice it until the pallet had been lifted and removed from the diked area. As the forklift operator backed out of the dike (down a ramp) the trail of leaking fluid became obvious to the driver, who reversed direction. The leaking drum was taken back into the diked area. Because the hole in the drum was at the bottom, practically the entire contents of the drum poured and pooled into the diked area.

When I arrived at the scene, I found workers in regular cotton coveralls sloshing around in the spill trying to absorb it. The plant shift supervisor/ on-scene commander was quick to ask, "This spill has tritium in it. Do these workers need any special protection?" It only took me a few seconds to notice the sign on the pole barn, "Flammable Storage," the label on the drum ("flammable"), and the readily available drum inventory sheet stating a dangerously low flash point for the liquid (and of course, a very small, practically irrelevant radionuclide content). I said, "Yes, a fire truck, SCBAs (self-contained breathing apparatus), and flash suits. Forget the tritium, these folks are standing in a pool of fuel!" You can imagine the look on the commander's face. The on-scene commander was distracted by a trace quantity of tritium listed on a piece of paper, when he should have focused on the obvious hazard—workers standing in flammable liquid.

First Responder Considerations

Accident Situations[2]

Most radiation-related accidents likely to be encountered by emergency medical personnel generally will involve radioactive materials or radiation-emitting devices being used in an industrial or institutional setting, or while being transported. However, there also may be incidents in which victims result from the accidental or deliberate misuse of radioactive materials.

Industrial accidents cover a range of situations, from activities within nuclear power plants, isotope production facilities, materials processing and handling facilities, and the widespread use of radiation-emitting measurement devices in manufacturing and construction.

Institutional accidents generally involve research laboratories, hospitals and other medical facilities, or academic facilities. Generally, the victim or patient is an individual who was directly involved in handling the material or operating the radiation-emitting device.

Transportation accidents may occur during the shipment of radioactive materials and waste. However, due to stringent and rigidly enforced regulations governing the packaging and labeling of radioactive material shipments, few of these incidents pose any serious threat to health and safety.

Here the victims or patients are usually vehicle operators, pedestrians, or occupants of other vehicles who are unlikely to come in contact with the radioactive material involved.

Commercial and private aircraft accidents also may involve radioactive materials, primarily radio-pharmaceuticals carried as cargo, or radioactive instrument components, but these seldom pose a serious exposure risk. Accidents involving military aircraft generally pose no increased risk since radioactive weapons elements are sealed and shielded, and protected against accidental detonation.

There have been several incidents worldwide in which radioactive materials were unknowingly released by individuals unaware that they were dealing with a hazardous substance. Improperly or illegally discarded radiation sources have been opened by scrap dealers and others, causing serious contamination and lethal exposure to a number of people.

EMS personnel responding in one of the preceding situations where radiation is involved or suspected need to keep in mind that the first priority remains the expedient delivery of appropriate emergency medical services to the patient, including transport to a hospital. Deal with the patient's medical condition first!

Critical Factor: Treat a patient's medical condition first.

In a first responder situation, ambulance, rescue, or medical services personnel will not know whether the patient is contaminated or exposed unless:

1. They are advised in advance by the party requesting assistance.
2. They are advised on arrival by other responders such as police or fire officials that radioactive materials are present at the scene.
3. They are advised by the patient that he or she is contaminated or was exposed.
4. They determine from their own observation of the accident site that contamination or exposure is a possibility (i.e., from visual signs, placards, or documents such as shipping papers).

Information regarding the source of the radiation, type of radioactive material involved, and length of time of exposure is valuable data that should be gathered at the scene if possible, but it does not alter the role of EMS personnel with respect to the handling and transport of the patient.

At the same time, it is important that EMS personnel remember the distinction between exposure and contamination. They should remember that there is little, if any, chance they will encounter a radiological incident that poses a serious threat to their own health and safety. While accidents involving small amounts of radioactive material may occur in industry or commerce at any time, incidents that involve high-level, dangerous amounts of radiation are extremely rare and almost never occur outside the surveillance of qualified experts. Laws and regulations governing the transport of high-level radioactive materials require packaging, escorts, and security to guard against anyone being accidentally exposed or contaminated.

The patient who has been exposed, but not contaminated, requires no special handling other than that appropriate to his illness or injury. Such a patient presents no radiological threat to medical personnel. The patient is not radioactive.

When contamination is known to be present, or suspected but unconfirmed, EMS personnel should take steps to minimize the spread of contamination to themselves and the transport vehicle.

Handling of contaminated injured involves the use of three basic principles: common sense, cleanliness, and good housekeeping. Let's examine some typical situations and consider the appropriate procedures when a radiologically contaminated or exposed individual is or might be involved.

The basic types of radiation accident victims are as follows:

External Radiation Exposure—A person exposed to even a lethal dose of radiation generally presents no hazard to the individuals around him. The patient is not radioactive, and is no different than the patient who has been exposed to diagnostic x-rays.

The only exception to this rule is the person who has been exposed to significantly high amounts of neutron radiation. Persons or objects subjected to neutron radiation may become radioactive themselves. Such activation is extremely rare and noted here for information purposes only.

External Contamination—The individual who has external contamination presents a different situation. Problems associated with this type of patient are similar to those encountered with chemical contamination. The presence of external contamination usually means the individual has come in contact with loose or unconfined radioactive material, such as a liquid or powder or airborne particles from a radioactive source. The objective should be containment to avoid spreading the contamination. Anyone or anything coming in contact with a person or an object that is contaminated by radioactivity must be considered as being contaminated until proven otherwise. Isolation techniques should be implemented to confine the contamination and protect personnel.

Internal Contamination—The patient who has been externally contaminated may also have received internal contamination by inhalation or ingestion. Internal contamination, however, is usually not a hazard to the individuals around the patient. The most common type of internal contamination involves the inhalation of airborne radioactive particles that are deposited in the lungs. Also possible is the absorption through the skin of radioactive liquids or the entry of radioactive material through an open wound. In all instances, there may be little or no residual surface or external contamination, but the patient may suffer the effects of exposure from the ingested or absorbed radioactive material.

Certain accident victims may be included in two or more of these categories. For example, an injured person may be contaminated both internally and externally with radioactive material. Such a patient should be handled with universal precautions, stabilized medically, appropriately decontaminated, and then evaluated for exposure by qualified medical experts.

External contamination may be eliminated or reduced by simply removing clothing and using conventional cleansing techniques on body surfaces, such as gentle washing and flushing that does not abrade the skin surface. Internal contamination, however, does not lend itself to conventional treatment using techniques normally associated with problems such as chemical poisoning, and must be evaluated by experts.

Terrorism Situations

Another potential source of contamination and/or exposure that must be considered involves the deliberate dispersal of radioactive material by terrorists. The Oklahoma City federal building and World Trade Center bombings, the subway poison gas attack in Japan, the use of chemical and biological agents during the Gulf War, and other incidents have heightened awareness regarding the potential for terrorist acts involving WMEs.

A WME incident in which chemical, biological, or radiological materials are used in conjunction with explosives or released environmentally under certain circumstances has the potential to cause significant numbers of casualties as well as create widespread public panic. Such situations require a very different approach to ensure appropriate steps are taken to protect medical service providers and facilities against unnecessary exposure.

Large-scale incidents, such as those involving catastrophic damage and mass casualties from occurrences such as tornadoes, earthquakes, or explosions, require a different approach than the small-scale incident in which only one or two victims are encountered. The concern here is with the type of contamination that may be present and its potential effect on the ability to sustain the viability of medical services. It is improbable that most natural disasters would release radioactive materials in sufficient quantities to cause mass casualties, but there are two man-made situations that could produce such results.

One is an accidental explosion that occurs within a facility where radioactive materials are used or stored. It is possible that a limited amount of radioactive material could be released as a direct result of the explosion or any ensuing fire. Local fire officials have access to information about where licensed radioactive materials are housed in their communities and can assist in determining when such a hazard might exist.

The other large-scale, man-made incident that potentially poses a serious risk is an act of terrorism. In any terrorist incident that produces mass casualties and extensive damage, the first consideration should be whether a chemical, biological, or radiological agent was involved. The presence of a hazardous material with the accompanying prospect of contamination and exposure drastically alters the approach that should be taken by medical service personnel.

When such involvement is known or suspected, the injured should be triaged, treated, monitored, and decontaminated to the extent possible, at the scene. The movement of contaminated or exposed patients to hospitals poses the substantial risk of contaminating transportation resources, treatment facilities, and staff, rendering these resources useless for the treatment of uncontaminated injured. EMS protocols should clearly outline the steps to be taken when there is notification of a terrorist incident possibly involving a hazardous material. Among the considerations are the following:

- Dispatching on-shift and off-shift emergency staff to establish on-scene triage and treatment capabilities. Most mass casualty plans provide personnel allocation schemes that can be adapted for this purpose.
- EMS providers and rescue services should not bring victims directly to hospitals unless specifically directed to do so.
- As early as possible after notification, contact state and federal support agencies for assistance with monitoring and decontamination functions and disposal of contaminated materials.
- It is also important to remember that any terrorist incident is a criminal act, and that while the first responsibility of medical service providers is to the patient, necessary interaction with law enforcement officials will be an integral part of the process from the outset. No physical evidence should be discarded without authorization from law enforcement officials, and all activities should be recorded as carefully and thoroughly as possible.

CHAPTER SUMMARY

Radiation is energy in the form of invisible electromagnetic waves or very small energetic particles. Radiation exposure differs from a traditional haz mat incident because the victims (or rescuers) do not have to come in contact with the source to be exposed.

Radiation exposures from a source outside the body are external exposures. An internal exposure is from a radioactive source within the body. Radioactive contamination is the presence of radioactive material where it is not desired. There is always a danger of contamination causing an internal exposure (usually by inhalation or accidental ingestion).

Exposure is the irradiation of any object. A RAD (radiation absorbed dose) is the unit of measure for radiation exposure. The unit of dose is a REM (Roentgen equivalent man). One Sievert equals 100 REM.

Special instruments that usually read in units per hour (R/hr) measure radiation. If an instrument reads 50 R/hr, and the exposure time is thirty minutes, the patient has received a dose of 25 REM. Radiation dosimeters should be worn by responders, because they measure exposure over a period of time (dose).

A common sense approach and a healthy respect for radiation are encouraged. There is wide debate over the health effects of low levels of radiation. High levels of radiation exposure cause severe health effects. These are called prompt or acute effects and include bone marrow destruction, nerve and digestive system damage, and death. The healthy respect, common sense approach means that radiation exposure should be minimized, but extraordinary measures to eliminate all exposures are not necessary.

Expedient emergency medical treatment is the first priority for responding to radiation exposed patients. Information related to the source, type of material, and length of exposure is valuable data if it can be gathered. EMS responders must remember the distinction between exposure and contamination. Contaminated patients must be decontaminated accordingly. Rescuers coming in contact with contaminated patients should also be considered contaminated.

In a mass casualty terrorism incident, the presence of a chemical, biological, or radiation agent should be considered. In suspected cases of radiological exposure, the patient must be triaged, treated, monitored, and decontaminated on the scene before transport. Transport of contaminated patients will result in contaminated transport units and contaminated hospitals. In any form of radiation incident, state and federal support agencies should be contacted.

CHAPTER QUESTIONS

1. How does a radiation event differ from a traditional haz mat incident?
2. List and define the three types of radiation.
3. Define the terms dose and exposure.
4. Discuss the common sense approach to radiation exposure.

    5. Discuss basic medical treatment procedures for each of the following radiation exposures:
- external radiation exposure
- external contamination
- internal exposure

    6. How does the prospect of radiation exposure in an incident affect EMS mass casualty protocols?

Simulation

There is a major international festival in your community or region with 50,000 attendees. There is a bomb detonation that generates 95 trauma casualties. An immediate assessment by the haz mat team reveals that the explosive device was combined with a radioactive material causing radiation exposure and contamination to 50 patients and 20 responders. Discuss the following questions in detail:

1. What EMS protocols and fire/rescue operational procedures are now in effect in your local agencies that address this scenario?
2. After studying this chapter, what protocols and operating procedures should be added to your community's response plan?
3. What role does hospital preparedness play in this incident? (Remember that many of these patients will self-present at hospitals or immediate care centers, circumventing the traditional EMS system.)
4. What are the contamination issues in this incident?
5. What are the state and federal support agencies available for assistance to your locale in a major radiation incident?

NOTES

[1]This anecdote is the personal experience of Craig Reed, Health Physicist, and is used with his permission.
[2]This section and the succeeding one are adapted and/or excerpted from *Radiation Accidents, March 1999, A Guide for Medical Professionals on Handling, Transporting, Evaluating and Treating Patients Accidentally Exposed to Radiation or Contaminated with Radioactive Materials*, first produced and published by the Illinois Department of Nuclear Safety.

10
Weapons of Mass Effect: Explosives

Hank T. Christen
Robert Walker

Source: Courtesy Liaison Agency, Inc.

Chapter Objectives

After reading this chapter, you will be able to accomplish the following objectives:

1. Recognize the significance of explosive devices in terrorism and tactical violence events.
2. Understand the categories of explosives and their characteristics.
3. Outline the basic elements in the explosive train.
4. Understand the basic initiating elements in explosive devices.
5. Outline the critical safety steps that must be utilized when operating in an environment where explosive devices are suspected or present.

Introduction

One of the first explosives was black powder. The Chinese invented black powder in A.D. 600. History has not recorded the first use of explosives for terrorism, but there is little doubt that soon after the invention of black powder, someone used it to blow up someone else.

Today there are many types of explosives designed for industrial use, military operations, and entertainment. All of these explosives are available to people through various means (legal and otherwise) for clandestine use. Some explosives can be made at home with common chemicals using "recipes" easily accessible to anyone seeking the information.

Explosive devices are effective for three basic reasons:

1. They create mass casualties and property destruction (a WMD).
2. Explosives are a major psychological weapon. An explosion instills terror and fear in survivors and the unaffected population (a WME).
3. Secondary explosive devices render a site unsecured, greatly complicating medical, rescue, and suppression efforts (WME).

Critical Factor: Explosives are very effective weapons for creating mass casualties and fear.

There are many historical examples of the terrorist use of explosives. Factions throughout Europe, the Middle East, Asia, and Africa have used long-term bombing campaigns. In the United States we have witnessed the horror of the World Trade Center and Oklahoma City explosions, but fail to realize that there are several detonations per week and numerous devices disarmed that do not get coverage beyond the local media.

There is every indication that emergency responders in the United States will see an increase in explosive terrorism and tactical violence. The Internet abounds with information about simple explosives and simple timing devices that can be made at home. Commercial explosives are readily available, and military explosives can be accessed in world black markets.

Explosive Physics

How do explosives work? How do explosives differ? What causes some explosive devices to fail? The answers to these questions fall under the general category of explosives physics (the science of explosives). Explosive physics is important to emergency responders. These are the scientific laws that kill you and your patients, and the laws that determine whether you will live through an event.

An explosive material is a substance that is capable of rapidly converting to a gas with an extreme increase in volume. This rapid increase in volume causes heat and noise, and a shock wave that travels outward from the detonation. Chemists and physicists like to point out that an explosion is not "instantaneous." Academically, they are correct because explosives require several nanoseconds to develop. To human observers, however, explosions are instantaneous. More importantly, significant injury and property damage also occurs "instantly."

The most damaging by-product of an explosion is the shock wave. The shock wave is an energy wave that originates at the source and travels outward in all directions. It behaves much like ripples on the surface of water when a pebble is dropped into a pond. The wave travels the course of least resistance, reflects off hard objects such as strong walls or buildings, and becomes concentrated in spaces such as hallways or areas between buildings. Shock waves can also be reflected back to the source.

The strength and characteristics of explosives are measured by the speed of the shock waves they produce. This is measured in feet per second (fps) or meters per second (mps). Their velocity of detonation determines the dividing line between low explosives and high explosives. A more precise and scientific definition is that a low explosive is one that deflagrates into the remaining unreacted explosive material, at less than the speed of sound. A high explosive is an explosive that detonates into the remaining explosive material faster than the speed of sound. A low explosive may become a high explosive by the way it is contained or initiated. Black powder, when burned in an open area, will not detonate. If you confine the powder in a container such as a pipe bomb, the outcome is very different. The same applies with the initiation of high explosives. When C4 is ignited it will burn without detonating. If you introduce a shock to C4 via a blasting cap, you get an explosive detonation.

Critical Factor: Explosives produce a very high speed, damaging shock wave.

Some explosives have a shock wave that produces a pushing effect. Detonation or deflagration that is slower than the speed of sound causes this push effect. Deflagration is a very rapid combustion that is less than the speed of sound. These explosives push obstacles and are commonly used for applications such as quarrying, strip-mining, or land clearing. Black powder, smokeless powder, and photoflash powders are examples of deflagrating or low explosives. A deflagrating effect or low explosive effect is similar to what you feel when a car is next to you with deep bass speakers at full volume.

High explosives have a sharp, shattering effect. This shattering effect is referred to as the explosive's brisance and is comparable to the high-pitched voice of an opera soprano that causes crystal glass to shatter. High explosives are very brisant, and produce shock waves greater than the speed of sound. For example, military explosives such as C4 produce a shock wave of 24,000 fps. High explosives (high brisance) have a very sharp and shattering effect. These explosives do extensive damage, causing severe injuries with a high percentage of fatalities.

The devastating effect of land mines is a product of brisance. The shock wave literally pulverizes bone and soft tissue in the lower extremities. In improvised explosive devices (IED) the shock wave causes severe barotrauma injuries, major internal organ damage, head injuries, and traumatic amputations. A lethal secondary effect is fragmentation. Concrete, glass, wood, and metal fragments are expelled at ballistic speeds. The effect causes multiple fatalities and critical injuries.

An explosive shock wave creates another effect called blast overpressure. Air in the vicinity of the explosion is compressed and expands creating a pressure higher than atmospheric pressure. Blast overpressure causes barotrauma damage in the form of air embolisms and damage to tethered organs. This pressure also causes severe structural damage to buildings. A blast pressure of 5 pounds per square inch (psi) does not sound high. However, on a standard door (80 by 30 inches), the total impact pressure is 12,000 pounds. On a wall that is 8 feet high and 10 feet long, the total pressure is 50,000 pounds.

In summary, the physics of explosives explain the effects that kill people and severely damage property. The most damaging by-product is a shock wave that travels very fast and is unseen. The shock wave causes fragmentation, blast overpressure, and barotrauma injuries.

Types of Explosives

Explosives are designed to detonate with maximum power when initiated, yet be extremely stable when stored or transported. One of the most widely known civilian explosives is dynamite. The invention of dynamite was a major breakthrough in explosive technology. The prime ingredient in dynamite is nitroglycerin (nitro), an extremely unstable liquid that detonates violently, with even minor shocks. In dynamite, the nitro is mixed with sawdust and other ingredients to stabilize the nitro.

Dynamite is a high explosive that generates a shock wave of 14,000 to 16,000 fps. It is readily available and legally procured in states that issue a blaster's permit. Quantities are stored on construction sites and are frequently stolen. Dynamite is also used on farms for digging, land clearing, and stump removal. Due to its availability, ease of use, stability, and explosive power, dynamite is a popular choice for IEDs.

Black powder and smokeless powder are also popular IED explosives. They can be easily purchased in small quantities in gun shops that cater to ammunition reloading hobbyists. These explosives are frequently used in pipe bombs. Black powder is a deflagrating explosive that detonates with extreme force when stored in a confined container. Pipe bombs were used in the Atlanta Olympics bombing and in many abortion clinic bombings.

Ammonium nitrate is another common civilian explosive. Ammonium nitrate fertilizer, when mixed with a catalyst, will detonate with violent force. This explosive is frequently used in agricultural operations, and was used in the Oklahoma City bombing.

Military explosives are extremely powerful, even in small quantities. The most well-known type is a plastic explosive called C4. It is soft and pliable, resembling a block of clay, and may be cut, shaped, packed, and burned without detonating. When detonated, C4 explodes violently and produces a very high-speed shock wave. Just two pounds of C4 can totally destroy a vehicle and kill its occupants.

C4 is not easily obtained, but is illegally available on the black market. Similar plastic explosives are available on foreign markets. A military plastic explosive called Semtex was used to make the IED that caused the Pan American crash in Lockerbie, Scotland. Other military explosives include TNT, Tritonal, RDX, and PETN.

All explosives (civilian and military) require an initial high-impact and concentrated shock to cause detonation. A small explosive device called an initiator produces this initiating shock. Initiators are a key step in a chain of events called the explosive train. The most common type of initiator is a blasting cap.

The first step in the explosive train is a source of energy to explode the initiator. This source is usually electrical, but can be from thermal, mechanical, or a combination of the three sources. The initiator contains a small amount of sensitive explosive, such as mercury fulminate. The detonation of the initiator produces a concentrated and intense shock that causes a high-order detonation of the primary explosive. The explosive train is diagramed as follows:

Initiating energy = initiator explosion = main explosive detonation

It is critical that all elements of the explosive train function properly. Any malfunction or separation of the elements breaks the explosive train resulting in a failed detonation.

Improvised Explosive Devices (IED)

An IED is any explosive device that is not a military weapon or commercially produced explosive device. In essence, IEDs are homemade. It is important to realize that homemade devices can vary from simple to highly sophisticated. Do not perceive IEDs as a high school product constructed from an Internet bomb recipe.

Approximately 80 to 90 percent of IEDs are made from smokeless powder or dynamite according to the Eglin Air Force Base, Florida, EOD School. Devices made from C4 or Semtex are rare, and usually lead investigators to suspect foreign sources.

A crucial element in an IED is a timing device. For many reasons, bomb makers do not want to be present when the device is initiated. Because of security and scope, this text does not cover timing devices in detail. Timers can be chemical, electrical, electronic, or mechanical. Simple timers include watches or alarm clocks that close an electrical circuit at a preset time. Electronic timers operate in a similar fashion, but are more reliable and precise. Some electronic timers or initiating devices can be activated from radio sig-

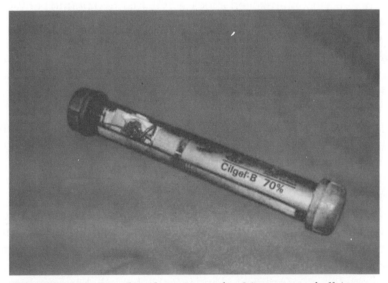

FIGURE 10-1 Pipe bombs account for 26 percent of all improvised explosive devices according to the FBI Bomb Data Center. The above photo is a confiscated pipe bomb constructed with a stick of Cilgel-B dynamite. *Source:* Courtesy R. Walker

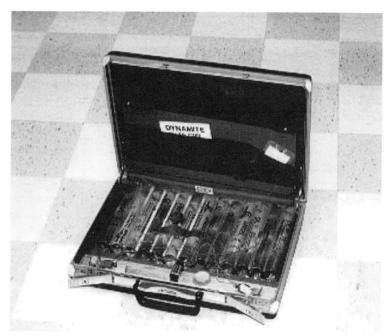

FIGURE 10-2 Briefcase rigged with dynamite and multiple detonation devices. *Source:* Courtesy R. Walker

nals from a remote site. In most cases, timers cause electrical energy to be routed from batteries to an initiator (usually an electric blasting cap).

Other devices have no timer, and are designed to detonate when people or emergency responders trigger the detonation. These devices are called booby traps. In simple devices, a trip wire or mechanical switch initiates the detonation. When the wire is touched or the device is tampered with, an explosion occurs. In more sophisticated devices, an invisible beam that is broken by people walking through it causes the detonation. Other high-tech booby traps include light, sound, or infrared triggering systems.

Critical Factor: Improvised explosive devices vary in the type of explosive, form of initiation, and degree of sophistication.

Chemical, Nuclear and Biological IEDs

An IED may be used to initiate a chemical, biological, or nuclear event. In these cases, the improvised explosive is used to scatter a chemical or biological pathogen or toxin. The history of such devices is scarce, but increased use of these devices is anticipated.

An especially "dirty" weapon is an improvised nuclear device (IND). An IND does not involve a nuclear explosion like a military nuclear weapon. In an IND, conventional explosives are used to scatter radioactive materials. The device is considered dirty because the radioactive contamination can render an area radioactively "hot" for thousands of years. Presently, there is no history of an IND incident, but the potential is there.

Critical Factor: Conventional explosives can be used to disperse chemical or radiological agents.

Secondary Devices

High threats to emergency responders are secondary devices (review Chapter 3, "Terrorism/Tactical Violence Incident Response Procedures"). Secondary devices are timed devices or booby traps that are designed and placed to kill emergency responders. The objective is to create an emergency event, such as a bombing or fire, that generates an emergency response. The secondary device explodes and may cause more injuries than the original event. In one of the Atlanta abortion clinic bombings, a secondary device in a dumpster exploded after EMS, fire, and law enforcement responders arrived. In the Columbine High School shooting, multiple devices scattered throughout the school greatly restricted the tactical operations of EMS units and SWAT teams.

Secondary devices can be used to create an entrapment situation. Beware of a situation that lures responders into narrow areas with only one escape route. A narrow, dead-end alley is a classic example. An incident such as a fire or explosion at the end of the alley is the initial event that causes emergency responders to enter the area. The secondary device (a booby trap or timed IED) is placed in the alley. When the IED detonates, there is only one narrow escape route that lies in the path of a concentrated shock wave.

A key to surviving an entrapment situation is to recognize the scenario by surveying the overall scene. A narrow focus (called tunnel vision) obscures the big picture. Look around! Do not concentrate on a small portion of the incident scene. Look for trip wires, suspicious packages, and objects that appear to be out of place. Be especially aware of dumpsters or abandoned vehicles. Question bystanders familiar with the area if possible. Try entering by an alternate route. In summary, look for a setup that could kill you.

Critical Factor: Beware of secondary devices!

FIGURE 10-3 Environmental awareness is critical to remaining safe at explosives or threat scenes. In the United Kingdom the general public is enlisted through a public awareness campaign and poster placement in areas of public assembly. *Source:* Courtesy P. Maniscalco

Safety Precautions

Many of the safety precautions for explosive devices were discussed in Chapter 4. Several safety steps bear repetition.

1. Avoid radio transmissions within at least fifty feet of a suspected device. Electromagnetic radiation (EMR) from radio transmissions can trigger an electric blasting cap or cause a sophisticated device to detonate.
2. Avoid smoking within fifty feet (or further) from a suspicious device.
3. Do not move, strike, or jar a suspicious item. Do not look in a suspicious container or attempt to open packages.

4. Memorize a clear description of suspicious items.
5. Establish a hot zone for 500 feet around small devices, and 1,000 feet around large devices or vehicles. (Large zones may not be practical in congested downtown areas). Maintain the required hot zone until bomb technicians advise otherwise.
6. Try to stay upwind from a device; explosions create toxic gases.
7. Take advantage of available cover such as terrain, buildings, or vehicles. Remember that shock waves bounce off surrounding obstacles.

Basic Search Techniques

Emergency responders often conduct primary searches or assist bomb experts in conducting a thorough search for explosive devices. Remember that emergency responders are not trained to clear an area of explosive devices; only bomb technicians perform this function.

Critical Factor: Bomb technicians are the only personnel qualified to clear an area or remove/disarm an explosive device.

In building searches, always search from the outside in. Building occupants are an excellent source of information, because they know what objects are supposed to be in a given location. Occupants can tell you that a trash basket has always been there, or that a paper bag is Joe's lunch. Likewise, they can point out that the innocent looking newspaper machine was never there before. In building interiors, custodians can assist in unlocking areas and pointing out obscure storage areas.

Search from the floor to the ceiling. Often objects that are not at eye level are unseen. Adopt the habit of making a floor level sweep, followed by an eye level sweep, and finally a high wall and ceiling sweep.

Begin vehicle searches from the outside to the inside (just like buildings). If the driver is present, assign one person to distract the driver from observing advanced search techniques. Leave the trunk and doors closed, and concentrate on the outside. Avoid touching the vehicle; touching can activate motion switches. Never open the vehicle until trained technicians

FIGURE 10-4 Suited in explosive protective ensemble EOD personnel will require special attention from on-scene EMS resources. EOD techs are prone to hyperthermia, dehydration, and in the event of device detonation during inspection, a whole blast of trauma related injuries. *Source:* Courtesy P. Maniscalco

arrive. If the driver is present, have the driver open doors, the trunk, and dash compartments.

Always emphasize the safety precautions previously discussed in this chapter. When in doubt, wait for bomb technicians. You can save lives by establishing a hot zone and exercising effective scene control.

CHAPTER SUMMARY

The use of explosives for terrorism goes back many centuries. Explosive devices are very effective weapons of mass effect. Bombings create mass fatalities and mass casualties. Bombs are also effective psychological weapons; they create fear in survivors and the community at large. Lastly, bombs can be used as secondary devices to kill or injure emergency responders.

An explosive is a material that converts to a gas almost instantly when detonated. This detonation creates a shock wave, which is a measure of the explosive power of a given material. In a low-order detonation, a shock wave travels through the remainder of the unexploded material at a speed less than the speed of sound. High order explosives create shock waves greater than the speed of sound. High explosives have a sharp, shattering effect called brisance.

There are many types of explosives, with black powder being the earliest type. When confined in a device such as a pipe bomb, it has considerable explosive force. The first commercial type of explosive was dynamite, which can produce a shock wave of 14,000 to 16,000 feet per second. Ammonium nitrate (fertilizer), when mixed with a catalyst, is a low-order explosive. Military explosives are extremely powerful. They have high brisance and create shock waves that are as high as 24,000 feet per second.

Improvised explosive devises (IEDs) are homemade weapons that contain an explosive material, a power source, and a timer. The explosives are usually dynamite or black powder. The timing devices can be chemical, electrical, or electronic. Special devices called booby traps contain a triggering mechanism such as a trip wire. These are secondary devices designed to injure emergency responders.

Secondary devices can be used to create an entrapment situation. Entrapment is a situation with a device in a narrow area, such as an alley, with no escape route. A key prevention step is to survey the entire scene before entry, and look for trip wires or other initiation devices.

Key safety steps in an unsecured area are:

- Avoid radio transmissions or smoking within 50 feet of a suspected device.
- Do not move, strike, or jar a suspicious item.
- Establish a hot zone 500 feet around a small device and 1,000 feet around a large device or vehicle. (These distances may not be practical in urban areas).

Emergency responders often assist in searching an area for suspicious devices. Bomb disposal experts are the only personnel that can clear an area or safely remove an explosive device.

CHAPTER QUESTIONS

1. Name three reasons why explosives are effective WMD/WME weapons.
2. Define and discuss the most damaging product of an explosion.
3. What is blast overpressure? What are the injury and damage effects of blast overpressure?
4. Name and briefly describe at least three types of explosives.
5. List four types of explosive timers.

6. List and discuss seven safety precautions relating to secondary explosive devices.
7. What is the role of emergency responders in a basic search for explosive devices?

Simulation

Research the previous year's history of explosive attacks in the United States. Ascertain trends in the types of explosives used, their effectiveness (casualties), and the primary motives for the attacks. What was the number of explosive detonations in the United States last year? (Note—sources can include publications, news articles, or Web sites for the ATF, FBI, Department of Justice, or other law enforcement sources.)

11
Mass Casualty Decontamination

Michael V. Malone

Source: Courtesy H. Christen

Chapter Objectives

After reading this chapter, you will be able to accomplish the following objectives:

1. List the three stages of decontamination.
2. Describe several methods used by fire departments for gross decontamination.
3. Recognize several considerations for setting up a decontamination area.
4. Define the decontamination requirements for hospital joint accreditation.
5. Understand the principles of mass casualty decontamination.
6. Recognize the decontamination requirements for various agents.
7. List features of biological agents that affect decontamination for biological agents.
8. Understand weather factors that affect decontamination.
9. Discuss considerations in local protocols for the establishment of triage procedures for contaminated patients.

Introduction

Decontamination is defined as the process of removing or neutralizing a hazard from the environment, property, or life form. According to the Institute of Medicine National Research Council, the purpose of decontamination is to prevent further harm and enhance the potential for full clinical recovery of persons or restoration of infrastructure exposed to a hazardous substance. This chapter will focus on mass casualty decontamination, and will discuss these areas:

- the traditional decontamination process used by fire departments and hazardous material response teams
- the decontamination capabilities of hospitals or health care facilities
- military types of decontamination
- methodology and principles applied to a mass casualty incident resulting from weapons of mass effect or an accidental release of a harmful substance
- containment procedures
- mass casualty decontamination, including decontamination requirements for victims with conventional injuries
- site selection, environmental, weather, and responder requirements during the decontamination process

Basic Principles of Decontamination

The management and treatment of contaminated casualties will vary with the situation and nature of the contaminant. Quick, versatile, effective, and large capacity decontamination is essential. Casualties must not be forced to wait at a central point for decontamination. Decontamination of casualties serves two purposes; it prevents their systems from absorbing additional contaminants, and protects health care providers and uncontaminated casualties from contamination. Review of after-action reports and video-tapes of the Tokyo subway incident in 1995 emphasizes this requirement.

Fire departments and hazardous materials response teams define the two types of decontamination as *technical decon* and *medical* or *patient decon.* Medical or patient decon is the process of cleaning injured or exposed individuals. This process, performed by haz mat teams, is far less common than technical decon. Personnel decon methods are traditionally established for haz mat team members and have not been applied in a mass casualty setting. The method of self-decontamination, or team decontamination, is considered a part of technical decontamination. Shower systems with provisions for capturing contaminated water runoff are commercially available and may provide a degree of victim privacy. This type of system does not have the capacity to treat a large number of casualties. The main limitations are availability of equipment and haz mat personnel.

Fire departments are equipped and structured for rapid and effective decontamination. Many fire departments have developed procedures that use existing low-tech equipment. A common practice is the use of two engines parked 20 feet apart with three ground ladders placed on top of the engines and spanning the gap between them. A fourth ladder is placed perpendicular on top of the other ladders at the midpoint of the span. This serves as a girder to support a hanging tarp creating two corridors (male and female). Tarps are stretched over the top of the ladders and down the middle of the corridor to create a male/female decontamination route. Hand lines and/or engine discharges with low pressure (60 psi) fog nozzles are utilized for water spray. The pumps are shut down and supplied by a distant engine to reduce noise in the decon area.

Fire departments with aerial devices utilize effective elevated deluge systems (low pressure fog) in conjunction with engine companies. Aerial ladders, extended horizontally, can also be used with tarps to create a decontamination corridor for privacy.

Stages of Decontamination

The process for decontamination of casualties involves three stages called *gross, secondary,* and *definitive* decontamination.

Gross decontamination:

1. Evacuate the casualties from the high-risk area. With limited personnel available to conduct work in the contaminated environment or hot zone, a method of triage needs to be established. First, decon those who can self-evacuate or evacuate with minimal assistance to decon sites, then start decontamination of those who require more assistance.
2. Remove the exposed person's clothing. The removal and disposal of clothing is estimated to remove 70 to 80 percent of the contaminant[1]; others estimate 90 to 95 percent.[2]
3. Perform a one-minute, head-to-toe rinse with water.

Secondary decontamination:

1. Perform a quick, full-body rinse with water.
2. Wash rapidly with a cleaning solution from head to toe. A fresh solution (0.5 percent) of sodium hypochlorite (HTH chorine) is an effective decontamination solution for persons exposed to chemical or biological contaminants. Undiluted household bleach is 5.0 percent sodium hypochlorite. Plain water has been found to be equally effective because of ease and rapidity of application. With certain biological agents, the sodium hypochlorite solution may require more than 10 minutes of contact. This is not possible in a mass casualty incident requiring rapid decontamination.
3. Rinse with water from head to toe.

Definitive decontamination:

1. Perform thorough head to toe wash until clean. Rinse thoroughly with water.
2. Dry victim and have them don clean clothes.

Critical Factor: The stages of decontamination are gross, secondary, and definitive.

Methods of Initial Decontamination

A first response fire company can perform gross decontamination by operating hose lines or master streams with fog nozzles at reduced pressure. The advantage of this is that it begins the process of removing a high percent-

FIGURE 11-1   Simulated patient undergoes gross decontamination process at chemical terrorism exercise in Newark, NJ. *Source:* Courtesy P. Maniscalco

age of the contaminant in the early stage of an incident. Methods to provide privacy and decontamination for non-ambulatory casualties must be addressed. Decontamination considerations include:

1. Prevailing weather conditions (temperature, precipitation, etc.), which affect site selection, willingness of the individual to undress, and the degree of decontamination required.
2. Wind direction.
3. Ground slope, surface material, and porosity (grass, gravel, asphalt, etc.).
4. Availability of water.
5. Availability of power and lighting.
6. Proximity to the incident.
7. Containment of runoff water if necessary or feasible. The department of Mechanical and Fluid Engineering at Leeds (U.K.) University has determined that if a chemical is diluted with water at the rate of approximately 2000:1, pollution of water courses will be significantly reduced[3] (Institute of Medicine National Research Council). Examples of containment devices or methods: children's wading pools, portable tanks used in rural fire fighting, hasty containment

pits formed by tarps laid over hard suction hoses or small ground ladders, diking with loose earth or sandbags covered with tarps.

8. Supplies including personnel protective equipment and industrial-strength garbage bags.

9. Clearly marked entry and exit points with the exit upwind, away from the incident area.

10. A staging area at the entry point for contaminated casualties. This is a point where casualties can be further triaged and given self-decontamination aids, such as spray bottles with a 0.5 percent solution of sodium hypochlorite solution of Fullers Earth.

11. Access to triage and other medical aid upon exit, if required.

12. Protection of personnel from adverse weather.

13. Privacy of personnel. (This will be a media intensive event; an example is B'nai B'rith, Washington, D.C., 1997).

14. Security and control from site setup to final cleanup of the site.

DECONTAMINATION TRIAGE

In a mass casualty event, decontamination of chemically exposed patients must be prioritized or triaged. The intent of this process is similar to the triaging of trauma patients in a conventional incident. The objective is to first decontaminate salvageable patients that are in immediate need of medical care. Patients that are dead or unsalvageable should not be immediately decontaminated. Patients that are ambulatory and non-symptomatic are the lowest decontamination priority. Again, the primary objective is to immediately decontaminate patients who are exposed, yet salvageable.

Critical Factor: First decontaminate victims who are severely exposed, yet salvageable.

The U.S. Army Soldier and Biological Chemical Command (SBCCOM) published a guide called *Guidelines for Mass Casualty Decontamination During a Terrorism Chemical Agent Incident* (January 2000). The SBCCOM guidelines suggest the following factors for assigning decontamination triage priorities:

1. casualties closest to the point of release
2. casualties reporting exposure to vapor or aerosol
3. casualties with liquid deposition on clothing or skin
4. casualties with serious medical signs/symptoms (shortness of breath, chest tightness, etc.)
5. casualties with conventional injuries

The major question in decontamination triage is the criteria for determining where or when not to treat/decontaminate a non-ambulatory patient that is symptomatic. Emergency response agencies must adopt a local protocol that should be based on the following issues:

- What is the nature of the incident? Severe exposure to nerve agents with major symptoms usually result in death.
- Are there high quantities of antidotes available? For example, nerve agents require very high doses of atropine and valium (for seizures).
- Are personnel available to move and treat mass numbers of non-ambulatory patients? A single non-ambulatory patient requires two to four responders.
- Ambulatory patients who are symptomatic or have been severely exposed should be immediately decontaminated.
- Ambulatory patients who are non-symptomatic should be moved to a treatment area for possible clothing removal and medical evaluation.
- Non-ambulatory patients should be evaluated in place while further prioritization for decontamination occurs (SBCCOM).
- Patients who are in respiratory arrest, grossly contaminated with a liquid nerve agent, having serious symptoms, or fail to respond to atropine injections should be considered deceased or expected to die ("expectant" in military terminology).
- In extreme cases, a patient in a hot zone may require immediate treatment prior to decontamination. Treatment usually consists of immediate antidote administration and airway maintenance. Clothing removal is the only expedient method of field decontamination, with decontamination by showering or flushing later, if appropriate.

HOSPITAL DECONTAMINATION STANDARDS

The Joint Commission on Accreditation of Healthcare Organizations (JCAHO) requires hospitals to be prepared to respond to disasters including hazardous materials accidents. The majority of hospitals that have decontamination capabilities utilize existing indoor infrastructure and do not have the ability to expand to accommodate mass casualties. Outside the standard universal protection procedures followed by the medical community, required protective equipment and trained personnel are limited in the hospital system.

A hospital standard practice for haz mat response is to call the fire department. Due to the stress placed on the response system mitigating the effects of a large incident, haz mat teams will not be available. The hospitals are at risk when the response system is stressed to the point that patients start self-referring or independent sources deliver patients to the hospital.

Critical Factor: Hospitals are required by joint accreditation standards to have decontamination procedures and equipment.

The military has identified two types of decontamination: personnel and equipment. Personnel decontamination has been divided into two subcategories: hasty and deliberate. Specialized units within the military (U.S. Marine Corps Chemical Biological Incident Response Force and the National Guard's Civil Support Teams) have further subdivided deliberate decontamination to encompass ambulatory and non-ambulatory personnel.

Hasty decontamination is primarily focused on the self-decontaminating individual using the M258A1 skin decontamination kit. This kit is designed for chemical decontamination and consists of wipes containing a solution that neutralizes most nerve and blister agents. Another type of kit,

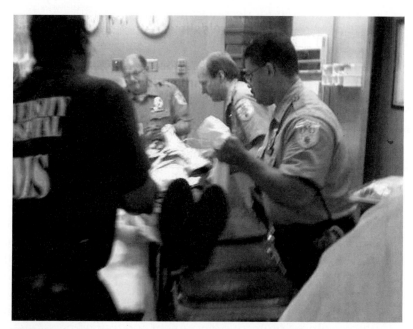

FIGURE 11-2   Hospitals must train personnel and equip facilities to conduct patient decontamination. There is no guarantee that all patients who arrive at your facility will be clean or have been through decontamination on-site at the emergency. Only 18.5%* of contaminated victims are treated at the scene of an exposure. The remainder seek out contamination treatment at medical facility/hospital. *Hazardous Substances Emergency Event Surveillance, Annual Report 1999. *Source:* Courtesy D. Gerard

the M291 decontamination kit, uses laminated fiber pads containing reactive resin, which neutralizes and removes the contaminant from a surface by mechanical and absorption methods. (These kits require user training, and are not usually available for civilian emergency response organizations.)

The procedure of removing and exchanging (donning and doffing) personnel protective clothing is also considered hasty decontamination. Deliberate decontamination is required when individuals are exposed to gross levels of contamination or for individuals who were not dressed in personnel protective clothing at time of contamination. The established process is to completely remove the individual's clothing, apply a decontamination solution (0.5 percent HTH or water) followed by a fresh water rinse, then use a chemical agent monitor (CAM) to detect the presence of nerve and blister agents or M-8 paper to validate the thoroughness of the decontamination process. At the end of this process the individual is provided new clothing and equipment. If the individual presents symptoms, he or she will be processed through the heath care system.

Decontamination Site Setup

The decontamination site should be established with the following considerations:

1. upwind from the source of contamination
2. on a downhill slope or flat ground with provisions made for water runoff
3. water availability
4. decontamination equipment availability
5. individual supplies
6. health care facilities
7. site security

Mass Casualty Decontamination

Specialized military units have developed personnel decontamination sites that can process large numbers of contaminated personnel, both ambulatory and non-ambulatory. These systems or sites are capable (agent dependent) of processing up to 200 ambulatory or 35 non-ambulatory personnel per hour depending upon the agent(s) involved. The sites are set up in a tent that incorporates a shower system that sprays a decontaminant followed by a rinse.

Step one of this process is removal of the patient's clothing. Ambulatory patients use a similar process that military personnel use

FIGURE 11-3 The USMC CBIRF establishing a decontamination line at an exercise in Tampa, Florida. *Source:* Courtesy P. Maniscalco

during their doffing procedures. Non-ambulatory casualties' clothing is cut off by decon specialists.

Step two is to place clothing into disposable bins, which are sealed.

Step three is to remove personal effects, tag them, and place them into plastic bags. Disposition of the personal effects will be determined later. These items may be crime scene evidence.

Step four is to apply a decontamination solution. For ambulatory casualties, this is done through a shower system. Non-ambulatory casualties are sponged down.

Step five is for individuals to use brushes to clean themselves, or be cleaned by a decon specialist. This step aids in the removal of the contaminant and allows for a three-minute contact time for the decontaminating solution.

Step six is a freshwater rinse.

Step seven is to monitor for the agent or contaminant. This is conducted using a CAM or M-8 paper for chemical agents, or using a radiation meter for radiation.

Step eight is to don dry clothing.

Step nine is medical monitoring. Individual documentation is developed.

Step ten provides for individuals' release or transport to a medical facility.

The non-ambulatory site uses the same steps; however, the casualty is moved along a series of rollers and cleaned by decon specialists. Care must also be taken at the non-ambulatory site to decontaminate the roller surface with a 5 percent solution of HTH between patients. These sites are self-contained, require a water source, and provide:

- heated water (if required; warm water opens the pores of the skin and could accelerate dermal exposure)
- water runoff capture
- decontamination solution
- protection from the elements
- privacy
- medical monitoring during the decontamination process
- post-decontamination checks
- clothing
- site control

The specialized decontamination assets just described are from pre-positioned military units and not usually available for rapid response to civilian incidents. These units are highly competent and professional, but limited by numbers and location. The military refers to them as low density, high demand assets. The U.S. Public Health Service has developed a similar capability resident in the Metropolitan Medical Response System (MMRS) and the National Disaster Medical Response Teams (NMRT).

Radiation Decontamination

Radiation injuries do not imply that the casualty will present a hazard to health care providers. Research has demonstrated that levels of intrinsic radiation present within the casualty from activation (after exposure to neutron and high-energy photon sources) are not life threatening. If monitoring for radiation is not available, decontamination for all casualties must be conducted. Removal of the casualty's clothing will reduce most of the contamination with a full body wash further reducing the contamination.

Wearing surgical attire or disposable garments such as those made of Tyvek will reduce the potential exposure of health care providers. Inhalation or ingestion of particles of radioactive material presents the greatest cross-contamination hazard. Care must be taken to capture runoff or retrieval of the material. Industrial vacuum cleaners are commonly used. The vacuum cleaner output air should be filtered with a HEPA filter to prevent rerelease of the material into the air.

FIGURE 11-4   North Carolina–Special Operations Response Teams (SORT) of the National Disaster Medical System (NDMS) participate in multiple training and response exercises yearly to maintain a heighten sense of readiness. NC–1 SORT is one of the best trained and equipped organizations in the world for response to these types of incidents and are deployable anywhere. *Source:* Courtesy P. Maniscalco

Decontamination Requirements for Various Agents

Decontamination requirements differ according to the type of chemical agent or material to which individuals have been exposed. Water is the accepted universal decontaminant. The importance of early decontamination cannot be overemphasized due to the mechanism of action with organophosphous compounds (nerve agents). Nerve agents may be absorbed through any surface of the body. Decontamination of the skin must be accomplished *quickly* to be fully effective. Liquid agents may be removed using Fullers' Earth. Persistent nerve agents pose the greatest threat to health care providers. Once a patient has been decontaminated or the agent is fully absorbed, no further risk of contamination exists.

> **Critical Factor:**   Early decontamination is critical for severe exposure to nerve agents.

Exposure to a vesicant (blister agent) is not always noticed immediately because of the latent effects. This may result in delayed decontamination or failure to decontaminate at all. Mucous membranes and eyes are too sensitive to be decontaminated with normal skin decontaminant solutions. Affected sensitive surfaces should be flushed with copious amounts of water or, if available, isotonic bicarbonate (1.26 percent) or saline (0.9 percent). Physical absorption, chemical inactivation, and mechanical removal should decontaminate skin. Chemical inactivation using chlorination is effective against mustard and lewisite, and ineffective against phosgene oxime. If water is used, it must be used in copious amounts. If the vesicant is not fully removed, the use of water will spread it.

Critical Factor: Vesicant contamination may not be immediately noticed.

Choking agents will not remain in liquid form very long due to their extremely volatile physical properties. Decontamination is not required except when used in very cold climates. Choking agents are readily soluble in organic solvents and fatty oils. In water, choking agents rapidly hydrolyze into hydrochloric acid and carbon dioxide.

Blood agents will not remain in liquid form very long due to their extremely volatile physical properties. Decontamination is not required.

In the case of incapacitants, total skin decontamination should be completed with soap and water at the earliest opportunity. Symptoms may appear as late as 36 hours after a percutaneous exposure, even if the individual is decontaminated within one hour of exposure.

Personnel exposed to riot control agents should be moved to fresh air, separated from other casualties, faced into the wind with eyes open, and told to breathe deeply. Clothes should be removed and washed to preclude additional exposure from embedded residue.

Biological Agents

Biological agents are unique in their ability to inflict large numbers of casualties over a wide area by virtually untraceable means. The difficulty in detecting a biological agent's presence prior to an outbreak, its potential to selectively target humans, animals, or plants, and the difficulty in protecting the population conspire to make management of casualties (including decontamination) or affected areas particularly difficult. The intrinsic features of biological agents that influence their potential use and establishment of management criteria include virulence, toxicity, pathogenicity, incubation period, transmissibility, lethality, and stability.

If a dermal exposure is suspected, it should be managed by decontamination at the earliest opportunity. Exposed areas should be cleansed using the appropriately diluted sodium hypochlorite solution (0.5 percent) or copious quantities of plain soap and water. The patient's clothing should also be removed as soon as possible.

Secondary contamination of medical personnel is a concern and can be avoided by strict adherence to universal medical precautions. Biological agents, for the most part, are highly susceptible to environmental conditions, and all but a few present a persistent hazard.

Anthrax is a very stable agent; however, in a non-aerosolized state it presents only a dermal (requiring breaks or cuts in the skin) or ingestion hazard. The strategy recommendations for potential exposure to anthrax are:

1. Gather personal information from the potentially exposed individual(s).
2. Explain the signs and symptoms of the disease.
3. Give them a point of contact to call if they show symptoms.
4. Send them home with the following instructions: remove clothing and place it in a plastic bag, securing it with a tie or tape. Shower and wash with soap for 15 minutes.
5. The individuals should be informed of the lab analysis results of the suspected agent as soon as possible. If results are positive, the correct medical protocol will be administered.

Effects of Weather on Decontamination

Weather effects will impact the manner in which an agent will act in the environment and will have an impact on decontamination requirements. A release of chemical agents or toxic industrial materials always has the potential to cause injuries to unprotected people proximal to the point of release. Strong wind, heavy rain, or temperatures below freezing may reduce effects. Weather will be of great importance for the respiratory risks expected at different distances from the point of release. Weather conditions will also influence the effect of ground contamination.

High wind velocity implies a short exposure time in a given area, reducing the number of casualties in an unprotected population. Low wind velocity will increase the exposure time, increasing the number of casualties, and may cause effects at a greater distance.

Critical Factor: Weather is an important determination in the effectiveness of a chemical attack.

To a high degree, the gas/aerosol concentration in the primary cloud depends on the air exchange or turbulence of the atmosphere. In clear weather, at night, the ground surface is cooled and inversion is formed (stable temperature stratification). Inversion leads to weak turbulence, resulting in the presence of a high concentration of material. Unstable temperature stratification occurs when the ground surface warms, resulting in increased turbulence. The effect is decreased concentration, particularly at increased distances from the point of release.

The concentration in the primary cloud may also decrease in cold weather, particularly at temperatures below −20°C (−4°F), due to a smaller amount of agent(s) evaporating during dispersal. However, this will increase ground contamination at the point of release. Precipitation also reduces concentration, but can increase ground contamination.

Low temperatures will increase the persistency of some agents. Some agents may cease to have an effect at very low temperatures due to their freezing point, however, they present a problem when temperatures increase or if they are brought into a warm environment.

Biological agents are potential weapons of mass destruction and generally have the following characteristics: odorless and tasteless, difficult to detect, and can be dispersed in an aerosol cloud, over very large downwind areas. Ideal weather conditions for dispersal include an inversion layer in the atmosphere, high relative humidity, and low wind speeds. Incubation periods can be as long as several days, therefore, wind speed and direction are a primary weather concern to determine the exposed population and predict the effects upon that population. Ultraviolet (UV) light has a detrimental effect on many biological agents, making periods of reduced natural sunlight the optimal time for release.

Most biological agents will not survive in extremely cold weather and it is difficult to aerosolize live biological agents in freezing temperatures. Toxins are less affected by cold weather, however, cold weather tends to provide a temperature inversion that prolongs the integrity of an aerosolized cloud.

CHAPTER SUMMARY

A common sense, well-informed approach to decontamination should be adopted. The following are additional considerations for decontamination operations in a mass casualty setting:

1. Establish a local protocol for decontamination triage.
2. Decontaminate as soon as possible to stop the absorption process.
3. Establish multiple decontamination corridors.

4. Establish security and control measures to contain contaminated casualties and prevent non-contaminated/nonresponders from entering the affected area.
5. Decontaminate only what is required.
6. Decontaminate as close to the point of contamination as possible (100 m or 328 ft outside, if the point of contamination was inside a building; 1 km (0.6 miles) for an outside release.
7. Involve the patient in the process, allowing as much self-decon as possible.
8. Use existing infrastructure.
9. Monitor the patients throughout the process.
10. Provide privacy if possible with use of tents, available facilities, and/or removal of the media.

Organizations that have potential requirements to provide decontamination support for a mass casualty incident should focus on existing inherent capabilities. With modifications and enhanced training, a good, thorough decontamination system can be effectively implemented.

CHAPTER QUESTIONS

1. List and discuss the three stages of decontamination.
2. Discuss at least five considerations for setting up a decontamination area.
3. Discuss lockdown procedures for controlling entry of contaminated patients at medical facilities.
4. Outline the 10 steps in mass casualty decontamination.
5. Outline triage procedures for mass casualty decontamination.
6. What factors determine the severity or effectiveness of a given biological agent?
7. How do the following weather elements influence the effects of a WME agent?
   • wind direction and speed
   • temperature
   • atmospheric stability

Simulations

1. Develop a mass decontamination procedure for your community. Consider training, equipment, protocols, and triage procedures.
2. Develop a mass decontamination plan for a medical facility. Consider security and lockdown, training, equipment, and control of contaminated vehicles.

NOTES

[1]Cox, R. D., *Annals of Emergency Medicine*, 23, 761–770, 1994.

[2]NATO Handbook on the Medical Aspects of NBC Operations, 1991.

[3]Institute of Medicine Research Council, *Improving Civilian Medical Response to Chemical or Biological Terrorist Incidents: Interim Report on Current Capabilities.* Washington, D.C.: National Academy Press, 1998.

12
Crime Scene Operations

Neal J. Dolan
Paul M. Maniscalco

Source: Courtesy P. Maniscalco

"Wherever he steps, whatever he touches, whatever he leaves, even uncon-sciously, will serve as silent evidence against him. Not only his fingerprints or his footprints, but his hair, the fibers of his clothing, the glass he breaks, the tool-marks he leaves, the paint he scratches, the blood or semen that he deposits or collects—all these and more bear mute witness against him. This is evidence that does not forget. It is not confused by the excitement of the moment. It is not absent because human witnesses are. It is factual evidence. Physical evidence cannot be wrong; it cannot be wholly absent. Only its interpretation can err. Only human failure to find it, study and understand it, can diminish its value."

Presiding Judge
Harris v. United States 331 US 145 (1947)

Chapter Objectives

After reading this chapter, you will be able to accomplish the following objectives:

1. Define a crime scene.
2. Recognize the value and importance of physical evidence.
3. Understand the evidence "theory of exchange."
4. Recognize the evidence classification of objects, body material, and impressions.
5. List key crime scene observations that should be made by initial responders.
6. Understand the key steps for emergency responders in preservation of evidence.
7. Understand the concept of "chain of custody."

Introduction

The significance of physical evidence at a crime scene cannot be overesti-mated. It is the only thing that will help prove who committed the crime. Proper training and technique are necessary to maintain the integrity and value of evidence. Each day emergency responders travel to emergency scenes to render aid. These emergency personnel shoulder a formidable burden to accomplish their mission and cause no further harm to the people or the scene. Crime scenes are exciting, chaotic, and dangerous places. They are replete with hidden clues that hold the answer to the ques-tion, "Who committed this crime?"

Emergency responders often carry out their duties in conflict with im-portant crime scene procedures. Emergency responders are focusing on the preservation of life and not the preservation of evidence. EMS responders may not be aware that the clothes they are cutting off and discarding from a shooting victim may contain valuable evidence to solve the crime. Firefighters may be employing legitimate fire fighting techniques at an explosion scene that are destroying evidence of who committed the offense.

It is possible to carry out the emergency responders' mission without creating more problems for the crime scene. This is best accomplished through training and awareness of potential crime scenes, and acting to minimize damage to the area and its contents.

The Crime Scene—Physical Evidence

A crime scene is any area in which a crime may have been committed. It is anywhere the criminal was during the commission of the crime and the egress from the scene. The exact dimensions of the scene will be determined by the nature and type of crime. For example, a shooting crime scene could be as large as the room or building where the victim was discovered. A terrorist incident could be several blocks or even miles in diameter. The Oklahoma City Federal Building incident had a 20-block perimeter established and a critical piece of evidence, the crankshaft from the rented Ryder truck, was found two blocks from the explosion site.

One key to uncovering the vast amount of information and physical evidence present at the crime scene is an awareness of what constitutes physical evidence. Evidence is something legally submitted to a competent tribunal as a means of ascertaining the truth in an alleged matter under investigation. Physical evidence is one form of evidence. It can be defined as anything that has been used, left, removed, altered, or contaminated during the commission of a crime by either the victim or the suspect.

Critical Factor: Emergency responders have a major role in assisting law enforcement agencies with identifying and preserving physical evidence.

The benefits of physical evidence are best summarized in the opening paragraph by the issuing judge in the case of *Harris v. U.S.* (1947). Physical evidence does not lie, forget or make mistakes. It has no emotional connection to anyone or anything. It is demonstrable in nature and not dependant on a witness. It is the only way to establish the elements of a crime.

In order to heighten the awareness of responders, it is necessary to explain how physical evidence evolves at the scene. Forensic scientists propose the *theory of exchange* to describe this process. Whenever two objects

FIGURE 12-1 Proper identification, preservation, and collection of physical evidence often can mean the difference between conviction or aquittal of a suspect. *Source:* Courtesy J. Pat Carter/AP/Wide World Photos

come in contact with each other, each will be altered or changed in some way. When a rapist comes in contact with a victim, numerous substances will be exchanged. The suspect or the victim could deposit or remove skin traces, blood, body fluids, carpet fibers, soil, and many other items. Bombing victims may have chemical traces on their clothing or fragments of evidence embedded in their bodies that may prove to be important in the investigation and prosecution of the perpetrators. These evidence sources have been invaluable to law enforcement investigators in the past, including high profile cases such as the bombing of a Pan Am airliner arriving in Honolulu, Hawaii, from Narjita, Japan, in 1982, as often taught by forensics expert Rick Hahn (retired FBI). Changes in the objects may be microscopic and require detailed examination to establish the variations. However, responders should be cognizant that exchanges will take place and are not always noticeable to the naked eye.

Physical evidence can be almost anything. Table 12-1 gives some examples of items that could be encountered by responders working at a potential or actual terrorism/crime scene.

Being aware of potential hazards at a scene is not new to EMS, firefighters, or haz mat responders. Scene safety, sizing up the scene, or just taking a minute to examine the environment of a scene can minimize the impact of costly mistakes of overzealous responders.

Table 12-1 Possible Items of Evidence at a Crime Scene

Objects	Body Material	Impressions
Weapons	Blood	Fingerprints
Tools	Semen	Tire traces
Firearms	Hair	Footprints
Displaced furniture	Tissue	Palm prints
Notes, letter, papers	Sputum	Tool marks
Matchbook	Urine	Bullet holes
Bullets	Feces	Newly damaged areas
Shell casings	Vomit	Dents and breaks
Cigarette butts		
Cigar butts		
Clothes		
Shoes		
Jewelry		
Bomb fragments		
Chemical containers		
Mechanical delivery systems		

Actions of First Arriving Units

Literature in the EMS, emergency management, and fire/rescue fields stresses initial scene evaluation. However, few texts elaborate on the importance of viewing an event as a crime scene and analyzing the hot zone accordingly.

First arriving units are usually overwhelmed. The scene is chaotic and dynamic. Any observations at this early stage are very important to law enforcement investigators.

Critical Factor: An initial scene evaluation should include a basic crime scene evaluation.

Several key observations are important for initial responders. There is not time to write anything; just remember key crime or evidence observations and report them to the incident manager or the law enforcement branch as soon as possible. Important observations include:

1. chemicals on the scene that would not normally be present
2. damage, debris fields, and fragmentation that indicate an explosion
3. suspicious people; people hiding or running
4. statements issued by bystanders
5. unusual odors on the scene
6. evidence of gunfire (shell casings, bullet holes, or gunshot wounds)
7. weapons in the area
8. suspicious casualties (patients may be terrorists)
9. multiple fires that appear to be from separate sources
10. suspicious devices

Evidence Preservation by Emergency Responders

What do we do with physical evidence when we find it? There are several answers. Observations should be written down on a notepad or electronic device as soon as practical. Admittedly, this step may take place hours later. Observations of weapons, suspects, or devices must be communicated to the incident manager or law enforcement branch immediately. Other observations should be conveyed when emergency response tasks are completed (put out the fire and treat the patients first).

There are several important rules in preserving physical evidence:

FIGURE 12-2 The Murrah Building bombing (April 1995) in Oklahoma City posed many issues for emergency responders with respect to evidence collection and recovery. *Source:* Courtesy P. Maniscalco

1. Do not touch or move evidence. Law enforcement knows how to photograph, document, package, and remove evidence.
2. If evidence must be moved for tactical reasons, note the original location and report it to law enforcement investigators.
3. If possible, photograph evidence removal during tactical operations. Forensic photographers will photograph normal evidence removal during an investigation.
4. Avoid evidence contamination caused by walking through the scene. Stretch a rope or scene tape into the crime scene area and instruct personnel to walk along the established path.
5. Minimize the number of personnel working in the area.
6. Check the soles of boots or shoes when personnel exit the crime scene because fragments or fibers may be imbedded.
7. Clothing or personal effects removed from victims are considered evidence. Ensure that law enforcement personnel practice biohazard safety procedures when examining "red-bagged" evidence or clothing.

Emergency responders can also become victims. A classic case was the Sarin gas attack in the Tokyo subway system. This incident was initially reported to be a normal call until emergency physicians realized they were dealing with a nerve agent as the cause of the sickness. The importance of preparation procedures and the use of protective equipment is paramount. Every day police, fire, and emergency medical services encounter dangerous situations during the normal course of performing their services. However, the potential for lethal hazards and long-term effects that result from terrorist incidents are much greater.[1]

Another unique hazard of terrorist incidents is the probability of a secondary device targeting the responders to the incident. Remember that the goal of the terrorist is to create chaos and fear, and what better way to accomplish this than to turn the responders into victims. Bombing incidents in Atlanta, Georgia, and Birmingham, Alabama, offer real examples of this scenario. In both cases, secondary devices were detonated. Emergency responders must be alert, aware, and suspicious of their surroundings when responding to incidents that have the potential to be terroristic in nature.

Crime Scene Analysis

The thorough analysis of a crime scene consists of the identification, preservation, and collection of physical evidence as well as the recording of testimonial evidence. Without adherence to this basic assertion during the initial stages of a crime scene investigation, the potential to disrupt the integrity of the evidence is great. Hawthorne states that this could result in the evidence being challenged by the defense in a court of law, which in turn could lead to dismissal of the charges or the finding of a lesser offense against the criminal defendant.[2] More often than not, the proper collection of physical evidence from a crime scene is the definitive portion in the resolution of a criminal offense. Admissible physical evidence has the potential to (1) establish that a crime has occurred; (2) place a suspect in contact with the victim and/or the scene; (3) establish identity of those associated with the crime; (4) exonerate the innocent; (5) corroborate the victim's testimony, and (6) cause a suspect to make admissions or confess.[3]

The crime scene should be approached as if it will be the only opportunity to gather the physical evidence present.[4] Attention should be initially directed toward observing and recording the information present at the crime scene, rather than taking action to solve the crime immediately.[5] Careful consideration should also be given to other case information or statements from witnesses or suspects.

This chapter was prepared with the intention of providing the reader with rudimentary principles of crime scene investigation, from the initial

FIGURE 12-3 Evidence preservation and recovery at a terrorism scene can involve thousands of pieces of physical evidence. With the TWA Flight 800 investigation federal investigators reassembled the fuselage from numerous pieces.
Source: Courtesy Mark Lennihan/AP/Wide World Photos

approach of the crime scene to final disposition of physical evidence found at the crime scene. Although the methods used to initially approach a crime scene are virtually universal in terms of application of use, it should be noted that at some point, the investigation may begin to take on unique characteristics that may be atypical or unorthodox in nature. Therefore, it is impossible to propose a single, step-by-step procedure that will ultimately resolve every type of crime scene[6] (Department of Justice, 2000).

However, regardless of the unique nature that a crime scene may take on, thorough crime scene analysis, effective interviews and interrogations and the use of common sense will make it less likely that evidence will be overlooked, improperly collected or preserved, or that mistakes will be made.[7]

A review of the literature reveals that a common set of generalized categories for crime scene procedures exists. These procedures are listed as follows:

1. protection of the crime scene
2. identification of evidence
3. documentation of evidence
4. collection of evidence
5. marking of evidence
6. packaging of evidence
7. transportation of evidence

Law Enforcement Responsibilities:

Upon initial arrival at the scene of a crime, the first responders are tasked with a grand responsibility. It is their task to set the foundation for what Hawthorne termed the process of analyzing a crime scene. The basic elements of the process are: (1) approach; (2) render medical aid; (3) identify additional victims or witnesses; (4) secure the crime scene and physical evidence, and (5) make appropriate notifications. While adhering to these principles, the first law enforcement responders will be certain to provide subsequent investigators and technicians with a sound foundation from which they can conduct a comprehensive analysis of the crime scene.

When approaching a crime scene, the first law enforcement responder must maintain professional composure regardless of the often overwhelming factors associated with the task to be completed.[8] He must be vigilant and able to recognize anything, whether it be animate or inanimate, which seems to have a connection to the crime committed. Furthermore, the relationship of items at rest as they relate to the position of other items present at the crime scene should be noted in terms of the distances and the angles that separate them. The first law enforcement responder must be objective in his initial approach of a crime scene, and resist the temptation to form conclusions as to what occurred.

Upon arrival at a crime scene, the paramount concern should be the preservation of human life and/or the prevention of additional injuries. The first law enforcement responders must be able to provide adequate first aid and/or request professional medical assistance. According to Hawthorne if law enforcement responders are providing medical assistance and the crime scene or physical evidence becomes contaminated, altered, or

> **Critical Factor:** The paramount concern at a crime scene is the preservation of life and injury prevention.

lost, that is a price that must be paid. The preservation of life outweighs the preservation of evidence at a crime scene.

After satisfying the immediate medical issues, the search for additional victims or witnesses should commence. This should be done for various reasons. These reasons include: (1) additional victims may require medical assistance and the need for additional medical personnel; (2) they may provide needed information that will aid the law enforcement responder in determining the extent of the crime, the crime scene, and any physical evidence; (3) they may also serve to corroborate what actually happened and provide needed information to establish the elements of the crime, suspect descriptions, vehicle descriptions, and avenue of escape. If there is more than one witness, the law enforcement responder should make arrangements to separate, and keep separated, those witnesses who have something to say about what they saw, in an attempt to prevent collaboration. There is always the possibility that the witnesses could have collaborated before the law enforcement responder's arrival, and the law enforcement responder should take all possible steps to ascertain if that was, in fact, what occurred. After obtaining all the facts from any additional victims and/or witnesses, the law enforcement responder now has knowledge that will enable him or her to implement the security of the scene and/or any physical evidence.[9]

It is the task of the first law enforcement responder to coordinate with emergency responders in properly identifying and securing the crime scene and its contents. The first law enforcement responder must continually question the scope of the crime scene and not limit the scope of his or her investigation. All possibilities must be considered regardless of their degree of improbability. Once the crime scene has been established, an account of personnel coming into and leaving the scene must be maintained through the use of a crime scene log. Such a log will lessen the possibility of unauthorized personnel entering into and contaminating the crime scene.[10] This log must be coordinated with the EMS/fire personnel accountability system.

The final step in what Hawthorne termed the process is making notification. This simply entails notifying supervisors as well as investigators or detectives who will be handling the case, and those people who will be ultimately responsible for documenting the scene and collecting the evidence.

The first law enforcement responders must be prepared to make split-second decisions on arrival at a crime scene. These decisions can have a lasting impact on victims, witnesses, the accusatory process, and even the community in which the crime occurred. For these reasons and others, the first law enforcement responders must be well trained in the significance of crime scene preservation, enabling the crime scene to be analyzed with as little disruption as possible. When this task is done properly, a successful investigation and conclusion of the case can be achieved.[11] Otherwise, all of the advanced technology and expertise at the disposal of law enforcement may potentially be rendered virtually useless.

Processing of Crime Scene/Physical Evidence

To achieve the maximum benefit from physical evidence, the investigator must not only be skilled in its identification, preservation, and collection, he or she must know how to handle and care for the evidence beyond the time of collection in order to preserve it for the development of leads, for laboratory examination, and/or for presentation in court. Such handling and care involves documenting and storing the evidence to retain the integrity of the item in its original condition (as nearly as possible), maintaining a chain of custody for the item to assure responsibility, and to ensuring its evidentiary value and its disposition when it is no longer of evidentiary value.[12]

The proper processing of a crime scene begins with properly documenting the evidence found within its boundaries. The investigator who first receives, recovers, or discovers physical evidence must be able to identify such evidence positively, at a later date, as being the specific article or item obtained in connection with a specific investigation.[13] This is best accomplished by utilizing various proven techniques of recording the nature of the scene and its contents as they are obtained or collected.[14] This process simply entails providing pertinent data about the evidence as it relates to a particular crime scene investigation.

Chain of Custody

In order for physical evidence collected from a crime scene to be considered admissible in a court of law, a valid chain of custody must be established.[15] The chain of custody, which ensures continuous accountability, is comprised of all those who have had custody of the evidence since its acquisition by a law enforcement agency. It begins when the item is collected and is maintained until its disposition. Each person in the chain of custody is responsible for the safekeeping and preservation of an item of evidence while it is under his or her control. Because of the sensitive nature of evidence, an evidence custodian often assumes responsibility for the item when it is not in use by the investigating officer or other competent authority involved in the investigation.[16]

Once the evidence from a crime scene has been properly identified, collected, and stored, it must be processed by a multitude of professionals who will analyze the evidence until its evidentiary value is no longer of use. It is at this point that the evidence may be considered for disposal. To determine when an item of evidence should be disposed of, the evidence custodian consults with the investigator who originally produced it, and any other investigator who has an official interest, to make sure the item is no longer needed as evidence.[17]

CHAPTER SUMMARY

Law enforcement history has shown that when mistakes are made, they predominantly occur during the initial stages of an investigation or at the crime scene.[18] More cases are lost or unresolved because the crime scene was not processed properly.

Furthermore, there are numerous incidents where police officers were careless and valuable evidence was not identified, not collected, or lost resulting in a poor follow-up by the investigating officers. Worse yet, this carelessness has, in some situations, lost the only evidence with which to prove or disprove that a crime was committed and to identify who the perpetrator might have been.

The critical nature of evidence simply cannot be ignored.[19] The responsibility of ensuring this does not happen belongs to all of those involved, from the first law enforcement responder to the investigators and technicians. Everyone within the system needs to know the importance of the crime scene and how it should be processed.[20]

CHAPTER QUESTIONS

1. Define a crime scene.
2. Discuss the theory of exchange in the evidence process.
3. What are the three major classifications of evidence? List several examples in each category.
4. List and discuss at least four crime scene observations that should be made by emergency responders.
5. List and discuss five key steps for first responder preservation of evidence.
6. Discuss the concept of chain of custody.

Simulation

Consult with at least two separate law enforcement agencies. Examine their training modules for evidence preservation and collection. Examine their standard operating procedures for evidence preservation and recovery. Based on your findings, write a comprehensive on-scene procedure for an emergency response agency relating to crime scene preservation.

NOTES

[1]Burke, Robert (2000). *Counter Terrorism for Emergency Responders.* Boca Raton, FL: Lewis Publishers.

[2]Hawthorne, Mark R. (1999). *First Unit Responders: A Guide to Physical Evidence Collection for Patrol Officers.* Boca Raton, FL: CRC Press.

[3]Fisher, Barry A. J. (2000). *Techniques of Crime Scene Investigation.* (6th ed.). Boca Raton, FL: CRC Press.

[4]Department of Justice (2000). *Crime Scene Investigation: A Guide for Law Enforcement.* Washington, DC: GPO.

[5]Hawthorne.

[6]Department of Justice.

[7]Adcock, James M. (19889). *Crime Scene Processing.* Dba JMA Forensics.

[8]Hawthorne.

[9]Ibid.

[10]Ibid.

[11]Ibid.

[12]Schultz, Donald O. (1977). *Crime Scene Investigation.* Upper Saddle River, NJ: Prentice Hall.

[13]Fox, Richard H., and Carl L. Cunningham (1973). *Crime Scene Search and Physical Evidence Handbook.* Washington, DC: U.S. Department of Justice.

[14]Schultz.

[15]Hawthorne.

[16]Schultz.

[17]Ibid.

[18]Adcock.

[19]Hawthorne.

[20]Adcock.

13
Technology and Emergency Response

Michael J. Hopmeier

Source: Courtesy R. Christen

Chapter Objectives

After reading this chapter, you will be able to accomplish the following objectives:

1. Recognize factors that have changed significantly in technology.
2. Understand the important linkage of technology with doctrine.
3. Recognize the limits of technology.
4. Have an awareness of technology use for personnel accountability.
5. Understand use of technology for situational awareness.
6. Understand technology applications that improve survivability.

Introduction

Any sufficiently advanced technology is indistinguishable from magic.

Arthur C. Clarke

Much discussion has taken place recently concerning technology as it relates to the new class of threats (specifically WMD), and applications for response to these threats. Many assessments are based on partially understood problems or partially understood technologies, and seldom reflect a true *system* approach. As a result, technology developers are often disappointed by the lack of acceptance of their ideas, and technology users rapidly become disillusioned about technology's ability to help them.

This chapter discusses issues associated with assessment of technology needs, and acquisition/purchase/design of technology solutions that help the first responder or public safety professional. The primary focus is on operators, not management, and as a result management issues are addressed only as they relate to field operations.

Background

New threats, especially WMD, have often been attributed to new technologies. With very few exceptions, this is not the case. Whether chemical or biological in nature, *new* threats are *not* the result of new technologies. In most cases the technology, and in fact its application to warfare, have been around for hundreds or thousands of years. From poisoning of wells with dead bodies to catapulting plague cadavers over the walls of castles, biological warfare has been used for over 1,000 years. Smoke, burning liquids, and even gases have been used for almost as long.

The exceptions to this are nuclear/radioactive threats. These threats are addressed directly, as there has been an enormous amount of writing, discussion, and preparation for these incidents over the past 50 years. In most cases it is not a need for new technologies or doctrine, but simply for re-learning what was implemented over the years of the Cold War. Our focus is on other classes of threats including chemical and biological, man-made as well as natural.

Several factors have changed significantly including:

- **Greater Recognition**—While many of these technologies have been used for many hundreds of years, we are only now becoming cognizant of them. It is only recently that they have entered the public and political consciousness, so it is easy to assume they never existed prior to current thought about them (the ostrich head-in-the-sand syndrome).
- **Greater Vulnerability**—For a variety of reasons, society is much more vulnerable to attacks of this nature than ever before, especially American society. From greater population densities to less robust infrastructure, society overall is less able to respond to *any* threat, let alone the catastrophic ones addressed in this book. Couple this with increasingly easier travel over longer distances and shorter times, as well as greater reliance on information technology with less understanding of it, and the vulnerability is obvious.
- **Greater Awareness**—With the increasing accessibility of information from the popular press and the Internet as well as other sources, the public has more and more information. Unfortunately, there is little ability to interpret this information and, as a result, even less understanding. This generally results in knee-jerk responses to the crisis *du jour*, and these responses are seldom correct.

Technology Considerations

This chapter will **not** address specific systems, manufacturers, or devices (they change rapidly and there are other sources of information on specific systems), but instead discuss generic considerations in choosing technologies or systems. For specific data concerning devices and manufacturers, refer to various standards and evaluation organizations.

Technology must be used to improve job performance.—Much of today's focus is on development, deployment, and marketing of technologies only for terrorist and WMD incidents. Some developers have lost sight of the fact that terrorist/WMD incidents seldom occur, but fires, haz mat accidents, explosions, and earthquakes occur much more frequently. As the old adage goes, "you fight as you train and you train as you fight." This is true in the public safety community. If technologies do not help in everyday operations and support the first responder on a regular basis, they will not be used (the training and support infrastructure needed to sustain them will never be absorbed). The end result is that vast quantities of effort and resources are expended as long as external funding is available, and thereafter fall into disuse, amounting to wasted effort. Technologies must be beneficial in raising overall preparedness, or not be pursued at all.

Technology without a plan is useless.—All too often it is assumed that a new technology (black box) will be a panacea and solve all problems. It will not. It is vital to have a way to use it when it arrives. Technology alone will never solve anything. Consider the issue of medical anti-shock trousers (MAST) in emergency medical services. The technology is sound, namely, squeezing the lower extremities to increase blood pressure. However, the protocols for the use of MAST trousers are still highly debated at medical conferences.

Training and sustainment.—When evaluating a technology, consider what is needed to maintain proficiency. If it is a long, arduous task to learn how to use a technology, and it will seldom be used, it will probably never be used at all. When a crisis develops that requires the technology, it will be unused because there is no time to relearn how to apply it. Consider computer software: We are all aware that Microsoft Windows has an enormous number of resources and capabilities, but few have mastered or recall all of them. As a result it is often easier to fall back on old ways of doing things (especially if we have a deadline to meet and it is coming up fast) than to try and figure out how to make the document look just right. The same thing happens with technology.

Maintenance, repair, and calibration.—It is often necessary to repair, adjust, or perform preventive maintenance on a technology, but this is not always factored into the cost of the new system. Consider the recent introduction of automatic external defibrillators (AED). Many underfunded volunteer fire departments received AEDs through grants, only to discover that the cost of the required yearly calibration and inspection was prohibitive. Also, many of today's highly accurate sensor and sampling devices require periodic calibration and adjustment that must also be factored into the support costs.

Understand the problem.—Often technologies are obtained to solve a perceived problem, but do little to address the actual problem. Consider, for example, personnel accountability. Exact location of personnel on the fire ground or at a crime scene may not be needed. It may be sufficient to know simply who is in the area of interest and be able to recall them, en masse, as needed. As a result, a simple inventory management system is better than a global positioning system (GPS) location system.

Recognize the limitations of technology.—As noted earlier, technology is not a panacea; it must be examined in the context of the entire problem. Make sure the limitations of technology are clearly recognized before use. As an example, uses of night vision goggles by police are a real improvement over simply having a flashlight. But low light level imaging can be spoofed or rendered inoperable through either intentional or unintentional means (shine a flashlight at the goggles or simply have someone turn on a porch light). All technologies have limitations; understand the limitations of the ones you plan to use.

Technology never replaces training and judgment.—Technology, no matter how good, can never replace proper training, skills, and good old-fashioned common sense. Always remember that the person on the ground is the ultimate tool and technology. Their training, preparedness, and ability to respond to and adapt to changing circumstances are paramount. Any technology, no matter how good, is only a tool and support method to extend the ability of the person on the ground. Technology must support the user; never make the user support the technology.

The user knows best and is the final arbiter.—Remember that no mater how many degrees or years of experience a technologist or scientist may have, he or she is not the one whose life is on the line and will depend on the technology. No matter how good a technology may look in a laboratory or during a demonstration, the true test is always in the field. When considering new technologies, always ask the question, "How will this help me do my job better?" If you don't know, don't buy it.

Strive for the acceptable solution, not the perfect one.—It is often very expensive to achieve an absolute, custom, and perfect solution. It is also nearly impossible. Instead, consider what would be acceptable (albeit not necessarily perfect), and consider optimizing the overall system and application of the technology. Can you make due with something less than perfect now, as opposed to waiting for perfection in a year or more? These questions must be asked and answered before embarking on a technology quest.

Ensure that a technology will be used.—If a technology is too time consuming or uncomfortable to be used, it won't be. Often the human factors associated with a technology are overlooked in the rush to get something into the market. If a technology is not comfortable, easy to use, and does not provide some capability that the user wants, it will be ignored.

Legacy system (hardware and procedural).—A legacy system is the existing system of hardware or procedures in a response organization. It is seldom successful to completely change the way something is done to fit a new technology. Instead, strive for *revolutionary* capabilities through *evolutionary* means. New systems must be compatible with existing technologies and legacy systems so that an upheaval is not required to introduce a new capability.

Technologies

Rather than providing a catalog of different technologies, this chapter is structured around expressed needs. Numerous studies have been done to define the needs of the emergency responder. The following is a compendium of the results. First the need is discussed briefly, and then potential technology solutions are addressed.

A major requirement in consequence response is personnel accountability. The questions of "where are you," "how are you," and "do you need help" are the most critical questions asked and answered during any operation. The ability to maintain accountability and knowledge of personnel assets on the ground or during an operation is paramount. Information such as individual location with sufficient accuracy to locate, identify, and render service if needed is crucial. The need to call for help in a crisis or the ability to render a determination as to whether help is required are all key. Furthermore, solutions must be economical, easy to use, and compatible with existing doctrine, protocol, and technologies within the emergency services.

Consider this issue in the context of inventory management or control. Industry manages its assets through use of various tags and points of entry/exit. Most of the classic solutions involve trying to track an individual, usually through some form of GPS-based system. This presents many problems, however. Among them is the difficulty of using GPS in built up areas as well as in buildings. What may make more sense is to use some form of inertial tracking and navigation system.

There are currently many technologies (generically referred as taggants) that can be used for tracking commodities and personnel. Consider using point of entry/exit control systems such as those used for inventory management in warehouses and supermarkets. These portals coupled with various forms of radio frequency (RF) identification tags, or even barcodes such as those used to identify automobiles going through tollbooths, could be very effective in this role.

Various attempts over the years have been made to create individual status/biotelemetry systems (the personnel status monitor (PSM) being a good example) that have all failed. Most have failed because of the designer's lack of knowledge as to what is actually needed. In general, physiologic status is *not* necessary to meet the needs of the incident commander in maintaining knowledge of his/her personnel and their status. Further, the need for physiologic monitoring creates the need for probes and sensors directly attached to the user. This is often uncomfortable, impractical, or both. As a result, the technology is seldom used accurately or effectively. Consider the minimum amount of technology and information needed to protect the user: systems that meet that need without getting overly complicated or complex.

Analysis and Evaluation of Emergency Response Data

While a large amount of data is currently being collected on emergency personnel and their operations, very little is being used to determine ways in which to improve effectiveness of the system. To determine which technologies and efforts should be pursued, and which of those have the highest potential payoff, it is vital that some methodology and analysis be created to assign benchmarks (also called metrics) to emergency personnel and system performance. As new technologies or techniques are brought online for evaluation and use, their various capabilities can be evaluated to determine what needs to be changed, dropped, pursued, or enhanced. This capability is vital to ensure that high value projects are pursued and marginal projects are discontinued.

The ability to assess the impact of new technologies on operations is paramount. The goals of a technology must first be defined, and then applied when a new technology or capability is assessed. To do this effectively, it is vital that the true problem be understood. Take the problem of firefighters being trapped in a zero visibility environment. The difficulty is not that they are unable to see, it is that they become disoriented, confused, and run out of air. Many die as a result of running out of air while trying to find their way out of a structure. In this case, knowing where they are is of little help without a detailed map of the structure. Therefore, the issue is trying to locate them in the building. One possible solution may be an inertial tracker that would let them simply walk back through the path they had taken, without knowing exactly where they are. This is an example of clearly understanding the problem, and the benchmarks (metrics) needed to assess the problem before attempting to obtain technology.

Improved Situation Awareness

Perhaps the most important aspect for any emergency responder is awareness of their surroundings (situational awareness). This is vital not only to allow the first responder to complete their mission, but to make sure that the team does the same in a safe manner. This awareness includes the ability to sense their surroundings, especially through improved sight in degraded conditions (thermal and acoustic imaging, image processing, radar vision, etc.). Individuals must know where they are situated, where their partners and other teams are located, and where major assets such as the fire truck, escape routes/safe zones, and supplies like oxygen and water are located. What type of atmosphere are they in (is it poisonous, too hot, contaminated)? What is the status of the building around them (will it collapse shortly, is a wall too hot, is an electrical short occurring nearby)? Are there signs of chemical, biological, or nuclear contamination, in the atmosphere around them, on them, on their patients, or partners? Of equal importance is how the information should be provided (heads-up displays, wrist worn indicators, voices, image displays)? What is the status of their equipment (amount of air, water, chemical agent, and power)?

Critical Factor: Responders must have situational awareness: an awareness of their surroundings.

Many of these questions have different answers depending on the environment in which they are asked, the community or department asking them, and the other resources available as infrastructure. The specific integrated systems, while unique, will still be able to be made from the same basic sets and systems of components. There are four areas/issues that need to be considered:

1. *Sensor technologies.* These are individual sensors that the responder can use such as temperature detectors, chemical sensors, thermal imagers, etc.
2. *Onboard/man information processing.* How is the data presented to the user? Heads-up displays? Voice presentation? Vibration? All at once, in individually controllable pieces, etc.?
3. *Externally available information transmission/display.* How much information goes back to the command post and incident commander? Back to headquarters? Back to the rest of the world? How much is recorded?
4. *How is information integrated into existing systems.* Is there an existing accountability system or mapping/situational awareness system that needs to be used? Are there already maps of city buildings available electronically and can they be accessed? Can this information be overlaid on an existing framework or incorporated into existing situational awareness systems?

Sensor technologies are expanding at an increasing rate. In the area of imagery, night vision scopes that function in the near infrared (IR) are available for less than $150.00. Uncooled thermal imagers (for example, new micro-bolometer technologies) are just starting to reach the mass market and will shortly provide thermal imaging and far IR capabilities for well under $1,000.00.

In the area of chemical/biological sensors, there is an enormous amount of work and effort going on. Much of the focus is on individually operated monitors and detectors designed for use by the first responder, which provide them with specific identification of individual agents. The first responder needs the ability to determine simply whether the environment is safe or not. The actual cause of the hazard is secondary and can be determined by other specialized units.

During the Tokyo sarin incident the "chemical detectors" used were parakeets. The responders knew upon arrival that they were dealing with some form of unknown chemical. However, they were not prepared to deal with any particular agent. Knowing exactly what agent they were dealing with at this level of response was of questionable value at best and useless at worst. Instead, what was needed was a simple safety detection device (safe/unsafe) with identification of the agent being performed at some other level of response (laboratory, hospital, etc.). The only real need is to determine whether the environment is safe to remove gear or whether the casualties have been properly decontaminated.

Until recently, chemical detectors consisted almost exclusively of carbon monoxide (CO) and lower/upper explosive limit detectors. Poisonous chemical agents were considered to only come from combustion products, and these products were correlated with CO concentrations. Therefore, when the CO concentration fell below a certain level, the other agents were considered to be below that level as well. Specialized gear, however, existed for use in specific, highly dangerous environments (chemical processing plants, rocket fueling areas, etc.). Specialized training went along with these areas.

Today that is no longer the case. In spite of the focus on chemical warfare agents, the real problem is, and continues to be, hazardous materials. Many chemical warfare agents are industrial chemicals transported throughout the country in bulk transporters (i.e., phosgene). As a result, it is necessary for technologies to be used in a variety of applications and circumstances, not just for terrorist incidents.

Tools and Technology to Improve Survivability

Improved hospital care.—Dealing with shock (hypo-perfusion), airway maintenance, and vital signs monitoring are all issues of concern that improve the survivability of the injured patient. Better ways to provide this prehospital care are of importance. These techniques and technologies need to be rugged, relatively easy to use, require minimal training and maintenance, and be cost effective.

Monitoring responder health conditions and performance.—The status of the emergency personnel during operations is vitally important. At a minimum, are they healthy, injured, or under unacceptable stress (i.e., heat exhaustion, dehydration, etc.)? Can this data provide a prediction as to their likely abilities and capability to continue operations? This information not only improves the safety of the individual emergency responder, but also their ability to perform their jobs. With sufficient information, improvements in training, equipment, procedures, and doctrine can be made.

Field performance predictive methods.—It is difficult to predict under non-stress conditions how a person will perform physically. Whether it is a masked or unknown cardiac condition or perhaps an allergy to specific chemicals that might be encountered, the need to be cognizant of these conditions is paramount. Issues such as buildup of chemical dosages (i.e., CO poisoning) that might need time to clear from a system, or longer-range impact (i.e., radiation) need to be monitored, accounted for, and used to help place and utilize emergency personnel to minimize danger to themselves and their colleagues.

Improved voice and data communications for intermodal use.—The transmission of data from the scene of an incident back to some central point, and vice versa, is vital. This data includes, but is not limited to, analog (i.e., voice) and digital (i.e., video) data. It might also include system status, maps, blueprints, and real-time access to databases. The communication system must be compatible with urban points of interference (such as buildings and other broadcasters), as well as rural environments (longer range, no repeaters). It must also be compatible with individual communications systems (battalion chief or incident commander's comm system). The system must be compatible with other equipment (i.e., trucks), as well as supporting legacy systems to minimize cost.

Improved voice and data communications for individuals.—Each individual on the scene must have the ability to communicate with the person next to him, every responder on the scene, or headquarters, and do so selectively and with ease. The capability to transmit voice, video, and status data is also important. The system must be compatible with the intermodal communications systems as well as individual equipment (i.e., self-contained breathing apparatus (SCBA), available batteries, water, smoke, etc.). It must function in urban areas (inside buildings with lots of electrical interference, including compatibility with other transmission sources such as radio stations, TV, etc.), and in rural areas (work at extended ranges in the woods, open fields, etc.). Additionally the system must support reception of data maps, graphics, personnel status, etc.

Measurement, prediction, and improvement of interpersonal skills.—Interpersonal and communication skills between emergency personnel and their various chains of command, and the community as a whole, should be monitored, measured, and improved through training. The ability to evaluate individual traits and capabilities to predict these skills can lead to improved operations and better interaction. This also provides a basis for focusing training and other resources where they can be the most effective. Issues concerning environment (room setups, facility design, etc.) will also be taken into account to make the emergency personnel as effective as possible.

Improved environmental data.—When engaging a fire or hazard, the personnel need to have as much information as possible about the environment around them, information such as the layout of the building and material stored in each room (i.e., are hazardous materials in danger of igniting and/or exploding). Are poisonous compounds or accelerants near by? What walls and floors are the strongest or weakest? Other information would include where the standpipes, power and gas lines, and structural components are located. What is the current structural status (is the building about to collapse, is a particular joist or main support weakened from heat or fire?)? What is the micrometeorology (local weather); will it rain, will the wind shift, is there lightning in the area? Real-time maps and geographic information are required to better locate the scene of an incident, or the best way to approach it (avoid mud, traffic congestion, small roads, etc).

Critical Factor: The data that relates to the surrounding environment is vital to emergency responders.

Improved threat assessment and intelligence.—Accurate information concerning likely threats and threat scenarios is vital to ensure proper training, equipment, and engagement of incidents. This information might range from terrorist threats to new forms of arson. Responses might include applying new methods or techniques of combating these incidents, to looking for specific early indications of a future incident that could be mitigated. Better methods of communicating information, analyzing its import, and determining actions based on this information is vital. This might include, but not be limited to, a central clearinghouse for information, better dissemination and communication of information, improved analysis and distribution, etc.

After-action analysis.—After an incident, it is vital to analyze and understand not only the cause, but also the effectiveness of the response. Better forensic analysis, record keeping, and logging of events correlated with the results of the response and the effects are important. The end goal is to reduce the effects of the response through mitigation and improved engagement techniques.

Remote analysis.—Many situations are so hazardous (biological, chemical, nuclear) that remote observation and initial analysis are the only safe means of determining the situation. Methods of remotely detecting contamination, heat sources, casualties, etc., are vital. Whether detection is by robotic or autonomous means, or through the use of remote sensing, this is crucial.

Robotics and autonomous devices.—Often, situations arise where a remote piloted or autonomous device may be needed. Requirements include the ability to maneuver through doors and up/down stairs; be self-powered and possibly self-directing; provide heavy lift (i.e., more than a person is capable of); and provide evacuation, breaching of barriers, auxiliary power, enhanced sensing, and other tasks that either augment or replace personnel.

Forced entry and breaching of barriers.—Depending on the incident, it may be necessary to breach security barriers, such as locked doors or chained gates, or if structural damage has occurred, to penetrate through debris. Power rams, hydraulic spreaders, pneumatic tools, etc., are all used; still better means are required. Barriers might include doors, bars and chains, walls, floors and ceilings, reinforced concrete walls, and pavement. Debris might include buckled beams, fallen concrete, crushed or damaged vehicles, timber, rock slides, etc. Technologies need to be safe and nonlethal (i.e., explosives generally create a greater hazard than not using them), man-portable, and require minimum support and training for their use.

Energy storage, control, transport and discharge.—Many missions require the application of external force or energy. Hydraulic tools, chain saws, and explosives are all examples. The capability to transport these energy sources to the scene and be able to apply the forces required is important. Current techniques include gas powered hydraulic pumps and air compressors, but these are often bulky, heavy, or unwieldy. Furthermore, because of the use of gas and the exhaust fumes, these techniques may not be suitable for all situations. New sources of easily applied power, compatible with the incident environment, are needed. These power sources must be compatible with exiting tools so that an entire new system of equipment and ancillary devices need not be purchased.

Chapter Summary

The so-called new threats of weapons technology are actually new or revised applications of old technology. Radioactive threats are an obvious exception.

There are several technology considerations:

- Technology must be used to improve job performance.
- Technology without doctrine is useless.
- Technology requires training and ongoing use.
- Technology requires maintenance, repair, and calibration.

- Technology has limitations.
- Technology never replaces training and judgment.

The user should be the final arbiter in technology considerations. The true test is in the field, not the laboratory. The technology solutions available will never be perfect due to human imperfection and cost. Instead, acceptable solutions should be the end state rather than perfect solutions.

New technology must be compatible with existing systems (legacy systems). It is not practical to change all existing technologies to be compatible with a new technology.

New technology requirements in the emergency response field include:

- tracking of resources and assets
- bio-monitoring with telemetry
- tracking of responders in a hot zone or building
- improved situational awareness
- sensor capability including biological, chemical, thermal, etc.
- survivability improvement
- improved voice and data communications
- improved environmental data
- improved threat assessment and intelligence

CHAPTER QUESTIONS

1. Discuss three factors relating to technology and threat vulnerability that have changed significantly.
2. What is the relationship between technology and doctrine?
3. Discuss at least three limits of technology.
4. How can technology be utilized for personnel accountability?
5. Discuss how technology can be used to enhance situational awareness.

Simulation

Select a major technology that has evolved in the emergency response or military arena in the past 20 years. Examples include noninvasive monitoring in EMS, water additives in fire operations, public health surveillance for disease agents, or DNA matching in law enforcement. Trace its evolution including early development to the present time, a success/failure analysis of products and manufacturers, costs, support, and training issues. Discuss how doctrine or operational procedures did or did not evolve with the technology. Conclude with a discussion of the outcome (endstate) of the technology, and a prediction for the future of the technology you selected.

Appendix A
Monitoring Devices

Paul M. Maniscalco
Eugene J. O'Neill

Source: Courtesy L. Robbins

Introduction

The importance of monitoring and detection equipment as part of the responder arsenal must never be underestimated. Our ability as responders to effectively detect and accurately identify chemical, biological and radiation substances at a terrorist incident must be our first priority. The quick identification of these substances aids in lessening the threat expansion factors to not only the civilian population but to the emergency response community also. By leveraging existing technology and equipment, we can implement these applications to enhance the safety and quality of a operations at these responses.

Detection and monitoring equipment can be divided into two categories: equipment and technology readily available on the civilian market and that which is available for use by the military.

Military Equipment

Chemical Detection Equipment

Utilization of chemical detection equipment is critical towards ensuring the safety of responders and to the accurate identification of the offending substance. Some military detection equipment and systems are available to the civilian responder community through programs such as the MMRS and the DOD/DOJ programs. These systems vary in their capacity, utility, and accuracy in the detection of various chemical threats, but for the most part they are deemed adequate for emergency response.

However, a limiting factor is that much of the military equipment requires users not only enter the contaminated environment, but also to spend up to 30 minute in the hazard zone to conduct testing. This time limit varies depending on the equipment being utilized.

Let's look at some of the hardware that is currently employed in this endeavor.

M8 Paper

M8 paper is issued in a book of 25 tan sheets of chemically treated, dye-impregnated paper, perforated for easy removal. A color comparison bar chart is printed on the inside of the front cover of the book. The book is heat-sealed in a polyethylene envelope. Detach the sheet of detector paper from the book and attach it to clothing or place on a surface so it can be exposed to drops of liquid splash of chemical agents. If colored spots appear, chemical agent is present. Put on protective mask. Compare colored spots with colors on inside cover to determine type of agent. The paper may also be

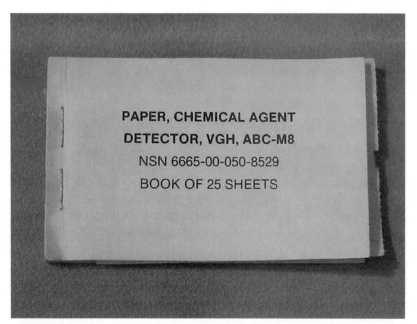

FIGURE A-1 M8 Chemical Agent Detector Paper. *Source:* Courtesy
P. Maniscalco

used to detect liquid contamination by placing the paper in contact with the
suspected surface.[1]

M8 paper is similar in function to litmus paper. If the test strips are exposed to a chemical agent, a color change occurs within 30 seconds, alerting the user. Dependent on the color change, the user can detect whether a
substance is a G or V nerve agent or a blister agent. It is limited to use only
on liquid; petroleum distillates, pesticides, and ethylene glycol can produce
false positive reading.

M9 Tape

M9 Chemical Detection Tape is a portable, expendable single roll of paper
that comes with mylar-adhesive backed and coated tape. It is 9.1 meters
long and 5.1 centimeters wide. It is packed in a cardboard dispenser with a
serrated edge. The paper contains a suspension of agent-sensitive dye in a
paper matrix. The paper is colored in page green with insoluble pigments.
The self-adhesive paper attaches to most surfaces, and a moisture-proof re-
sealable bag is provided to store the dispenser after removal from its original package.[2]

M9 tape is used by the military as a chemical detector on vehicles,
equipment, and chemical protective outerwear. It is similar in structure and
function to M8 paper but has an adhesive backing so that it can be affixed
to objects. It is capable of detecting both the G and V nerve agents in liquid

form and blister agents in liquid form also. Like M8 paper, it can only detect the presence of the agent, not their concentrations; also, false positives can be produced by exposure to petroleum distillates, pesticides, and ethylene glycol. It is recommended that chemical protective outerwear be utilized when using M9 paper; the reagent dye is a potential carcinogen.

The M256A1 Detection Kit

The M256A1 Chemical Agent Detector Kit is a portable, disposable chemical agent detector kit that can detect and identify nerve, blister, or blood agent vapors. It is typically used to determine when it is safe to unmask after a chemical agent attack. A test disk contains a glass ampoule with compounds that react with an agent to give a color change.

Each kit consists of 12 disposable sampler-detectors, one booklet of M8 paper, and a set of instruction cards attached by a lanyard to a plastic carrying case. The case is made from molded, high-impact plastic and has a nylon carrying strap and a nylon belt attachment. Each sampler-detector contains a square, impregnated spot for blister agents, a circular test spot for blood agents, a star test spot for nerve agents, and a lewisite-detecting tablet and rubbing tab. The test spots are made of standard laboratory filter paper. There are eight glass ampoules, six containing reagents for testing and two in an attached chemical heater. When the ampoules are

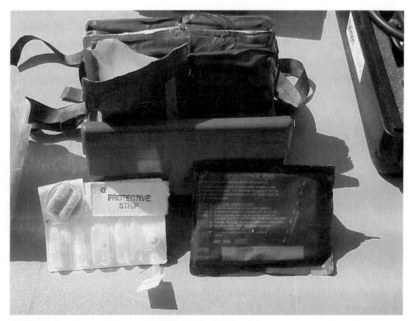

FIGURE A-2   M256A1 Chemical Agent Detector Kit. *Source:* Courtesy H. Christen

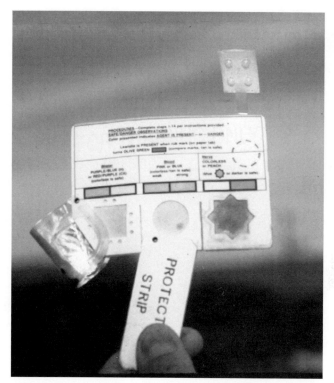

FIGURE A-3 M256A1 Sampler-Detector Found in the Kit. *Source:* Courtesy United States Government

crushed between the fingers, formed channels in the plastic sheets direct the flow of liquid reagent to wet the test spots. Each test spot or detecting tablet develops a distinctive color, which indicates whether a chemical agent is or is not present in the air.[3]

Unlike the M8 paper and M9 tape, it can be used for both liquid and vapor sampling. It also is capable of registering relativity low concentrations of agents that could escape detection by the M8 and M9 detectors.

A disadvantage of this system is the fact that complete environmental sampling may take up to 30 minutes; in the civilian response community, this may be unacceptable due to SCBA limitations.

M18A2 Chemical Agent Detection Kit

The M18A2 Chemical Agent Detector Kit uses both detector tubes and paper tickets to detect and classify dangerous concentrations of lethal chemical agents in the air, as well as liquid chemical agent contamination on exposed surfaces. Agents detected are CK, H, HN-1, HN-3, CX, AC, CG, L, ethyl dichlorsarsine, methyl dichloroarsine, the G-series nerve agents, and VX.

FIGURE A-4 M18A2 Chemical Agent Detector Kit. *Source:* Courtesy P. Maniscalco

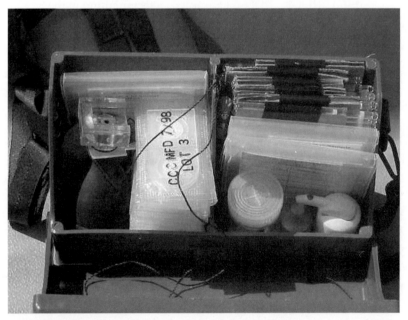

FIGURE A-5 Open M18A2 Chemical Agent Detector Kit. *Source:* Courtesy P. Maniscalco

The M18A2 kit is a portable, expendable item capable of surface and vapor analyses. The M18A2 kit is designed primarily for detecting dangerous concentrations of vapors, aerosols, and liquid droplets of chemical agents. The kit's capability provides for the sampling of unknown NBC agents. The presence of a chemical agent is detected by distinctive color changes. If a chemical agent is suspected but cannot be detected with the kit, vapor samples can be collected in sampling tubes for forwarding to a laboratory for identification. The kit contains detector tubes, detector tickets, M8 paper, and instruction cards.[4]

Chemical Agent Monitors (CAMs)

These devices use a microprocessor chip to identify the presence of the V and G series nerve agents and blister agents. In its original incarnation, the CAM had to be manually switched between detection and nerve and blister agents; also, no agent concentration values were available. Newer models are capable of switching between nerve agent detection and blister agent detection automatically, are able to reflect the type of detected agent, and reflect its relative concentration in a rudimentary system of LED bars. The number of bars that light indicate the concentration of the agent. In its latest incarnation, the improved chemical agent monitor (ICAM), it can differentiate between not only the nerve and blister agents but actually identify the agent specifically.

One of the disadvantages to the CAM in all its forms is its sensitivity to pesticides, which will produce a false positive for nerve agents. It can also take up to 15 minutes to clear itself and reset if the detection chamber becomes saturated with agent.

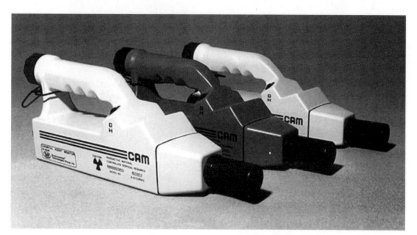

FIGURE A-6 Handheld Chemical Agent Monitors. *Source:* Courtesy of Environmental Technologies Group, Inc. (ETG) Baltimore, Maryland, USA

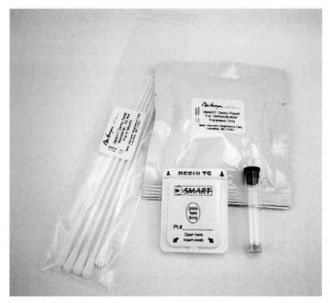

FIGURE A-7   SMART Ticket Kit. *Source:* Courtesy Environmental Technologies Group, Inc. (ETG) Baltimore, Maryland, USA

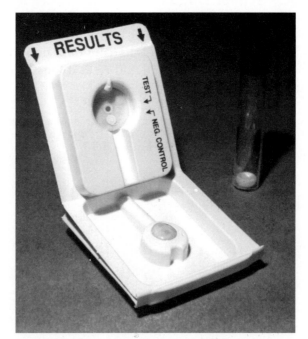

FIGURE A-8   SMART Ticket Opened. *Source:* Courtesy Environmental Technologies Group, Inc. (ETG) Baltimore, Maryland, USA

Biological Detection

Detection of biological agents with field testing equipment is difficult at best. Even under the best conditions, results obtained by field detection devices require confirmation by more detective and traditional methods such as microbiological assay, immunoassay, and genome detection.

Biological SMART Tickets

SMART tickets are a handheld point detection system developed for early detection of biological warfare agents. When utilized properly, they are capable of detecting eight different biological agents: Y. Pestis, F. Tularensis, B. Anthracis, V. Chorerae, SEB, Ricin, botulinium toxins, and brucella. These kits are produced by the Navy Medical Research Institute and are filtering down to the civilian response community.

The primary technology employed is antigen capture chromatography, in which agent-specific color changes are used to identify positive or negative results usually within a 15-minute time window. It can only be used as a screening tool and is not capable of giving concentrations of a detected agent. It is still, however, a useful component in early detection of the presence of a biological warfare agent.

Electro Chemical Devices

These devices function by using two internal electrodes. One electrode establishes and maintains a constant electrical potential; the second measures the electrochemical potential produced when gases are drawn into the chemical bath through a semipermeable membrane. The electrodes in conjunction with the chemical bath produce an electronic gradient that is measured and amplified, and then displays as a percentage of the gas being tested for. Its advantages are in the fact that it is a combination device, meaning that it can be used for testing for two or more gases. Disadvantages include inaccuracies caused by contamination of the chemical bath, and it is sensitive to atmosphere changes.

Sound Acoustic Wave—Mini Chemical Agent Detector (SAW Mini Cad)

The SAW Mini Cad is a portable chemical agent detector. Although its use parameters parallel those of the military and civilian CAMs, it is capable of detecting much lower concentrations of agents. It is also capable of the simultaneous detection of both nerve and blister agents without a mode transition and is less susceptible to false positive readings than other military or civilized detectors.

FIGURE A-9   Microsensor Systems Inc. SAW Mini Cad MKII. *Source: Courtesy P. Maniscalco*

The primary technology used in the SAW is a pair of surface acoustic wave microsensors. These are piezo electric crystals that detect the mass of chemical vapors absorbed into chemically selective coatings on the sensor's surface. The rate of absorption causes a change in the resonant frequency of the sensor, which is in turn read by an internal microcomputer to determine the presence and concentrations of the agent or agents in question.

Benefits of the SAW Mini Cad include a relatively short detection time (usually under 60 seconds), an overall lightweight package (0.5 Kg), a resistance to false positive readings, and a minimal interference by meteorological conditions such as humidity or ambient temperature.

The M31E1 Biological Integrated Detection System (BIDS)

The BIDS system is a vehicle-mounted biological detection platform, and at present it is only available for military use. It can be used for large area monitoring and early warning of biological agent release or use. Essentially it is a laboratory on wheels, and it can test for the presence of biologicals by particle sizing, bioluminescence, flow cytometry, immunoassays, and mass spectrometry. Through these methods, it is capable of identifying specific biological agents with a high index of confidence and producing safe samples for later laboratory testing.

Civilian Detection Equipment

Combustible Gas Indicators (CGI)

In a CGI, the primary sensing is referred to as a Wheatstone Bridge circuit. This consists of two filaments: One filament acts as the actual sensor component while the other acts a compensatory. It functions by detecting the resistance differences between the filaments, thus producing a yes or no answer to the presence of combustible gases. Although accurate in its detection capabilities, it can be adversely affected by high humidity, exposure to high-voltage sources (will produce false negatives and skewed readings), and contamination by heavy metals and sulfur compounds; and it will produce false high readings in the presence of oxidant gases.

Photo Ionization Devices (PID)

Photo ionization devices are useful in the initial reconnaissance of agents. They function by bombarding gases or vapors with ultraviolet light. This ionizes the substances, which are then analyzed for their ionization potential and are thereby unidentified by their known IP. PIDs are useful for the identification of the actual concentration of a known agent and are accurate for readings between 0.1 and 2,000 PPM. However, they are of little use if

FIGURE A-10 CGI—Mine Safety Appliance Explosimeter.
Source: Courtesy of Mine Safety Appliances Pittsburgh, Pennsylvania, USA

FIGURE A-11   Portable Photo Ionization Device (PID). *Source:* Courtesy of HNU Systems, Inc.

the identity of an agent is not known, because the apparatus uses known IPs to produce a concentration. Interference gases, airborne particulate, condensation, atmospheric changes, and a lack of regular maintenance and cleaning can also adversely affect them.

Multi Substance Detection Equipment

Currently available in the civilian marketplace are several Multi Substance Detectors. The primary benefit to the purchase and use of this type of detector is its versatility. Several relatively small, ergonomically designed devices capable of detecting the "traditional" blister and chemical agents are available; these devices can also detect the presence of riot control agents such as mace and pepper spray.

One such detector, the APD 2000, manufactured by the Environmental Technologies group, is capable of not only continuous simultaneous monitoring of any blister, nerve, or riot agents, but it will automatically "clear down" after an alarm and continue to monitor through its spectrum of indicated agents.

The benefits of these detectors include portability (<3 Kg), readily available power (many function on C cell alkaline batteries), short detection times (<30 seconds), and easily readable visual displays with accompanying audible alarms. They are relatively insensitive to meteorological extremes or humidity changes.

FIGURE A-12 Some devices are configured for multipurpose, such as this Mine Safety Appliance Five Star® Alarm—A Combination Instrument (multi-sensor) and Combustible Gas Indicator. *Source:* Courtesy of Mine Safety Appliances Pittsburgh, Pennsylvania, USA

FIGURE A-13 Portable handheld chemical detector situated in storage/carry case. APD 2000. *Source:* Courtesy of Environmental Technologies Group, Inc.

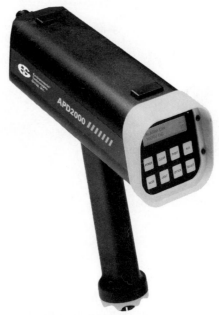

FIGURE A-14 APD 2000. *Source:*
Courtesy of Environmental Technologies Group, Inc.

Flame Ionization Devices (FID)

These are similar in function to PIDs but use a hydrogen flame to ionize molecules and then identify them. FIDs like PIDs are useful in pinpointing the exact concentration of a known substance but lack the ability to identify an unknown, because the apparatus has to be specifically calibrated to the substance that is being measured. It is capable of measuring concentrations in the range of 0.1 to 1,000 PPM. It cannot be used to measure inorganic compounds and requires operation by skilled technicians.

Colormetric Tubes

Essentially a civilian counterpart to the M256A1 kit, it consists of a collection of sealed glass tubes, each containing a substance-specific reagent strip. This reagent strip reacts when a prescribed amount of air is aspirated through the tube using a high-volume syringe.

Although a seemingly obvious choice for civilian detection equipment and available for detection of a wide array of chemical warfare agents (G and V nerve agents, blister agents, choking agents, and riot control agents), it should be kept in mind that these devices may have a margin of error as high as 50 percent. They also may not detect agents in low concentrations, as there

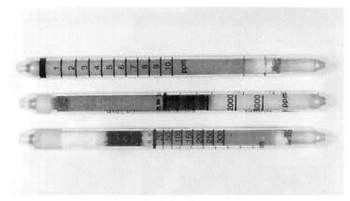

FIGURE A-15 Dräger Colormetric Tubes. *Source:* Courtesy of Dräger Safety Technology

FIGURE A-16 Dräger Gas Detector Pumps with Colormetric Tubes inserted. Power and manual pumps shown. *Source:* Courtesy of Dräger Safety Technology

FIGURE A-17 Dräger Multi Colormetric Tube Extension with Manual Pump. *Source:* Courtesy of Dräger Safety Technology

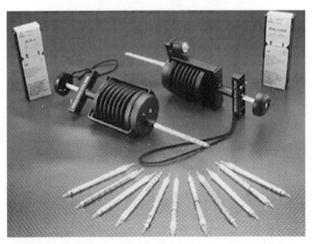

FIGURE A-18 MSA Kwik-Draw Detector Tube Pumps. *Source:* Courtesy of Mine Safety Appliances, Pittsburgh, Pennsylvania, USA

is not enough to react with the reagent strip. In addition, they can be adversely affected by age, temperature, humidity, and storage conditions.

Radiation Detection

A subject that cannot be overlooked in response to a terrorist incident is radiation detection. Although the actual detonation of a nuclear device is a low probability, it is not beyond the realm of possibility for a terrorist organization to utilize a "dirty bomb," a conventional explosive device used as a means of dissemination for radioactive materials. This could be used as a means to contaminate large numbers of civilian and response personnel. In addition to chemical and biological surveillance, radiation surveillance should also be conducted; there are wide means of accomplishing this goal. Traditional alpha, beta, and gamma detectors should all be used as a means of establishing a baseline and for point detection, as well as film badges (capable of measuring exposure to gamma, X ray, and neutron radiation) and personal dosimeters (capable of measuring alpha, beta, and gamma radiation) to gauge individual exposure. No response arsenal should be without radiation detection equipment.

Devices designed to measure radiation consist of a radiosensitive detector and a means of recording the effects of radiation on the detector. The detectors respond to radiation by producing various physical, measurable effects. Ionization is one of these effects; the ion pairs can be collected to give an electrical signal, which is related to the intensity of the radiation. Some detectors will emit light pulses in response to radiation, and by counting the pulses, the intensity of the radiation can be found; others store the radiation readings over prolonged time frames for review of the information later. All of these devices respond to the energy deposited in them by the radiation.

There are several kinds of radiation detectors that have been used over the years. With the emergence of the Geiger-Mueller tube that is filled with a gas, the term "Geiger Counter" emerged. Geiger counters are handheld devices designed to identify levels of radiation. As radiation particles pass through the tube, a voltage change is detected, a light flashes, and an audible clicking sound is generated. It shows you the radiation intensity at the time you are measuring it.

Dosimeters are designed to measure absorbed radiation doses and the quantity of gamma radiation by monitoring the amount of radiation that enters the dosimeter. They are about the size of a large pen or cigar. Prior to measuring the gamma radiation, the dosimeter's self-contained ion chamber and fiber voltmeter must be zeroed out by charging the device to approximately 165 volts DC. The operator can peer into the device through the view port to interpret the readings.

Some examples are:

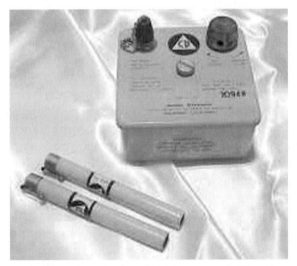

FIGURE A-19 Pocket Dosimeters and Charging Base. *Source:* Courtesy United States Government—FEMA

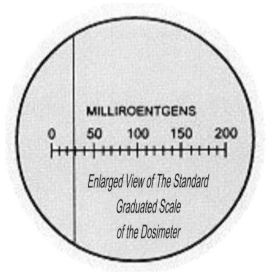

FIGURE A-20 An Enlarged Standard View of the Dosimeter Graph as Would Be Seen Through the Viewpoint. *Source:* Courtesy United States Government—FEMA

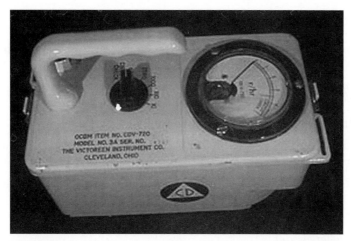

FIGURE A-21 Top View of a Victoreen 720 Radiation Measurement Equipment. *Source:* Courtesy Inovision Radiation Measurements Cleveland, Ohio, USA

FIGURE A-22 Bottom View of a Victoreen 720 Radiation Measurement Equipment. *Source:* Courtesy Inovision Radiation Measurements Cleveland, Ohio, USA

We encourage you to refer to the radiation chapter and to conduct independent research to determine which devices are the most appropriate for your organization and their operation.

Summary

Although this is a brief overview of the equipment currently utilized in the field, we encourage you to conduct additional research on the equipment and the threat you wish to address. In the reference section, we have provided some links to assist you with this endeavor.

Recommended Reference Sources

Defense Against Weapons of Mass Destruction Responder Awareness and EMS Technician Course Materials—DoD/DoJ Domestic Preparedness Program

SBCCOM
http://www.sbccom.apgea.army.mil

Nuclear, Biological, and Chemical Defense—Contamination avoidance
www.sbccom.army.mil/products/nbc.htm

Hazardous Materials Field Guide—Armando Bevelacqua and Richard Stilp Delmar Publishing, *www.firesci.com*

Jane's Chem-Bio Handbook
www.janes.com

CW Protective Equipment—An overview of respiratory and body protection
www.opcw.nl/chemhaz/equip.htm

US EPA Chemical Emergency Preparedness Program—Counterterrorism
www.epa.gov/swercepp/cntr-ter.html

National Domestic Preparedness Office (NDPO)
www.ndpo.gov

Environmental Technologies Group, Inc. (ETG)
http://www.envtech.com

Mine Safety Appliances
http://www.msanet.com

Microsensor Systems Inc.
http://www.microsensorsystems.com

Mitretek Systems
http://www.mitretek.org/mission/envene/chemical/detection.html

Dräger Safety Technology
http://www.drager.com

NOTES

[1] M8 Chemical Agent Detector Paper: http://www.sbccom.apgea.army.mil/products/m8.htm

[2] M9 Chemical Detection Paper: http://www.sbccom.apgea.army.mil/products/m9.htm

[3] M256A1 Chemical Agent Detector Kit: http://www.sbccom.apgea.army.mil/products/m256a1.htm

[4] M18A2 Chemical Agent Detection Kit: http://dp.sbccom.army.mil/fs/equip_m18a2.html

Appendix B
A Basic Review of Personal Protection Equipment (PPE)

Gerald F. Dickens EMT-D (retired)
Hazardous Materials Technician
Logistics Coordinator
NYC-EMS Special Operations Division

Introduction

Selection and utilization of personal protection equipment (PPE) is a critical task that will mean the difference between a safe operation and one that can cause catastrophic injuries or death to responders entering a contaminated area. This *brief overview* is designed as a primer to assist the emergency responder with understanding the basics of PPE and the process required to identify which level and protective device is best suited for an environment being entered. PPE is generally divided into two categories: respiratory protection and chemical protective garments. Let's take a look.

Respiratory Protection

Prolonged, unprotected exposure to toxic gases, vapors, and particulates can cause serious health complications including long-term medical woes or even death. Many chemical substances will damage or destroy portions of the respiratory tract, and they are also absorbed directly into the bloodstream from the lungs. The respiratory system can be protected by avoiding or limiting exposure to these substances.

Self-contained breathing apparatus (SCBA) or supplied-air breathing apparatus (SABA) systems can help decrease exposures. When these systems are not available, respirators can provide protection, but only for specific products. The respirator devices that are currently available can provide filtering of gases, vapors, and particulates or supply clean breathing air to the user.

The Occupational Safety and Health Administration (OSHA) regulates the use of respirators under their regulation 29 CFR 1910.134, which delineates the requirements for the use of approved respirators; proper selection and individual fit testing of users [1]. You can also reference the National Fire Protection Association (NFPA) standards for the professional standards that NFPA has promulgated.

Respiratory Hazard

Normal atmosphere contains 78 percent nitrogen, 21 percent oxygen, .9 percent inert gases, and .04 percent carbon dioxide. Human life depends on oxygen to exist; even small decreases have dramatic consequences. A list of decreasing percentages for oxygen and the effects on the body due to decreased amounts of oxygen follows.

OXYGEN %	EFFECTS (1)
16–21%	Nothing abnormal.
12–16%	Loss of peripheral vision, increased breathing, accelerated heartbeat, altered mental status and coordination.
10–12%	Very poor judgment, very poor coordination, muscular exertion causes fatigue that may cause permanent heart damage, intermittent expirations.
6–10%	Nausea, vomiting, unable to perform vigorous movement or unable to move at all, unconsciousness, followed by death.
<6%	Spasmodic breathing, convulsive movements, death in minutes.

The various regulations and standards dealing with respirator use recommend that oxygen levels ranging from 16 to 19.5 percent be an indication of an oxygen deficient environment. Such numbers take into account individual responses, errors in measurement, and other safety considerations. When responding to a hazardous materials emergency, 19.5 percent oxygen should be considered the "Lowest—Safe" working concentration. When the working environment will be below 19.5 percent available oxygen, a supplied air source (SCBA or SABA) must be employed to ensure the safety of the responders.

Respiratory Protection Devices

The basic function of a respirator is to reduce the risk of respiratory injury due to inhaling airborne contaminants. Respirators are made up of two main parts: (1) the part that supplies or purifies the air and, (2) the face piece, covering the nose and mouth or entire face, to seal out contaminants. The first part defines the class of respirator; the second determines the measure of protection given by that respirator.

Classes of Respirators

Respirators are divided into two major classifications depending on their use:

1. *Air-Purifying Respirators (APRs)* remove contaminants by passing air through a purifying element. There are two subclasses: (a) particulate respirators that use a mechanical filter element and (b) gas and vapor respirators that use chemical sorbent contained in a cartridge or canister.

   It is important to realize the limitations on the use of respirators. These devices are specific for certain types of contaminants, so it is very important to identify the substance involved. There are maximum concentration limits and a maximum use limit (MUL) for the respirator, so it is very important to know the concentration of the contaminant. Air-purifying respirators only clean the air; the ambient concentration of oxygen must be sufficient (=/>19.5 percent) for the person using the respirator.

2. *Atmosphere-Supplying Respirators (ASRs)* provide a source of clean breathable air. This type of respirator provides breathable air to the user in two ways, connected to a stationary source through a long hose or from a container worn by the user. The first type is called a supplied-air breathing apparatus (SABA) and the second type is a self-contained breathing apparatus commonly known as an SCBA.

These devices can be used in most instances, regardless of the type of contaminant or the concentration of oxygen. However, the contaminant concentration limits vary for the different types of ASRs. The user must be aware of the limitations of their respirator.

Fit Testing and Protection

The degree of respiratory protection provided to the wearer also depends on the proper fit of the face piece/mask. It does not matter how clean the supplied air is or if you selected the correct APR; if you don't have a proper fit (leak-free seal), you will have little protection from the contaminants. There are three basic designs of face pieces for protection.

1. Quarter-Mask—fits over the bridge of the nose, along the cheek, and across the *top* of the chin. It is held in place using a two- or four-point suspension system. Limited protection is afforded the user due to the ease of its being dislodged, causing a breach in the seal.

2. Half-Mask—fits over the bridge of the nose, along the cheek, and *under* the chin. It is held in place by a four-point suspension system. Greater protection is afforded the user due to maintaining a better seal, and it is less likely to be dislodged.

3. Full-Face Piece—fits across the forehead, down over the temples and cheeks, and under the chin. It is held in place usually by a head harness using a five- or six-point suspension system. The full-face mask offers the greatest protection because of the secure fit, making it easier to maintain a proper seal. Using a full-face piece also gives the user additional protection by covering the eyes.

All personnel required to wear respirators must successfully pass a fit-test designed to check the integrity of the seal. This is an essential part of a safe operation.[2]

Critical Factor: Not all respirators fit everyone.

Each individual must find out which manufacturer's masks they can properly wear. At best, any given respirator will fit 60 percent of the working population. With the large number of respirators available, at least one type should be found to fit an individual. The use of respirators is prohibited when certain factors prevent a good face piece-to-face seal. Some examples are facial hair (any face hair that interferes with a good seal), skullcaps, long hair, make-up, or temple pieces on eyeglasses.

There are two types of fit-tests, quantitative and qualitative. Because quantitative tests are expensive and tedious, qualitative tests are most often used to check respirator fit.

The *quantitative fit-test* is an analytical determination of the concentration of a test agent inside the face piece compared to that outside the mask. This concentration ratio is called the *assigned protection factor (APF)* and is a measure of the relative protection offered by a respirator. For example, if the ambient concentration of the test agent is 1,000 and the concentration inside the face piece is 10 PPM, the respirator gives the tested person an APF of 100.

A *qualitative fit-test* is a subjective test where an irritant or aroma is used to determine if there is a good face piece-to-face seal. If the test subject does not respond (smell, taste, cough, etc.) to the test agent, they can wear the tested respirator with the APF for that type of face piece.

A protection factor is used to determine the maximum use limit (MUL) of a successfully fit-tested respirator. The MUL is the highest concentration, not exceeding IDLH concentration, of a specific contaminant in which a respirator can be worn. For example, MUL = APF × TLV if a contaminant has a TLV-TWA of 10 PPM. Then the MUL for any half-mask respirator is 100 PPM; the MUL for a full face piece APR or demand SCBA is 1,000 PPM.[3] If the ambient concentration is greater than 1,000 PPM, then a pressure demand

SCBA is required. Fit testing and assigned protection factors are only two of the several considerations for selecting the proper respirator.

Respirator User Requirements

The American National Standards Institute (ANSI) has prepared the *American National Standard Practices for Respiratory Protection* and updates it periodically. ANSI Z88.2-1980 has been issued as a voluntary standard and addresses all phases of respiratory use and is highly recommended as a guide to respiratory protection.[4] The most recent version is Z88.2-1992.

The Occupational Safety and Health Administration (OSHA) refers to 29 CFR Part 1910.134 as the source of respiratory protection regulations. Section C of 29 CFR 1910.134 requires the employer to develop and implement a written respiratory protection program with required worksite-specific procedures and elements for required respirator use.

Chemical Protective Clothing (CPC)

Before using any protective equipment, personnel must be trained in compliance with all applicable OSHA, state, and local standards. Special protective clothing is required to ensure your safety in a situation involving a release of hazardous materials; this includes a release of nuclear, biological, or chemical (NBC) agents. In addition to the aforementioned, you can also reference the NFPA standards for guidance, particularly NFPA 471, 472, and 473.

Protective Garment Selection

When selecting protective garments for employment in a hazardous environment, it is not only important to select the right suit, but it is also important for the wearer to know the limitations and condition of the suit they are wearing.

- No fabric is impermeable forever.
- Suit should be strong enough so it does not tear easily.
- Use an appropriate fitting suit (too small or too large have obvious limitations).
- When donning the suit, all potential problem areas (wrist, ankles, mask if not encapsulated) should be properly secured with duct tape or chemical resistant tape.

Under the provisions of 29 CFR Part 1910.120, Appendix B, Part A, dated March 06, 1989, there are four levels of protective clothing. They are defined as Levels A, B, C, and D with Level A being the highest level of protection and Level D providing the least amount of protection.

Personal Protection

Level A

It consists of a fully encapsulated, gas and vapor proof, chemical resistant suit and self-contained breathing apparatus (SCBA). This level of protection should be worn when the highest level of skin, respiratory, eye and mucous membrane protection is needed. The following list includes Level A equipment:

- Reusable or disposable fully encapsulated chemical resistant suit (tested and certified against CB threats)
- Testing equipment for fully encapsulated suits
- Spare bottles for SCBA and appropriate service/repair kits
- Chemical resistant gloves, including thermal, as appropriate to the hazard
- Personal cooling system, vest or full suit with support equipment
- Hardhat
- Inner chemical/biological resistant garment (fire resistant optional)
- Inner gloves
- Chemical resistant tape
- Chemical resistant boots, steel or fiberglass toe and shank
- Outer booties
- Two-way local communications (secure preferred)
- Personnel accountability system (to account for missing personnel)

Level B

This level consists of a splash resistant chemical suit, encapsulated or non-encapsulated, and an SCBA. This level of protection should be used when the highest level of respiratory protection is required, but a lesser level of skin and eye protection is sufficient. The following list includes Level B equipment:

- Splash resistant chemical clothing, encapsulated or nonencapsulated
- Splash resistant hood
- Spare bottles for SCBA and appropriate service/repair kits
- Chemical resistant gloves, including thermal, as appropriate to the hazard
- Personal cooling system, vest or full suit with support equipment
- Hardhat
- Inner chemical/biological resistant garment (fire resistant optional)
- Inner gloves

- Chemical resistant tape
- Chemical resistant boots, steel or fiberglass toe and shank
- Outer booties
- Two-way local communications (secure preferred)
- Personnel accountability system (to account for missing personnel)

Level C

This level consists of a splash resistant chemical suit, with the same level of skin protection as Level B, and an air-purifying respirator. This level of protection should be used when the concentration(s) and type(s) of airborne substance(s) are *known* and the criteria for using air-purifying respirators are met. The following list includes Level C equipment:

- Splash resistant chemical clothing
- Splash resistant hood
- Air permeable or semipermeable chemical resistant clothing
- Full face air-purifying respirators with appropriate cartridges
- Chemical resistant gloves, including thermal, as appropriate to the hazard
- Personal cooling system, vest or full suit with support equipment
- Hardhat
- Inner chemical/biological resistant garment (fire resistant optional)
- Inner gloves
- Chemical resistant tape
- Chemical resistant boots, steel or fiberglass toe and shank
- Outer booties
- Two-way local communications (secure preferred)
- Personnel accountability system (to account for missing personnel)
- Extrication gear

Level D

This level consists primarily of a standard work uniform. This level of protection should be used when *no* respiratory protection and *minimal* skin protection is required. *Also the atmosphere contains no known hazard, and work functions preclude splashes, immersion, or the potential for unexpected inhalation of, or contact with, hazardous levels of any chemicals.*

- Coveralls
- Safety boots or shoes
- Safety glasses
- Hardhat
- Gloves
- Emergency escape breathing apparatus
- Face shield

> **Critical Factor:** Level A protection is the *ONLY* level to be worn when entering an area containing an unknown or questionable contaminant.

Standard Firefighter Protective Ensemble with SCBA

Standard NFPA compliant turnout gear with an SCBA will provide adequate protection against inhalation exposure and *limited* protection against small splashes, but will not protect against skin absorption from chemical agent vapors in some instances.[5]

> **Critical Factor:**   Although there has been some research conducted by the SBCCOM Domestic Preparedness program regarding turnout gear use, they determined that turnout gear with self-contained breathing apparatus (SCBA) provides less protection than Level A suits, but will allow short exposures. Through scientific testing, they have determined that taping cuffs and openings and face pieces can accomplish enhanced protection to the responder. Configurations of turnout gear with SCBA, listed in order of increased protection, include:
>
> • Standard (no use of duct tape)
> • Self-taped
> • Buddy-taped
> • Turnout gear over Tyvek undergarment
>
> Saving *live victims* while minimizing risk of harm to the rescuers is the rescue mission.
>
> This brief overview is intended to assist the reader with identifying the limitations of turnout gear and serve as a catalyst for them to seek more in-depth information. That information can be downloaded from the SBCCOM Internet site. The file in PDF format is entitled "Guidelines for Incident Commander's Use of Firefighter Protective Ensemble (FFPE) with Self-Contained Breathing Apparatus (SCBA) for Rescue Operations During a Terrorist Chemical Agent Incident."

Recommended Initial Levels of Protection

You must always enter a contaminated or a suspected contaminated area with the level of protection that will best ensure your survival. Therefore, knowledge of the contaminant is of the utmost importance, which in turn will allow you to select the appropriate level of protection. This will increase the survivability rate of your personnel. During the process of equipment selection, be prudent; you do not want to overburden people with equipment that may be nice to have, but hinders the job due to weight, bulk, or a heat factor. You do not want to become a victim, either as a result of the hazard, improper protection, or exhaustion.

Listed here are minimum recommended levels of protection for chemical-biological threats facing emergency responders:

Agent/Category	Minimum Level of Protection[6]
Unknown	Level A
Blister	Level A (i)
Nerve	Level A (ii)
Choking	Level B
Blood	Level B
Biological	HEPA Filter
Radiological	HEPA Filter

(i) Sufficient vapor will cause blisters.
(ii) High concentrations may result in nerve agent poisoning.

Critical Factor: Remember the SBCCOM research yielded two critical findings:

"Standard turnout gear with SCBA provides a first responder with sufficient protection from nerve agent vapor hazards inside interior or downwind areas of the hot zone to allow 30 minutes' rescue time for known live victims.[3]"

"Self-taped turnout gear with SCBA provides sufficient protection in an unknown nerve agent environment for a 3-minute reconnaissance to search for living victims (or a 2-minute reconnaissance if HD is suspected).[3]"

Ensure the safety of your members by verifying that the PPE selected is appropriate for the environment and will sustain the best level of protection while being exposed to hazardous substances.

Summary

The appropriate training, selection, and employment of personal protective equipment are critical to the safety of your members who will be operating in hostile environments. All too frequently we are forced into situations where we will take calculated risks in order to save lives or achieve our desired mission goals. With the heightened threat of weapons of mass destruction, terrorism, the dangers to our members have been increased exponentially, with the risks of failure being catastrophic. Ensure that your organization has taken the proper time to review the matters of protection so that all members will be "as safe as reasonably possible" when responding to acts of terrorism.

Remember, "A good day is going home at the end of your shift. A great day is going home in one piece."

NOTES

[1]29 CFR 1910.134. *Respiratory Protection.*

[2]29 CFR 1910.134 (c). *Respiratory Protection Program,* 1st Qtr 2000.

[3]*Emergency Response to Hazardous Materials Incidents.* Student Manual, United States Environmental Protection Agency, 165.15, February, 1990.

Borak, Jonathan M.D., Callan, Michael, and Abbott, William. *Hazardous Materials Exposure, Emergency Response and Patient Care, Emergency Response and Patient Care.* Brady, 1991.

[4]ANSI Z88.2-1002. *American National Standard Practices for Respiratory Protection.* American National Standards Institute.

29 CFR 1910.120. *Hazardous Waste Operations and Emergency Response.*

29 CFR 1910.120. *App B Standard Title: General Description and Discussion of the Levels of Protection and Protective Gear.*

[5]*Guidelines for Incident Commander's Use of Firefighter Protective Ensemble (FFPE) with Self-Contained Breathing Apparatus (SCBA) for Rescue Operations During a Terrorist Chemical Agent Incident.* Prepared by U.S. Army SBCCOM, Domestic Preparedness, Chemical Team, August, 1999.

Chemical Protective Clothing for Law Enforcement Patrol Officers and Emergency Medical Services When Responding to Terrorism with Chemical Weapons. Chemical Weapons Improved Response Program, Domestic Preparedness, U.S. Army Soldier and Biological Chemical Command, November, 1999.

[6]*Defense Against Weapons of Mass Destruction.* Responder Operations Course, Domestic Preparedness.

Defense Against Weapons of Mass Destruction. Technician—EMS Course, Domestic Preparedness, Developed by Science Applications International Corporation, Edgewood, Maryland; Prepared for U.S. Army Chemical and Biological Defense Command, Aberdeen Proving Ground, Maryland, August, 1997.

New Jersey HAZMAT Emergency Response. Level 3 Hazardous Materials Technician Course, Presented by NJ State Police Office of Emergency Management, November, 1993.

Appendix C
Chemical Stockpile Emergency Preparedness Program: How to Use Auto-Injectors; When to Use Auto-Injectors

HOW TO USE AUTO-INJECTORS

The recommended procedure is to inject the contents of the auto-injector into the muscles of an auterolateral thigh (through pocket). Proceed as follows:

1. Remove safety cap (yellow on atropine; gray on 2-PAM Cl; both in clip on Mark I). Do not touch the colored end of the injector after removing the safety cap, since the injector can and will function into the fingers or hand if any pressure is applied to this end of the injector.

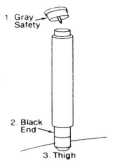

2. Hold injector as you would a pen. Place colored end (green on atropine, black on 2-PAM Cl) on thickest part of thigh and press hard until injector functions. Pressure automatically activates the spring, which plunges the needle into the muscle and simultaneously forces fluid through it into the muscle tissues.

3. Hold firmly in place for ten seconds, then remove. Massage the area of injection.

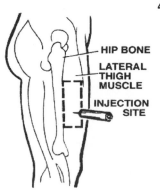

4. After each auto-injector has been activated, the empty container should be disposed of properly. It cannot be refilled nor can the protruding needle be retracted. It should be disposed of in a "sharps" container in accordance with rules for handling medical wastes and possible blood-borne pathogens.

   Dosage should be noted on a triage tag or written on the chest or forehead of the patient.

IMPORTANT: Physicians and/or other medical personnel assisting evacuated victims of nerve agent exposure should avoid exposing themselves to cross-contamination by ensuring they do not come in contact with the patients' clothing.

CHEMICAL STOCKPILE EMERGENCY PREPAREDNESS PROGRAM

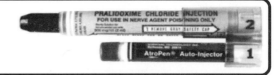

WHEN TO USE
AUTO-INJECTORS

Use only after the following events have occurred:
- Emergency medical personnel have donned personal protective equipment subsequent to recognizing existence of chemical agent hazard in area

- Some or all of signs and symptoms of nerve agent poisoning listed are present:
  —unexplained runny nose
  —tightness of chest with difficulty in breathing
  —pinpointed pupils of the eye (miosis)
  —blurred vision
  —drooling, excessive sweating
  —nausea, vomiting, and abdominal cramps
  —involuntary urination and defecation
  —jerking, twitching, and staggering
  —headache, drowsiness, coma, convulsions
  —stoppage of breathing

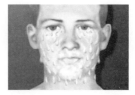

Treatment
- Immediately administer one atropine auto-injector (2 mg), followed by one 2-PAM Cl auto-injector (600 mg).
- Atropine should be given first; followed immediately by 2-PAM Cl.
- If nerve agent signs or symptoms are still present after 5–10 minutes (depending on severity), repeat injections.
- If signs or symptoms still exist after an additional 10 minutes, repeat injections for a third time.
- If signs or symptoms remain after third set of injections, DO NOT give any more antidotes but seek medical help immediately.

If severe signs and symptoms are present:
- Administer all three auto-injector kits (atropine and 2-PAM Cl) in rapid succession; then medical help should be sought.
- Remove secretions, maintain a patient airway and, if necessary, use artificial ventilation.
- Morphine, theophylline, aminophylline, or succincylcholine should not be used with 2-PAM Cl. Avoid reserpine or phenothiazine-type tranquilizers.
- 2-PAM Cl is most effective if administered immediately after exposure. Less effective if given more than 6 hours after termination of exposure.

Appendix D
CSEPP Recommended Guidelines for Antidote Treatment for Nerve Agent Exposure

- Laws regulating the use of controlled drugs differ from state to state. In some states atropine or 2-PAM Chloride can only be administered under the direction of a physician. You should be familiar with the laws and local protocols governing drug administration in emergency situations in your state.
- A MARK I kit contains two auto-injectors: unit 1 contains 2 mg of atropine, and unit 2 contains 600 mg of 2-PAM Chloride.
- You should identify at least 2 signs and symptoms of nerve agent poisoning before beginning treatment. Doses may be repeated as clinically indicated. Atropine treatment should be repeated until the patient is atropinized. Incremental 2-PAM Chloride dosages may be repeated until the maximum dose based on body weight is achieved.
- If an adult patient shows mild signs of miosis and rhinorrhea after vapor exposure, experts generally recommend observation only.
- A slow IV should be administered over a 20–30 minute period in 250 ml of normal saline or 250 ml 5% D/W solution.

Signs and Symptoms

If an adult patient shows mild signs of miosis and rhinorrhea after vapor exposure, experts generally recommend observation only.

Mild Signs and Symptoms:

Miosis (pinpoint pupil)
Blurry vision

Chest tightness
Rhinorrhea (runny nose)
Lacrimation (tearing)

Moderate Signs and Symptoms:

Above signs/symptoms plus
Significant respiratory distress
GI effects
Muscle weakness
Fasciculations
Excessive lacrimation
Nausea; vomiting; diarrhea; cramps

Severe Signs and Symptoms

Above signs/symptoms plus
Convulsions
Respiratory failure
Loss of consciousness

Treatment of Adults Exposed to Nerve Agent Vapor

Mild Signs and Symptoms:

Atropine:
2 mg IV or IM (1 auto-injector)
Repeat doses at 5 to 10 minute intervals until patient is atropinized.
2-PAM Chloride:
1 g by slow IV or 600 mg IM (1 auto-injector)

Moderate Signs and Symptoms:

Atropine:
4 mg IV or IM (2 auto-injectors)
Repeat doses at 5 to 10 minute intervals until patient is atropinized.
2-PAM Chloride:
1 g by slow IV (repeat hourly as needed for up to a total of 3 g in 3 hours)
or 1200 mg IM (2 auto-injectors)

Severe Signs and Symptoms:

Atropine:
6 mg IV or IM (3 auto-injectors)
Once hypoxemia is reversed, an additional 2 mg IV at 3 to 5 minute intervals may be required to support airways.

2-PAM Chloride:
>1 to 2 g by slow IV (repeat hourly as needed for up to a total of 3 g in 3 hours) or 1800 mg IM (3 auto-injector)

Diazepam:
>10 mg IM or 5 mg IV (repeat doses as required)

Treatment for Direct Contact Exposure to Nerve Agent: Adults

A person exposed to liquid nerve agent should be treated according to the signs and symptoms as a person exposed to nerve agent vapor. Due to the slower uptake, however, onset of symptoms may be delayed for 1 to 2 hours and some symptoms may not appear until after 6 hours.

Mild Signs and Symptoms:

Onset of sweating and muscle fasciculation at site of exposure 1 to 2 hours after exposure should be treated with:
Atropine:
>2 mg IM (1 auto-injector)

2-PAM Chloride:
>600 mg IM (1 auto-injector) or 1 g slow IV

Moderate Signs and Symptoms:

Onset of GI symptoms more than 6 hours after exposure should be treated with:
Atropine:
>2 mg IM (1 auto-injector)

2-PAM Chloride:
>600 mg IM (1 auto-injector)

Severe Signs and Symptoms:

Same as for vapor exposure.
Atropine:
>6 mg IM (3 auto-injector)

2-PAM Chloride:
>1 to 2 g by slow IV (repeat hourly as necessary for up to 3 g in 3 hrs)

Treatment of Adolescents, Children, and Infants

Treatment varies depending on age and body weight of child or adolescent. The adult-size atropine and 2-PAM Chloride auto-injectors should never be given to infants.

Atropine: (depends on age)

Repeat doses for all age groups as clinically indicated until patient is atropinized.

Less than 2 years:	0.5 mg IV or IM
2 to 10 years:	1 mg IV or IM
Over 10 years or adolescent:	2 mg IV or IM (1 auto-injector)

2-PAM Chloride: (depends on body weight)

Less than 50 lbs. (22.7 kg):	15 mg per kg of body weight by slow IV
Over 50 lbs.:	600 mg IM (1 auto-injector)
	Repeat doses at hourly intervals as clinically indicated (no more than twice).

Diazepam: (depends on age)

Infants over 30 days to children age 5 years:
0.2 mg to 0.5 mg per kg of body weight slowly every 2 to 5 minutes, up to maximum total dose of 5 mg IV or IM
Children over 5 years:
1 mg every 2 to 5 minutes, up to maximum total dose of 10 mg

Appendix E
Characteristics
of Radiation[1]

Radiation is energy traveling in the form of particles or waves in bundles of energy called photons. Some everyday examples are microwaves used to cook food, radio waves for radio and television, light, and x-rays used in medicine.

Radioactivity is a natural and spontaneous process by which the unstable atoms of an element emit or radiate excess energy in the form of particles or waves. These emissions are collectively called ionizing radiations. Depending on how the nucleus loses this excess energy, either a lower energy atom of the same form will result, or a completely different nucleus and atom can be formed.

Ionization is a particular characteristic of the radiation produced when radioactive elements decay. These radiations are of such high energy that when they interact with materials, they can remove electrons from the atoms in the material. This effect is the reason why ionizing radiation is hazardous to health, and provides the means by which radiation can be detected.

A typical model of the atom is called the Bohr Model, in honor of Niels Bohr who proposed the structure in 1913. The Bohr atom (Fig. E-1) consists of a central nucleus composed of neutrons and protons, which is surrounded by electrons that "orbit" around the nucleus.

Protons carry a positive charge of one and have a mass of about 1 atomic mass unit or amu (1 amu = 1.7×10^{-27} kg, a very, very small number). Neutrons are electrical neutral and also have a mass of about 1 amu. In contrast, electrons carry a negative charge and have mass of only 0.00055 amu. The number of protons in a nucleus determines the element of the atom. For example, the number of protons in uranium is 92 and the number in neon is 10. The proton number is often referred to as Z.

Atoms with different numbers of protons are called elements, and are arranged in the periodic table with increasing Z. Atoms in nature are

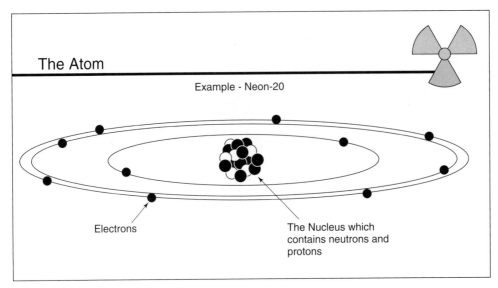

FIGURE E-1   Model of a Typical Atom

electrically neutral so the number of electrons orbiting the nucleus equals the number of protons in the nucleus.

Neutrons make up the remaining mass of the nucleus and provide a means to "glue" the protons in place. Without neutrons, the nucleus would split apart because the positive protons would repel each other. Elements can have nuclei with different numbers of neutrons in them. For example hydrogen, which normally only has one proton in the nucleus, can have a neutron added to its nucleus to form deuterium, or have two neutrons added to create tritium, which is radioactive. Atoms of the same element that vary in neutron number are called isotopes. Some elements have many stable isotopes (tin has 10), while others have only one or two. We express isotopes with the nomenclature neon-20 or $^{20}Ne_{10}$, with 20 representing the total number of neutrons and protons in the atom, often referred to as A, and 10 representing the number of protons (Z).

Radionuclides can be arranged by A and Z in the chart of the nuclides.

Alpha decay (Fig. E-2) is a radioactive process in which a particle with two neutrons and two protons is ejected from the nucleus of a radioactive atom. The particle is identical to the nucleus of a helium atom.

Alpha decay only occurs in very heavy elements such as uranium, thorium, and radium. The nuclei of these atoms are very "neutron rich" (i.e., have a lot more neutrons in their nucleus than they do protons), which makes emission of the alpha particle possible.

After an atom ejects an alpha particle, a new parent atom is formed that has two less neutrons and two less protons. Thus, when uranium-238

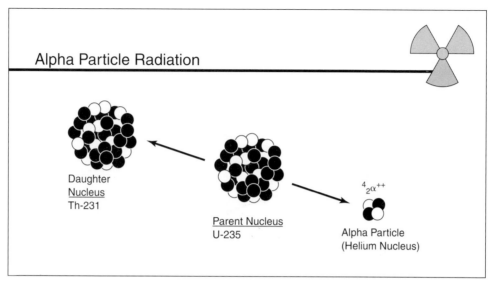

FIGURE E-2 Alpha Particle Radiation

(which has a Z of 92) decays by alpha emission, thorium-234 is created (which has a Z of 90).

Because alpha particles contain two protons, they have a positive charge of two. Further, alpha particles are very heavy and very energetic compared to other common types of radiation. These characteristics allow alpha particles to interact readily with materials they encounter, including air, causing many ionizations in a very short distance. Typical alpha particles will travel no more than a few centimeters in air, and are stopped by a sheet of paper.

Beta decay (Fig. E-4) is a radioactive process in which an electron is emitted from the nucleus of a radioactive atom, along with an unusual particle called an antineutrino. The neutrino is an almost massless particle that carries away some of the energy from the decay process. Because this electron is from the nucleus of the atom, it is called a beta particle to distinguish it from the electrons, which orbit the atom.

Like alpha decay, beta decay occurs in isotopes that are "neutron rich" (i.e., have a lot more neutrons in their nucleus than they do protons). Atoms that undergo beta decay are located below the line of stable elements on the chart of the nuclides, and are typically produced in nuclear reactors and cyclotrons. When a nucleus ejects a beta particle, one of the neutrons in the nucleus is transformed into a proton. Since the number of protons in the nucleus has changed, a new daughter atom is formed that has one less neutron, but one more proton than the parent. For example, when Rhenium-187 decays (which has a Z of 75) by beta decay, Osmium-187 is created (which has a Z of 76).

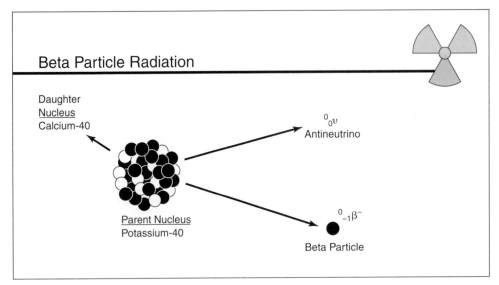

FIGURE E-3 Beta Particle Radiation

Beta particles have a single negative charge and weigh only a small fraction of a neutron or proton. As a result, beta particles interact less readily with material than alpha particles. Depending on the beta particle energy, (which depends on the radioactive atom and how much energy the antineutrino carries away), beta particles will travel up to several meters in air, and are stopped by thin layers of metal or plastic. Antineutrinos have very little mass and no charge. Its been said that a neutrino has more than a 50 percent chance of traveling through a light-year of solid lead without being stopped.

After a decay reaction, the nucleus is often in an "excited" state. This means that the decay has produced a nucleus that still has excess energy to get rid of. Rather than emitting another beta or alpha particle, this energy is lost by emitting a pulse of electromagnetic radiation called a gamma ray (Fig. E-4). The gamma ray is identical in nature to light or microwaves, but of very high energy.

Like all forms of electromagnetic radiation, the gamma ray has no mass and no charge. Gamma rays interact with material by colliding with the electrons in the shells of atoms. They lose their energy slowly in material, being able to travel significant distances before stopping. Depending on their initial energy, gamma rays can travel from one to hundreds of meters in air and can easily go right through people.

It is important to note that most alpha and beta emitters also emit gamma rays as part of their decay process. However, there is no such thing as a "pure" gamma emitter. Important gamma emitters including Tech-

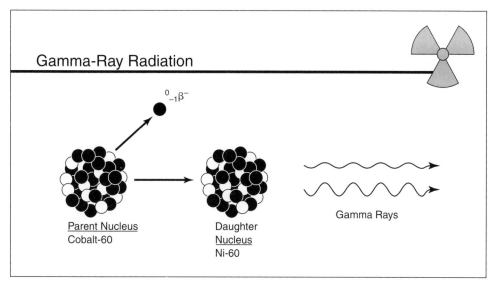

FIGURE E-4   Gamma Ray Radiation

netium-99^m, which is used in nuclear medicine, and Cesium-137, which is used for calibration of nuclear instruments.

Over a century ago in 1895, Roentgen discovered the first example of ionizing radiation, x-rays. The key to Roentgen's discovery was a device called a Crooke's tube, which was a glass envelope under high vacuum, with a wire element at one end forming the cathode, and a heavy copper target at the other end forming the anode. When a high voltage was applied to the electrodes, electrons formed at the cathode would be pulled toward the anode and strike the copper with very high energy. Roentgen discovered that very penetrating radiations were produced from the anode, which he called x-rays (Fig. E-5).

X-ray production occurs whenever electrons of high energy strike a heavy metal target, like tungsten or copper. When electrons hit this material, some of the electrons will approach the nucleus of the metal atoms where they are deflected because of their opposite charges (electrons are negative and the nucleus is positive, so the electrons are attracted to the nucleus). This deflection causes the energy of the electron to decrease, and this decrease in energy then results in the formation of an x-ray.

Medical x-ray machines in hospitals use the same principle as the Crooke's tube to produce x-rays. The most common x-ray machines use tungsten as their cathode, and have very precise electronics so the amount and energy of the x-ray produced is optimum for making images of bones and tissues in the body.

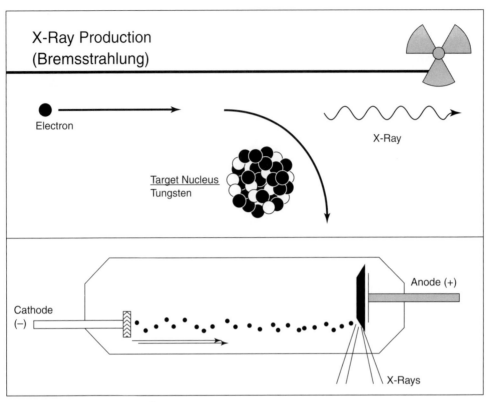

FIGURE E-5 X-Ray Production

Properties of Radiation

Different radiations have different properties, as summarized in Table E-1:

Table E-1 Radiation Properties

Radiation	Type of Radiation	Mass (AMU)	Charge	Shielding Material
Alpha	Particle	4	+2	Paper, skin, clothes
Beta	Particle	1/1836	±1	Plastic, glass, light metals
Gamma	Electromagnetic wave	0	0	Dense metal, concrete, Earth
Neutrons	Particle	1	0	Water, concrete, polyethylene, oil

In summary, the most common types of radiation include alpha particles, beta and positron particles, gamma and x-rays, and neutrons. Alpha particles are heavy and doubly charged, which causes them to lose their energy very quickly in matter. They can be shielded by a sheet of paper or the

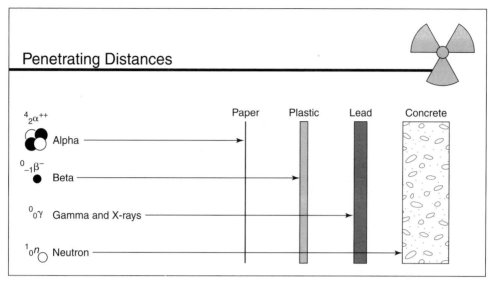

FIGURE E-6   Penetrating Distances

surface layer of our skin. Alpha particles are only considered hazardous to a person's health if an alpha-emitting material is ingested or inhaled. Beta and positron particles are much smaller and only have one charge, which causes them to interact more slowly with material. They are effectively shielded by thin layers of metal or plastic and are again only considered hazardous if a beta emitter is ingested or inhaled.

Gamma emitters are associated with alpha, beta, and positron decay, X-rays are produced either when electrons change orbits within an atom, or electrons from an external source are deflected around the nucleus of an atom. Both are forms of high energy electromagnetic radiation, which interact lightly with matter. X-rays and gamma rays are best shielded by thick layers of lead or other dense material and are hazardous to people when they are external to the body.

Neutrons are neutral particles with approximately the same mass as a proton. Because they are neutral, they react only weakly with material. They are an external hazard best shielded by thick layers of concrete.

Half-life (Fig. E-7) is the time required for the quantity of a radioactive material to be reduced to one-half its original value.

All radionuclides have a particular half-life, some of which are very long, while other are extremely short. For example, uranium-238 has such a long half-life, 4.5×10^9 years, that only a small fraction has decayed since the Earth was formed. In contrast, carbon-11 has a half-life of only 20 minutes. Since this nuclide has medical applications, it has to be created where it is being used so that enough will be present to conduct medical studies.

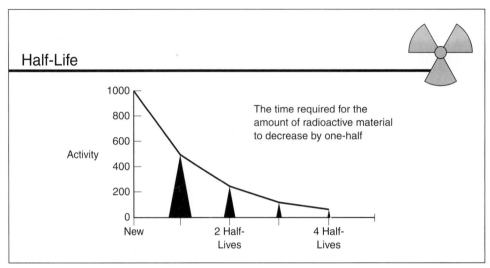

FIGURE E-7   Half-Life

When given a certain amount of radioactive material, it is customary to refer to the quantity based on its activity rather than its mass (Fig. E-8). The activity is simply the number of disintegrations or transformations the quantity of material undergoes in a given period of time.

The two most common units of activity are the curie and the Becquerel. The curie is named after Pierre Curie for his and his wife Marie's discovery of radium. One curie is equal to 3.7×10^{10} disintegrations per second. A newer unit of activity is the Becquerel named for Henry Becquerel, who is credited with the discovery of radioactivity. One Becquerel is equal to one disintegration per second.

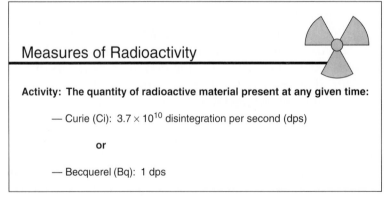

FIGURE E-8   Measures of Radioactivity

Radiation Units

— Roentgen: A unit for measuring the amount of gamma or X-rays in air.
— RAD: A unit for measuring absorbed energy from radiation.
— REM: A unit for measuring biological damage from radiation.

FIGURE E-9   Radiation Units

It is obvious that the curie is a very large amount of activity and the Becquerel is a very small amount. To make discussion of common amounts of radioactivity more convenient, we often talk in terms of milli- and micro-curies or kilo- and mega-Becquerels.

Radiation is often measured in one of the three units noted in Figure E-9, depending on what is being measured and why. In international units, these would be Coulombs/kg for Roentgen, Grays for RADS, and Seiverts for REM.

NOTES

[1]Provided by the University of Michigan chapter of the Health Physics Society, www.umich.edu.

Appendix F
Survival Strategies

Dennis Krebs

Upon completion of this chapter, the reader will be able to:

- Identify the need to train personnel in a variety of survival skills ranging from basic domestic arguments to acts of terrorism.
- Demonstrate the proper means to secure weapons found on patients and describe what objects might provide cover or concealment if a person is discharging a firearm at a scene.
- Identify the measures required in the proper selection and care of body armor.
- Describe the strategies employed during a civil disturbance.
- Demonstrate the proper techniques utilized by emergency responders in preserving evidence at a crime scene.

In 1982 the first survival skills program debuted in the fire and EMS services. At that time such training was often thought to be overkill and completely unnecessary to the provision of service to one's community. Certainly no one would attempt to harm an EMT, paramedic, or firefighter. Yet, in most metropolitan areas emergency responders were increasingly becoming the victim of violent assaults. Even today fire and EMS personnel routinely face violent situations often with tragic consequences.

Los Angeles, CA: Scott Miller of the Los Angeles City Fire Department was shot in the neck, severing his carotid artery, while responding to an incident during the riots of April 1992.[1]

Philadelphia, PA: On Christmas Day 1994 two medics were slashed by a knife wielding man who had feigned injury in an attempt to steal their drug bag.[2]

Denver, CO: A Denver firefighter was shot in the neck and killed on January 31, 1993, after climbing a ladder to a second-story window to check on a man who had reportedly committed suicide.[3]

Washington, DC: A firefighter was shot on December 9, 1996, while attempting to stabilize a gunshot wound victim. Two men rushed the ambulance firing approximately eight rounds into the unit.[4]

Toledo, OH: Firefighter David Bilius was injured by flying glass secondary to a shotgun blast to the windshield of his ambulance, and Lt. Jeffrey Cook was struck by a second shotgun blast on February 13, 1998. The incident left a total of two dead and four injured.[5]

Respected leaders of the emergency services maintained their view that most communities did not face a crime problem and thus teaching their personnel to "sneak up to a scene" was foolish. Emergency responders themselves would often state, "My job is to treat people; whether knives are being thrown, bullets are flying, my job is to treat people." They were not comfortable with awaiting the scene to be secured by law enforcement when responding to a shooting, stabbing, or trauma-related suicide.

Early survival programs were very basic and often focused on merely changing attitudes. Students were given the ability to identify a potentially dangerous situation and avoid it if possible. Since many communities across the nation are rural and police may have an extended arrival time, sometimes in excess of one hour, students were given a variety of safety tactics. At each step of involvement at a scene they needed to have the capability to extract themselves should violence erupt. Emergency responders quickly realized that if they were injured, they became part of the problem and not part of the solution. There were clear roles and responsibilities that needed to be emphasized. Law enforcement would be responsible for security at any scene; fire and EMS were to be responsible for treating injuries and mitigating any other hazards.

Since 1982, it is obvious that society has changed. Violence, especially among our nation's youth, is pervasive. Many leaders in our communities long for a return to the time when concern was raised over "Saturday Night Specials," the cheaply made and cheaply purchased handguns many states made illegal a number of years ago. Today, we are concerned about the armed bank robbers carrying military style weapons with thousands of rounds of ammunition, or the clandestine drug lab that has claymore mines dispersed as booby traps.[6] The problem is no longer that of the big city. Small rural communities are experiencing problems as well. The drug abuse problem, and ancillary crimes associated with it, is touching every community. Gangs are migrating from the Los Angeles and Chicago areas and moving into "smalltown USA." Names like Black Gangster Disciples, El Rukns, Vice Lords, Latin Kings and Tiny Rascal Gang are becoming common even in the smallest of communities. The complacency that "it never happens here" is a dangerous mind-set.

With the change in our society there also came a change in the delivery of fire and emergency medical services. Basic survival skills continue to be a necessary part of a responder's education. Teaching the EMT, firefighter or paramedic how to effectively deal with the intoxicated driver slumped

FIGURE F-1   SWAT team personnel carrying a victim in a stokes basket. *Source:* Courtesy D. Krebs

over the steering wheel of a car is still appropriate. Providing basic information on what objects will stop a bullet and what objects merely conceal a person from view is still relevant. Our roles, however, have changed and possibly expanded. Who could have anticipated that firefighters in a number of jurisdictions across the country would wear body armor as a matter of routine? Did anyone believe in 1982 that paramedics would be assigned to SWAT teams? Finally, who could have imagined that the emergency services would be called upon to deal with a mass casualty incident the size of the World Trade Center bombing, Oklahoma City bombing or the bombing at the Olympics in Atlanta? Though our roles and responsibilities might be separate fire, EMS and law enforcement are finding themselves operating at scenes side by side, each relying on the capabilities of the other to bring the incident to a successful conclusion.

As we briefly examine various survival concepts, keep in mind that your basic responsibility, whether your role be that of police office, EMT, firefighter, or paramedic, is to complete your tour of duty and return home safely to your family at the end of each day.

There is a wide dispersion of incident types that might turn violent. For that reason it is imperative that law enforcement, fire and EMS begin to work together and coordinate activities. However, these agencies need to work together prior to an emergency. This can be done in a variety of ways. One of the more effective ways to stimulate a cooperative working relationship between the emergency services is through training. Police, fire

FIGURE F-2   An armored personnel carrier with "Rescue" on the side and Star of Life. *Source:* Courtesy D. Krebs

and paramedic entrance level training should highlight the roles and responsibilities of associated emergency services. If we become familiar with each other's needs, duties on an incident can be more easily coordinated. Regular meetings between agencies allows for discussion of problems. For instance, if police continually position their vehicles in front of structure fires blocking access for fire equipment, this problem can be discussed openly and remedied at roll call. If EMS personnel unnecessarily trample through crime scenes destroying evidence, a training class can be discussed and implemented. Through preparation and preplanning, agencies can develop a cohesive document outlining their response to demonstrations, rallies and other social gatherings that may not, at least initially, be considered a threat to public safety. Intelligence information that might not otherwise find its way to your agency may now have an avenue for limited dissemination. If police are conducting an investigation into an illegal fireworks manufacturer operating from a house in your district, such information would be invaluable to companies responding to that address for a reported explosion. Interagency teamwork on a small scale can only benefit operations at a major disaster. Certainly, sensitive information must be guarded. At times this sensitivity may dictate that only the agency head is informed of a possible risk. When writing standard operating procedures,

FIGURE F-3 Medics receive briefing prior to deploying with SWAT team. *Source: Courtesy D. Krebs*

the department may have other agencies examine the document to ensure it does not negatively impact their operations. Each agency might provide constructive comment that may have been overlooked during the initial draft. At critical times during a civil disturbance, bombing or mass casualty incident resulting from a shooting, having a familiar representative available to respond to your needs is imperative.

Avoiding the Violent Confrontation

Emergency services personnel respond to a variety of calls that can quickly turn violent. Shootings, stabbings, domestic arguments, and barroom brawls are but a few of the incidents to which we routinely respond. Often times our arrival prior to law enforcement places us in an unsafe environment. Especially on shootings, stabbings and some suicides, personnel should stage approximately one block from the scene and await it to be secured by law enforcement. Our routine of quick arrival on the scene with lights flashing and sirens blaring may not always be beneficial. The entire neighborhood, including those nosey neighbors down the street, is now approaching to provide unsolicited help and expertise. Should the injury at this address have resulted from a domestic squabble, you can expect the scene to be quite sterile since you announced your arrival with sufficient fanfare to wake the dead. Consider shutting down emergency equipment, lights, and siren approximately one block away. Go in quiet. You will be amazed at what can be seen and heard when the occupants are not aware of your arrival.

By now it should be common knowledge that responders should not stand in front of doors. But many do not realize that standing on the hinged side of the door is nearly as dangerous as standing in front of the door. When approaching a house or apartment, stand to the doorknob side of the door. Take a second to listen. Are the occupants arguing? Is there mention of a weapon? Should it sound too dangerous, back away and call for assistance. If there are no outward signs of trouble, knock on the door and announce yourself, "EMS, did you call for help?"

Once inside the residence, let the occupant lead the way. This practice not only gets you to your patient more expeditiously but also provides you a shield when violence erupts. Arriving at your patient's side, once again identify yourself. Remember, someone whose level of consciousness has been altered by drugs, alcohol, or mental instability may, due to your uniform, mistake you for the police. Operating at overdoses, suicides, shootings, domestic arguments, stabbings, and so on, is dangerous. Parties involved may decide to start or continue a previous violent confrontation in your presence. Check the area for weapons, even if the police are present. Two sets of eyes are better than one. Quickly glance around the room looking for the scissors, vase, steak knife, iron or any other object within reach that might be used as a weapon. Secure those objects by unobtrusively moving them to a safe location. Sliding the object under a couch or in the drawer of an end table is quite simple.

People often give clues that they are about to become violent. Clenched fists or teeth, fidgeting in a chair, or pacing back and forth as well as abusive language are good signs of impending violence. Try to diffuse the situation prior to its escalation. Should violence erupt, do not stand between fighting parties. User your voice, body language and eye contact in an attempt to separate the combatants. By manipulating the parties to turn away from each other, somewhat back-to-back, you may be able to control the situation. One of the greatest survival tools you have is your mouth. Treat people with the same amount of respect that you would like to have shown to you. The speed of your voice, tone of your voice, and certainly your vocabulary can mean the difference between a situation escalating to a violent confrontation or it being diffused. Condescending words like "sport," "buddy," or "pal" should be avoided. Terms such as "gramps," "old man," or "pops" are not only demeaning to your patient but family members as well. Certainly, if you were employed by a large successful business, you would conduct yourself appropriately with customers and colleagues, ensuring that all facets of your personal presentation were appropriate. Why do we not then consider the emergency that all facets of your personal presentation were appropriate. Why do we not then consider the emergency services as a large successful business and provide our customers with the same level of consideration. This type of presentation is not always successful. There will be those instances where it may be appropriate to use more of an authoritative tone. You must still, however, act professionally. If at any time you feel your safety is threatened and police are not already on the scene, back out and call for assistance.

Weapons

Weapons, especially firearms, are a part of our society. When responding to calls for assistance, we must be ever wary that our patient or bystanders might be armed. When discussing the topic of weapons, many emergency responders think of the 9mm handgun or sawed-off shotgun. Most firefighters, EMTs and paramedics are unaware of the vast array of commercially made and make-shift weapons that are readily available on the street. How many of us would think that the small 3″ toy at the end of a keychain could fire a single .38 caliber round, or that the ballpoint pen in your patient's shirt pocket can open into a butterfly knife? How many of us place a female's purse on the stretcher between her legs? Do you realize the

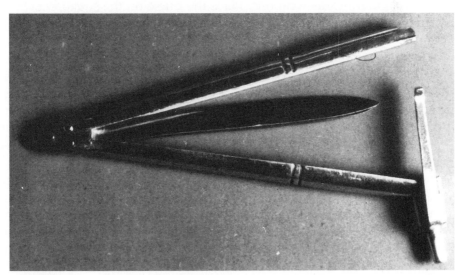

FIGURE F-4 This ballpoint pen opens into a deadly butterfly knife. Many such objects can find their way through airport metal detectors. *Source:* Courtesy D. Krebs

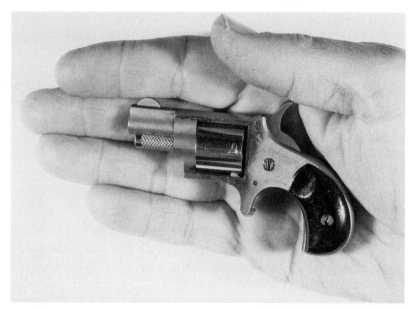

FIGURE F-5 A .22 calibre revolver fits in the palm of the hand. *Source:* Courtesy D. Krebs

number of weapons that are readily available in that purse? (Don't count the credit cards!) Females are often taught in self-defense classes to use keys, nail files, and hairpins as weapons. Place the purse on the squad bench where your patient can see it but does not have quick access to the bag.

Most of us have been taught to conduct a primary and secondary survey, the secondary survey being a head-to-toe examination for lumps, bumps, and deformities. During this exam we should not only focus on the detection of injuries but weapons as well. Suppose while conducting this secondary survey you feel a bulge at your patient's waist line. Your suspicion is raised because it feels like a firearm. Your first option is to let a police officer disarm this person. Doing so reduces the possibility of a confrontation with the patient; police officers are trained to disarm people, we are not. You also remove yourself from most legal issues such as chain of custody. Again, if at all possible, let police do the disarming. There will be those situations where law enforcement has an extended arrival time or the situation dictates immediate action. But, are you sure this armed person is not a police officer? Does it matter? Should you ask, "Hey, are you a cop?" If you are concerned that the person carrying this firearm might be a law enforcement official, there are two simple questions that can be asked. "Are you on the job?" and "What agency are you with?" To a police officer, the questions are obvious. To the normal citizen, these questions will have little meaning. Our reason for asking such questions is clear. Police officers go through extensive training to ensure that no one takes their weapon. Attempting to disarm this patient who just happens to be a police officer could be disastrous. Even in a semiconscious state, officers have been known to fight for control of their weapon. Should this person respond that he is a police officer, your next question should be, "Can I secure your weapon for you?" Usually the officer will prefer another police officer to take control of the weapon. Another decision must be made. Do you feel comfortable that this person is a law enforcement official, and do you feel comfortable providing medical care even though he is armed? The answer is not simple and can only be made by on-scene personnel evaluating the situation as it presents itself.

One quick and simple method of controlling the situation where your patient is armed is to take both radial pulses. You are not attempting to control the person's hands! That is the responsibility of law enforcement. The fact that the person cannot pull a weapon with anything other than his hands is merely an added benefit of conducting a good patient assessment. At this point your partner can remove and secure the weapon. "Secure" does not mean on the coffee table or the roof of the car. The weapon should be locked in a glove box or drug locker within your unit. If you are in a building, it can be secured in an aid kit. **Never attempt to unload a firearm.** Firearms are routinely altered. The pound per square inch pressure needed to pull the trigger may have been reduced. Some semiautomatic pistols are carried "cocked and locked" where the hammer is pulled back with only the safety engaged to prevent firing. Lacking any knowledge of this weapon's history, you are safest to merely pick the firearm up by the handgrips and secure it in a safe location. You do not place a pencil down the barrel, and you never pick it up by the trigger guard as some have suggested.

The first arriving police officer should be appraised of the situation. Turn the weapon over and have them sign your report that custody of the weapon was transferred. In addition, your report should reflect the facts of the incident, where the weapon was found and a description of the weapon, i.e., a black revolver, a 6-inch knife with wood handle. Remember, if possible, allow the police to disarm a patient.

Cover and Concealment

This is actually a relatively easy concept to understand. It's mid-afternoon and you are dispatched to the "unknown type medical emergency." Arriving on the scene, you grab the appropriate equipment and walk toward a large office building. As you do so, the crack of gunfire is heard from the lobby and multiple rounds scream past your head. What to do? If you contemplate that thought for much longer than two seconds, you could be dead! Thus, cover, objects that are difficult or impossible for bullets to penetrate, is just what the doctor ordered. Simply put, this is the biggest, heaviest object you can *immediately* find. Such objects include large concrete planters, trees, brick or concrete block walls, dumpsters, utility poles or the motor block and wheels of your unit. Fire hydrants, although they may not have a pleasant aroma, provide excellent cover. Many large-caliber rounds will barely mar the paint. Indoors, there is little that will provide adequate protection. Walls are generally made of only metal or wood studs and thin drywall where most rounds can easily penetrate. Since most furniture is lightweight, your options are limited to the heavy wood desk or refrigerator. In a hospital, objects like the portable x-ray machine may provide limited protection. Remember, the object that can stop a round from a .22 caliber pistol may not stop the round from a military style rifle.

Indoors you will most often rely on concealment, protection from being seen. If the suspect cannot see you, he may not fire. Concealment can be provided by darkness, shadows, bushes, or even the overstuffed chair. The wall constructed of studs and drywall, although not adequate protection in stopping a bullet, may provide excellent concealment. Emergency responders should be aware that the reflective striping attached to many of the garments worn on emergency scenes not only allows you to be seen by motorists but does wonders to assist a gunman in target acquisition. If the shooting starts, remove anything that has reflective striping. Although concealment can protect you from being seen, it will not protect you from being heard; remain as quiet as possible. Once again, concealment will not stop a bullet.

When arriving at a scene, take note of the objects that might be utilized as cover or concealment: the tree in the front yard, the car in the driveway, or the shadow at the corner of the house. In time, your conscious effort to survey the area will occur subconsciously and become part of your overall size-up of every scene.

Soft Body Armor

With the increased concern for the welfare of fire and EMS personnel in the United States, some departments are examining the use of soft body armor as a means of providing protection to their personnel. The Los Angeles City Fire Department undoubtedly utilizes body armor more widely than any other department in the country, with garments being assigned to each engine company, truck company, ambulance, and battalion chief's vehicle. Each EMS member is also provided their own body armor.

One of the widest misconceptions is that soft body armor is bulletproof. Body armor is most widely made of Kevlar and, depending on which threat level is utilized, provides varying degrees of bullet and fragment

FIGURE F-6 Brochure advertising firefighter and EMT ballistic protection. *Source:* Courtesy Second Chance Body Armor, Inc., Central Lake, Michigan

protection.[7] This material was introduced in 1972 by E. I. Du Pont De Nemours and Company. One ounce of Kevlar is 5 times stronger than an equal amount of steel and will not melt, is flame resistant, and stretch resistant as well.[8] Changing technology has allowed body armor to become more lightweight and flexible.

Kevlar, as in most other articles of clothing, is woven to form the vest. When a bullet strikes the vest, the energy of the round is dissipated throughout the weave of the garment. Some bruising, lacerations or internal injuries

Threat Level Chart

Threat Level	Ballistic Protection
I	.22 caliber long rifle handgun rounds .38 caliber special projectiles
IIA	.22 caliber long rifle handgun rounds .38 caliber special projectiles .357 Magnum 158 grain striking at 1259 ft/sec 9 mm 124 grain striking at 1090 ft/sec
II	.22 caliber long rifle handgun rounds .38 caliber special projectiles .357 Magnum 158 grain striking at 1250 ft/sec .357 Magnum 158 grain striking at 1395 ft/sec 9 mm 124 grain striking at 1090 ft/sec 9 mm 124 grain striking at 1175 ft/sec
IIIA	.22 caliber long rifle handgun rounds .38 caliber special projectiles .357 Magnum 158 grain striking at 1250 ft/sec .357 Magnum 158 grain striking at 1395 ft/sec 9 mm 124 grain striking at 1090 ft/sec 9 mm 124 grain striking at 1175 ft/sec .977 124 grain striking at 1400 ft/sec .44 Magnum semi wad cutter striking at 1400 ft/sec
III	.22 caliber long rifle handgun rounds .38 caliber special projectiles .357 Magnum 158 grain striking at 1250 ft/sec .357 Magnum 158 grain striking at 1395 ft/sec 9 mm 124 grain striking at 1090 ft/sec 9 mm 124 grain striking at 1175 ft/sec .977 124 grain striking at 1400 ft/sec .44 Magnum semi wad cutter striking at 1400 ft/sec 7.62 mm (308 Winchester)
IV	.22 caliber long rifle handgun rounds .38 caliber special projectiles .357 Magnum 158 grain striking at 1250 ft/sec .357 Magnum 158 grain striking at 1395 ft/sec 9 mm 124 grain striking at 1090 ft/sec 9 mm 124 grain striking at 1175 ft/sec .977 124 grain striking at 1400 ft/sec .44 Magnum semi wad cutter striking at 1400 ft/sec 7.62 mm (308 Winchester) 30.06 projectiles

may occur after being struck, and a medical examination should always be conducted to rule out injury.

Body armor is classified into six levels of protection. Most fire and EMS providers have elected to utilize Threat Level II or Threat Level IIIA. Check with your local police agency for assistance in determining which level is appropriate for the area. Whenever purchasing body armor, check the garment to be sure that it meets the standard set forth by the National Institute of Justice Standard No. 0101.03.

Body armor needs to be worn to be effective. Simply placing it in a compartment like breathing apparatus, only to be used when there is visible smoke, or in this case when the situation becomes violent, defeats the garment's usefulness. Seldom will you be capable of predicting when violence is about to erupt. Los Angeles City Fire Department has strict guidelines when the vest must be worn. Although it can be worn at any time the provider feels threatened, the vest must be worn on shootings, stabbings, and domestic violence cases. Should your armor be of the type that is worn under a uniform shirt, the garment would be donned prior to the beginning of the shift. The other style is worn like a vest, over the uniform, and should be donned prior to your arrival at the scene. This second type should also be concealed by wearing a jacket or turnout coat over the vest. Most armed suspects will attempt to shoot at the torso unless its obvious that you are wearing a protective device.

Initially, soft body armor will feel bulky and uncomfortable, especially during summer months. However, the protection gained far outweighs the discomfort.

Your soft body armor should be inspected routinely for signs of abrasion, fraying of the fabric, or unraveling. In addition, the user should watch for permanent folds or wrinkles, which might indicate unusual wear. Although the manufacturer will provide specific instructions on caring for the garment, the following general guidelines remain useful:

- Hand wash the body armor in cold water with a mild detergent.
- Thoroughly rinse and remove *all* detergent. (An improperly rinsed vest can develop reduced stopping power.)
- Never use bleach or any product containing bleach.
- Never dry clean. (Check manufacturer's specifications.)
- Never machine wash or machine dry.
- Always drip-dry indoors.
- Return the vest to the manufacturer for repairs.[9]

Soft body armor should always be replaced after it has stopped a bullet. Due to normal wear, Du Pont also suggests that armor be replaced after 5 years.

Most departments are not sufficiently funded to provide this type of protection to each member. Many individual firefighters, paramedics, and EMTs are purchasing their own body armor. Departments should have pre-plans noting where sufficient protection can be obtained in case of a large-scale emergency such as a civil disturbance. In today's society no one from fire or EMS should be sent into such a situation without some type of ballistic protection. Local National Guard or Army Reserve units may be capable of providing some assistance. Ballistic armor may also be available through state and federal surplus dealers. One must be aware that the military utilizes "flak jackets." These garments are meant to protect against fragmentation devices like hand grenades, not projectiles, and are not subjected to the same standard as law enforcement type armor. They will, however, provide limited protection against certain types of rounds. It is suggested that you obtain a "flak jacket" and request that your local police agency test the vest to ensure it will meet your local needs.

Appendix G*
Integrating Criminal Investigation into Major EMS Scenes

David Cid
Paul M. Maniscalco

The use of weapons of mass destruction and the possibility of bioweapon deployment add urgency to the need for effective coordination between crisis and consequence managers. A critical dimension of successful collaboration is mutual appreciation for each other's mission requirements. This cross-cultural awareness can provide the basis for effective interaction during a crisis, when mutual understanding and goodwill carry the day.

Presidential Decision Directive #39, issued June 21, 1995, assigns specific roles and responsibilities for responding to terrorism acts within the United States or within its interests abroad. Within the context of this directive, the FBI has the lead role for crisis management and the Federal Emergency Management Agency the lead for consequence management.

Crisis management, defined as law enforcement response, focuses on the criminal aspects of the incident.[1] Consequence management, defined as the response to the disaster, focuses on alleviating damage, loss, hardship, or suffering.[2]

Response Coordination

Should an act of terrorism occur, the law enforcement professionals who arrive on scene are, as a group, highly motivated, outcome-oriented professionals comfortable in a high-stress environment. In short, they are much like their consequence management counterparts. When two such groups

*This appendix first appeared as an article in *JEMS,* January 1999, pp. 68–69.

meet in common understanding and cooperation, a synergy occurs, supporting the resolution of issues and leading to a successful outcome.

To ensure success, it is helpful for today's emergency responders to develop an understanding of law enforcement concerns and the factors that drive them.

Major Case Dynamics

Every criminal investigation has unique factors and dimensions. A study of major case dynamics indicates that information and evidence *essential* to a solution is most often collected, or is available for collection, within the first 48 hours after the crime.

Over time, the elements may carry away or alter physical evidence, and rescuers or others with a legitimate need to access the scene may inadvertently lose, damage, or destroy the evidence. Personal accounts and witness statements are equally fragile because they depend on memory. These recollections may dim or alter over time, especially regarding an event that garners media attention.

Excepting the individual whose injuries or emotional condition renders them incapable of responding appropriately to questions, the more timely the interview, the better the information and intelligence derived.

A second, important consideration adding to the urgency of timely investigative action is the concern that the individuals responsible for the crime may commit another before being apprehended. This means that law enforcement will require access to the scene and relevant people—victims, witnesses, and responders—immediately or as soon as is practical. The need for effective coordination in this setting should be apparent.

Therefore, local emergency response systems should develop a predefined protocol to ensure smooth implementation of this process. Application of the incident management system (IMS) with the assignment of a law enforcement liaison will ensure that emergency response organizations involved with the incident will have a conduit for information sharing.

By having a policy in place that facilitates information exchange, emergency response organizations play a key role in assisting law enforcement to achieve its mission.

The Era of Microscopic Evidence

On December 21, 1988, at 6:56 P.M., radar lost Pan Am Flight 103. Residents of Lockerbie, Scotland, reported a thunder-like sound coming from a flaming object falling from the sky. Several minutes later, pieces of the aircraft and the bodies of its passengers began striking the earth.[3] So began one of the most challenging and intensive crime scene investigations in history.

Debris was strewn along two trails, one of which extended more than 130 kilometers. Two major portions of the wreckage fell on Lockerbie, killing 11 people on the ground. The total dead, including passengers, came to 270. For weeks, a cadre of law enforcement officers combed this vast area searching for evidence as small as a penny. Ultimately, a reconstruction of the aircraft and its contents led to the indictment of two men currently enjoying protection by the Libyan government.

This incredible feat of forensic investigation was made possible by the technological advances in crime scene examination and evidence collection during the past decade. Microscopy and computer-driven analytical models of every imaginable kind render the smallest items of evidence valuable. Especially in settings such as this, critical evidence can be carried away on the shoes and clothing of victims and responders or altered by the movement of people or material during rescue.

Clearly, consistent with our mutual public safety missions, the emergency response community can lend assistance with the investigation via a number of methods, first by developing policies and operation protocols that bring attention to the necessity of taking all reasonable measures to ensure crime scene preservation. Next, the emergency response community can encourage cooperation between emergency service and law enforcement organizations to construct and deliver an overview of what emergency service members should remain cognizant of at a terrorist scene and what events or materials they encounter that require immediate law enforcement attention. Finally, we should incorporate this protocol and training into local emergency exercises and test it.

Information, Intelligence, and Testimony

Information is the lifeblood of an investigation. The skilled investigator acts much like an astute clinician: able to view an array of signs and symptoms and define patterns that lead to the identification of a disease process. As the clinician searches for a first cause in disease, the investigator searches for the actors in a crime and reconstructs their activities. Often, a fact not immediately recognizable as important becomes critical to the success of an investigation as the process matures; therefore, investigators must have access to information in the broadest sense. Make no prejudgments regarding what to share.

When an individual provides probative information, we call it testimony. Testimony is evidence other than physical or documentary items, which, like all other evidence, remains subject to the court's scrutiny. Because of this scrutiny, investigators are compulsive documentarians. This obsessive attention to detail may appear inappropriate during a crisis, but we cannot overstate its importance. Investigators need timely and substantial access to rescuers and victims.

Our judicial system is part of the bedrock of the democratic process. A system of law governed by oversight from a disinterested judiciary and played out in an adversarial setting between the prosecution and the defense ensures the greatest possible degree of fairness: It acts as the arena in which the investigator and related work are tested. Imagine if your every action were subjected to the scrutiny of a cadre of professionals whose obligation under the law was to discredit all you did.

Investigators arrive at the crime scene with this perspective—and carry it throughout the investigative process. Investigators, therefore, find it necessary to thoroughly establish the basis for any conclusions in a way that will firmly convince a jury, withstand the attacks of the defense, and fall within the framework of rules and procedures acceptable to the court. This complex exercise in the best of circumstances becomes doubly difficult during a crisis. An understanding of these concerns by the consequence management community is essential to an effective collaboration.

Conclusion

Emergency service (consequence) and law enforcement (crisis) professionals must develop an appreciation of each other's roles and responsibilities. This relationship begins with dialogue that leads to understanding. Tabletop and field exercises should be ongoing. Continuing interaction fosters relationships that engender trust. Don't meet your counterparts for the first time at the scene of an act of terrorism.

Notes

[1] *Emergency Response to Terrorism.* U.S. Department of Justice, Office of Justice Programs, Bureau of Justice Assistance & Federal Emergency Management Agency, U.S. Fire Administration, National Fire Academy. 59; 1997.

[2] Ibid.

[3] Aircraft Accident Report No. 2/90 (EWC1094, UK Air Accidents Branch).

Appendix H
Tactical Ultraviolence

James P. Denney

Background

The United States is constantly in a state of limited warfare. The limited war is crime based, and the violence associated with it is covered with a thin sheath of rationality and societal acceptance (one expects criminals to be violent). The violence is usually a secondary act that is instrumental to, or a by-product of, criminal activity. Examples include takeover-style bank robberies where the immediate use of violence places the perpetrator in control of the environment, or the use of violence in the protection of ones drug market.

Working in contemporary emergency services has been likened to working in a limited war environment where resources respond into combat zones for the purpose of neutralizing the aggressor through capture or overriding force, retrieval of the wounded, and evacuation of the vulnerable population. Providers are often heard referring to particular incidents as being "like a war zone" and their experience analogous to the conditions of war they see on the nightly news.

Increasingly violent populations, coupled with a rise in criminal violence and easily obtainable modern weaponry, have combined to present emergency service workers with conditions usually reserved for veteran combatants. Fully automatic military-style weapons and explosive devices are not uncommon in the field today and may be encountered by a responding emergency service organization unprepared to manage the potential impact or consequence generated by these weapons.

Following the 1992 Los Angeles civil disturbance, a manual intended to heighten the awareness of emergency service workers regarding gangs, violence, and tactical response was developed for first responders. Material for the manual was provided by local and national law enforcement agencies and dealt primarily with working in the gang environment. Interest has been maintained on these issues since then, and public discussion of their impact on emergency service providers has been conducted.

Emerging Threat

A disturbing phenomenon has recently been noted that, following analysis, has resulted in the identification of an emerging threat to emergency service providers within the culture of criminal violence. The phenomenon has been identified as *tactical ultraviolence.*

Definition

Tactical ultraviolence is defined as the predetermined use of maximum violence in order to achieve control of the immediate environment, regardless of victim cooperation, level of environmental threat to the perpetrator, or the need to evade law enforcement or capture that results in physical or psychological injury or death to the victim(s).

Rationale

The rationale for employing tactical ultraviolence is to exert predatory control over the immediate environment through the creation of chaos and the infliction of terror, trauma, and death on presenting targets.

Tactical ultraviolence is never spontaneous. It is not intended to fit in the socially acceptable definition of rational violence, nor can it be associated with senseless acts of violence. Tactical ultraviolence differs from senseless violence in that it has a specific purpose (manifest violence) and is a component of criminal enterprise, whether for political purposes or personal gain; the purpose may include the achievement of the act itself or compel law-enforcement–assisted suicide.

Tactical ultraviolence may or may not be victim precipitated and may be initiated at the onset of criminal activity (it may be the sole activity), at any point during the activity, or following conclusion of the activity and is inclusive of all targets within range without regard to age, gender, or ethnicity.

Incidence

Since 1992, a steady increase in the use of tactical ultraviolence during the commission of crimes has been noted, even though there has been a concurrent drop in the overall violent crime rate according to the Department of Justice. However, today a single violent crime may involve one or several perpetrators and may result in several if not hundreds of victims.

The most common expression of tactical ultraviolence is the utilization of improvised explosive devices, designed for maximum impact, placed in high-volume public areas where broad patterns of injury can be achieved. Recent examples of tactical ultraviolence include:

- The Columbine High School incident
- The Oklahoma City bombing
- The 1996 Olympic bombing
- Clinic bombings in Atlanta (which were accompanied by secondary devices that targeted responding public safety agencies)
- The Empire State Building shooting in New York
- The botched bank robbery in Los Angeles
- The bank shootings in Detroit

Public Safety Concerns

Public safety personnel consistently operate in environments characterized by extensive damage, human injury, and limited resources. The most frustrating problem encountered during these events is the lack of accurate, timely information. Lack of information creates an air of uncertainty concerning what has happened or is likely to happen, coupled with a strong urge to take some action "before it is too late." The problem is deciding which action to take. Personnel who know what to do, when and how to do it, and who to report it to during an incident are invaluable. The need for clear management objectives, directed activity, and precise communication is never more evident than during these operations.

The implication is clear: Public safety risk managers must reassess current response strategies. Consideration must be given to existing vulnerability and the attendant consequences associated with exposure to acts of tactical ultraviolence or terrorism. Managers must develop exposure-reduction methodologies that include the use of ballistic garments, police escorts during tactical operations, and an awareness of potential secondary antipersonnel devices at the scene of bombings and incorporate them into their tactical operation plans.

A separate but equal concern is the significant volume of resources committed to incidents involving tactical ultraviolence that may result in a short-term reduction in the level of community protection. This includes medical community resources and capability.

Media Concerns

Consideration must also be given to the impact of a convergent press corps on these incidents, many of whom arrive before adequate public safety resources are on scene. In most cases, critical resources are diverted from public safety duties in order to control access to potentially hazardous environments by the press corps. Ironically, both the public and the press expect the emergency to be successfully mitigated and abated by capable emergency managers in minimal time regardless of incident type.

In regions of media concentration, such as large metropolitan areas, air space utilization by an airborne press corps may present a risk multiplier to ground resources. Therefore, the application of tactical air space restrictions should be considered as a precautionary measure until the environment is stabilized.

An additional concern is that the media may provide real-time intelligence to perpetrators in a manner similar to that experienced during Desert Storm. The "CNN Effect" includes the broadcasting of tactical information to the viewing audience and is a source of intelligence utilized internationally by the threat medium. The media may inadvertently provide information regarding the emergency extraction of public safety personnel, civilian evacuation, and SWAT team tactics directly to the viewing audience and to targeted perpetrators, thereby increasing provider vulnerability.

Appendix I[1]
FEMA Backgrounder: Terrorism

Emergency Information

1. Most terrorist incidents in the United States have been bombing attacks, involving detonated and undetonated explosive devices, tear gas and pipe and fire bombs.
2. The effects of terrorism can vary significantly from loss of life and injuries to property damage and disruptions in services such as electricity, water supply, public transportation and communications.
3. One way governments attempt to reduce our vulnerability to terrorist incidents is by increasing security at airports and other public facilities. The U.S. government also works with other countries to limit the sources of support for terrorism.

What Is Terrorism?

Terrorism is the use of force or violence against persons or property in violation of the criminal laws of the United States for purposes of intimidation, coercion or ransom. Terrorists often use threats to create fear among the public, to try to convince citizens that their government is powerless to prevent terrorism, and to get immediate publicity for their causes.

The Federal Bureau of Investigation (FBI) categorizes terrorism in the United States as one of two types—domestic terrorism or international terrorism.

Domestic terrorism involves groups or individuals whose terrorist activities are directed at elements of our government or population without foreign direction.

International terrorism involves groups or individuals whose terrorist activities are foreign-based and/or directed by countries or groups outside the United States or whose activities transcend national boundaries.

Biological and Chemical Weapons

Biological agents are infectious microbes or toxins used to produce illness or death in people, animals, or plants. Biological agents can be dispersed as aerosols or airborne particles. Terrorists may use biological agents to contaminate food or water because they are extremely difficult to detect. Chemical agents kill or incapacitate people, destroy livestock or ravage crops. Some chemical agents are odorless and tasteless and are difficult to detect. They can have an immediate effect (a few seconds to a few minutes) or a delayed effect (several hours to several days).

Biological and chemical weapons have been used primarily to terrorize an unprotected civilian population and not as a weapon of war. This is because of fear of retaliation and the likelihood that the agent would contaminate the battlefield for a long period of time. The Persian Gulf War in 1991 and other confrontations in the Middle East were causes for concern in the United States regarding the possibility of chemical or biological warfare. While no incidents occurred, there remains a concern that such weapons could be involved in an accident or be used by terrorists.

Facts about Terrorism

- On February 29, 1993, a bombing in the parking garage of the World Trade Center in New York City resulted in the deaths of five people and thousands of injuries. The bomb left a crater 200 by 100 feet wide and five stories deep. The World Trade Center is the second largest building in the world and houses 100,000 workers and visitors each day.
- The Department of Defense estimates that as many as 26 nations may possess chemical agents and/or weapons and an additional 12 may be seeking to develop them.
- In recent years the largest number of terrorist strikes have occurred in the western states and Puerto Rico. Attacks in Puerto Rico accounted for about 60 percent of all terrorist incidents between 1983 and 1991 that occurred on United States territory.
- The Central Intelligence Agency reports that at least 10 countries are believed to possess or be conducting research on biological agents for weaponization.

Terrorism in the United States

- In the United States, most terrorist incidents have involved small extremist groups who use terrorism to achieve a designated objective. Local, state and federal law enforcement officials monitor suspected terrorist groups and try to prevent or protect against a suspected attack.

Additionally, the U.S. government works with other countries to limit the sources of support for terrorism.

- A terrorist attack can take several forms, depending on the technological means available to the terrorist, the nature of the political issue motivating the attack, and the points of weakness of the terrorist's target. Bombings are the most frequently used terrorist method in the United States. Other possibilities include an attack at transportation facilities, an attack against utilities or other public services or an incident involving chemical or biological agents.

- Terrorist incidents in this country have included bombings of the World Trade Center in New York City, the United States Capitol Building in Washington, D.C., and Mobil Oil corporate headquarters in New York City.

NOTES

[1]www.fema.gov/library/terror.htm

Appendix J
Federal Agencies with Responsibilities for WMD and Terrorism Issues Charts (2001)

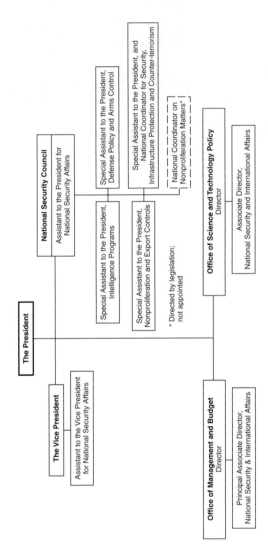

FIGURE J-1

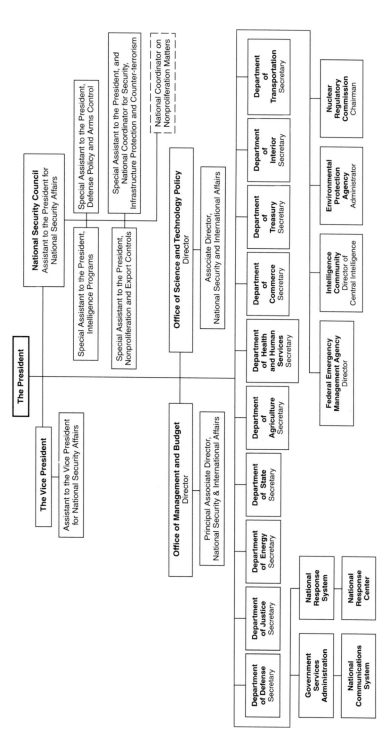

FIGURE J-2

357

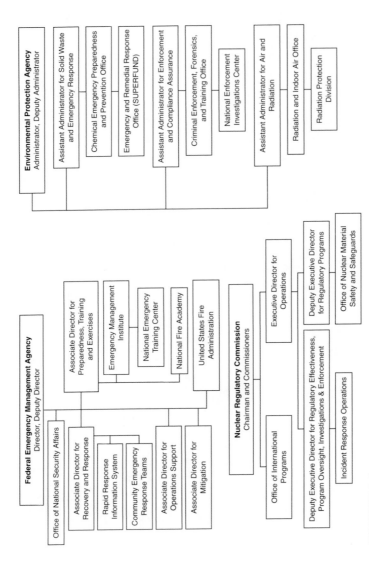

Environmental Protection Agency
Administrator, Deputy Administrator

- Assistant Administrator for Solid Waste and Emergency Response
  - Chemical Emergency Preparedness and Prevention Office
  - Emergency and Remedial Response Office (SUPERFUND)
- Assistant Administrator for Enforcement and Compliance Assurance
  - Criminal Enforcement, Forensics, and Training Office
  - National Enforcement Investigations Center
- Assistant Administrator for Air and Radiation
  - Radiation and Indoor Air Office
    - Radiation Protection Division

Federal Emergency Management Agency
Director, Deputy Director

- Office of National Security Affairs
- Associate Director for Recovery and Response
  - Rapid Response Information System
  - Community Emergency Response Teams
- Associate Director for Operations Support
- Associate Director for Mitigation
- Associate Director for Preparedness, Training and Exercises
  - Emergency Management Institute
  - National Emergency Training Center
  - National Fire Academy
  - United States Fire Administration

Nuclear Regulatory Commission
Chairman and Commissioners

- Office of International Programs
- Deputy Executive Director for Regulatory Effectiveness, Program Oversight, Investigations & Enforcement
  - Incident Response Operations
- Executive Director for Operations
- Deputy Executive Director for Regulatory Programs
  - Office of Nuclear Material Safety and Safeguards

FIGURE J-3

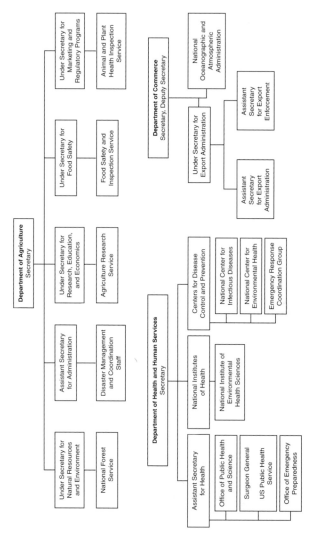

FIGURE J-4

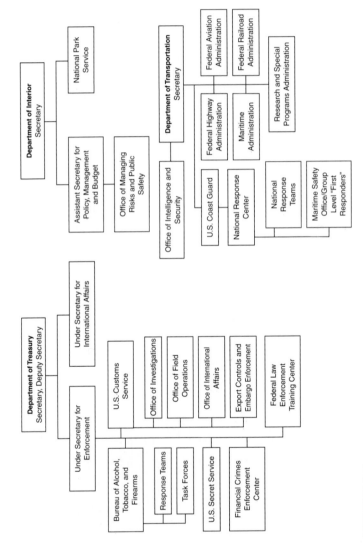

FIGURE J-5

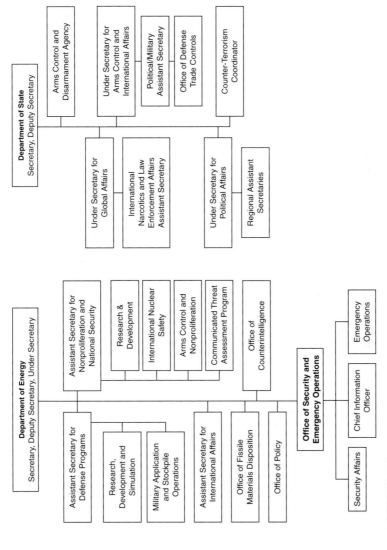

FIGURE J-6

361

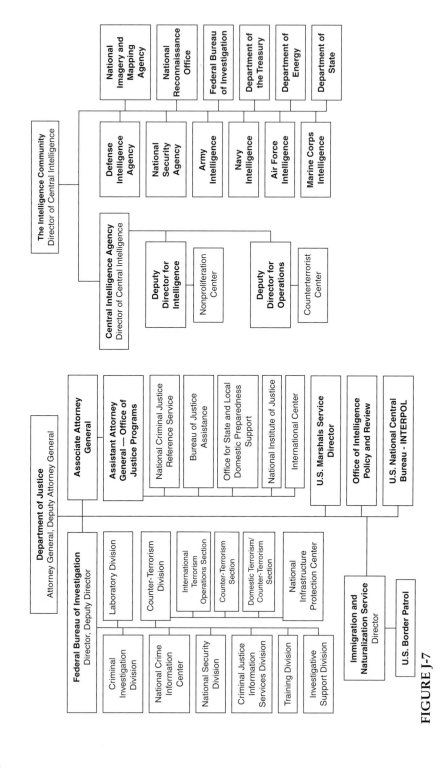

FIGURE J-7

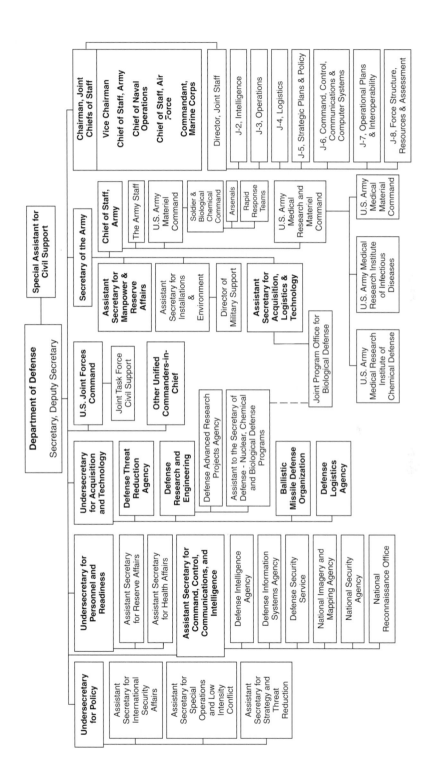

FIGURE J-8

Appendix K
HMEP Special Topics— Terrorism and Illicit Use of Hazardous Materials

Responders to incidents involving terrorism may encounter unusual chemicals or biological agents or unusual uses of those hazardous materials that have not been addressed thoroughly in current hazardous materials training. For example, nuclear response training for first responders has traditionally been for major catastrophes (i.e., nuclear war and power plant emergencies), and not for small isolated terrorist events. As a second example, the high risk chemical and biological agents that might be involved in terrorist incidents may require unusual protocols and procedures for patient decontamination and treatment that are not addressed in current EMS training. As an additional example, some of the materials that may be involved have unusual dispersal characteristics that responders may not be trained to accommodate when determining of safe perimeters and public protection/evacuation requirements at the incident.

Current training for community emergency planning and preparedness strategies and existing response plans use risk predictions based upon known vulnerabilities and hazard identifications, such as commodity flow studies, fixed facility storage of material, etc. This allows responders to plan for the response prior to an emergency and to assess whether the response capability and resources in the area are sufficient to meet potential emergencies. However, terrorist and other illicit acts involving hazardous materials may occur in untraditional locations that are not normally thought of as high risk hazardous materials locations, such as public gathering places or remote transportation areas. As a result, current protocols for allocating response resources and preparing for hazardous materials emergencies may not allow sufficient response capability for terrorist-related hazardous materials emergencies.

Finally, hazardous materials emergencies involving terrorism or other illicit use of hazardous materials may involve additional and unusual risks to responders beyond those presented by the hazardous materials themselves. Public sector responders may be at additional risk due to secondary releases targeted at responders, primary releases that intentionally create extremely high risk rescue situations, and even to primary releases targeted at public response facilities.

The Challenge to Public Sector Response and Planning Organizations

Public sector response and planning organizations should examine all facets of their response system to ensure preparedness for response to incidents of terrorism and illicit use of hazardous materials. This review should include existing plans, operating procedures, equipment, training and exercises.

Plans should include:

- Consistency and interface with plans from all levels of government, specifically the Federal Response Plan (FRP) and the FRP Terrorism Annex;
- Presidential Decision Directive 39, specifically examining responsibility for crisis management and consequence management in their community;
- Unified command operations with all levels of government; and
- Thorough, in-depth plans for response to mass casualty chemical incidents.

Operating procedures should include:

- Command post operations including command post security, responder accountability, and on-site responder identification;
- Protection against secondary explosive devices and other secondary events;
- Responsibility for and support to crime scene operations, evidence collection and chain of custody; and
- Emergency decontamination at mass casualty chemical incidents.

Equipment should be evaluated to ensure appropriate protection and detection of nuclear, chemical and biological agents (NBC). Existing training, including annual refresher training, for all responders should be enhanced to include competencies for response to incidents involving terrorism or other illicit use of hazardous materials. Finally, agencies should identify a person or persons within their organization as their point of contact for issues regarding terrorism and the illicit use of hazardous materials. These persons should interface with appropriate response agencies to include EMS, fire, haz mat, and law enforcement.

Training Strategies

Training for public sector employees who respond to hazardous materials emergencies at the Awareness, Operations, Technician, EMS, and Incident Commander levels should include thorough instruction to prepare those responders to safely and efficiently respond to hazardous materials emergencies involving terrorism or other illicit use of hazardous materials.

For some metropolitan areas, the Department of Defense, the Department of Justice, and the United States Public Health Service are developing programs to provide in depth training and logistical support to assist public sector response organizations in preparing local responders to better prepare for terrorist-related hazardous materials emergencies. However, for most state, tribal, territory and local training systems outside these high risk metropolitan areas, training for response to such incidents should be addressed as additional competencies within current hazardous materials responder curriculums and training delivery systems.

This additional hazardous materials response training can be accomplished either through additional courses or through enhancement of current hazardous materials courses. Those training systems who have sufficient resources to do so and who would like to add additional courses to their curriculums should be advised that several courses will soon be available in this training area. Training programs are currently under development by several federal training providers, such as the National Fire Academy, and several state training organizations, such as the Virginia Department of Emergency Services, that address these competencies. Information regarding these and other programs and their availability for use will be provided as soon as available to HMEP grantees under separate cover through the HMEP response course assessment and catalog mechanism.

For many training providers, insufficient resources and limited access to responder training time many render impractical the use of additional, supplemental responder training courses addressing terrorism competencies. In that case, training providers may wish to consider addressing the needed training through modification to and enhancement of existing courses within their curriculums. As training providers develop updated modules and training resource materials for use in updating existing courses, information on these materials will be provided to HMEP grantees when available.

Responder Competencies

The National Fire Protection Association has issued a proposed tentative interim amendment to NFPA 472 that articulates additional responder competencies at the Awareness, Operations, Technician and Incident Commander levels for response to hazardous materials emergencies involving terrorism or other illicit use of hazardous materials. These competencies have been included in the recommended portions of the response training guidelines, *Guidelines for Public Sector Hazardous Materials Training,* and are also displayed later.

The domestic terrorism sub-committee for NFPA Standard 473, EMS Responder Levels 1 and 2, is also preparing supplemental competencies for EMS Level 1 and Level 2 responders to terrorist hazardous materials incidents. These competencies have been drafted and will be undergoing extensive national review and comment prior to their implementation. These draft competencies are displayed below for reference but will not be added to the recommended portions of the response training guidelines, Guidelines for Public Sector Hazardous Materials Training, until the competencies have been finalized by NFPA.

Note that the following are intended to supplement existing competencies for response as presented in NFPA 472 and 473, 1997 edition, and should be reviewed in context with the full set of competencies for each response level in order to properly depict the complete responder competency requirements for response to hazardous materials incidents involving terrorism or other illicit use of hazardous materials.

First Responder at the Awareness Level

Analyzing the Incident: Detecting the Presence of Hazardous Materials

(new) 2-2.1.13
Identify types of locations that may become targets for criminal or terrorist activity using hazardous materials.
*The following are some examples of locations:
  a. Public assembly
  b. Public buildings
  c. Mass transit systems
  d. Places with high economic impact
  e. Telecommunications facilities
  f. Places with historical or symbolic significance

(new) 2-2.1.14
Identify at least four indicators of possible criminal or terrorist activity involving hazardous materials.
*The following are some examples of indicators:
  a. Hazardous materials or lab equipment that is not relevant to the occupancy
  b. Intentional release of hazardous materials
  c. Unexplained patterns of sudden onset illnesses or deaths
  d. Unusual odors or tastes
  e. Unexplained signs of skin, eye, or airway irritation
  f. Unusual security, locks, bars on windows, covered windows, and barbed wire
  g. Unexplained vapor clouds, mists, and plumes
  h. Patients twitching, tightness in chest, sweating, pin-point pupils (miosos), runny nose (rhinorrhea), and nasal vomiting.

Analyzing the Incident: Initiating Protective Actions

(new) 2-4.1.6
Identify the specific actions necessary when an incident is suspected to involve criminal or terrorist activity.
*The following are some examples of action:
  a. Communicate the suspicion during the notification process
  b. Isolate potentially exposed people
  c. Document the initial observation
     (Recommended by authors but not included with NFPA T.I.A.)
  d. Attempt to preserve evidence while performing operational duties

Analyzing the Incident: Initiating the Notification Process

(add to) 2-4.2
Given either a facility or transportation scenario of hazardous materials, with or without criminal or terrorist activities, the first responder at the awareness level shall identify the appropriate initial notifications to be made and how to make them, consistent with the local emergency response plan or the organization's standard operating procedures.

First Responder at the Operations Level

Analyzing the Incident: Surveying the Hazardous Materials Incident

(new) 3-2.1.6
Identify at least three additional hazards that could be associated with an incident involving criminal or terrorist activity.

*The following are some examples of hazards:
   a. Secondary events intended to incapacitate emergency responders
   b. Armed resistance
   c. Use of weapons
   d. Booby traps
   e. Secondary contamination from handling patients
       (Recommended by authors but not included with NFPA T.I.A.)
   f. Hostage barricade situations

Analyzing the Incident: Collecting Hazard and Response Information

(new) 3-2.2.6
Identify the type of assistance provided by the federal defense authorities, such as the Defense Logistics agency and the U.S. Army Operations Center, with respect to criminal or terrorist activities involving hazardous materials.

(new) 3-2.2.6.1
Identify the procedure for contacting federal authorities as specified in the local emergency response plan (ERP) or the organization's standard operating procedure (SOP).

Analyzing the Incident: Predicting the Behavior of a Material and its Container

(new) 3-2.3.9
Given the following types of warfare agents, identify the corresponding DOT hazard class and division:
   a. Nerve agents
   b. Vesicants (blister agents)
   c. Blood agents
   d. Choking agents
   e. Irritants (riot control agents)
   f. Biological agents and toxins

*Some examples are as follows:

		DOT Hazard Class
a.	Nerve agents	
	Tuban (GA)	6.1
	Sarin (GB)	6.1
	Soman (GD)	6.1
	V agent (VX)	6.1
b.	Vesicants (blister agents)	
	Mustard (H)	6.1
	Distilled mustard (HD)	6.1

Nitrogen mustard (HN)	6.1	
Lewisite (L)	6.1	
c. Blood agents		
Hydrogen cyanide (AC)	6.1	
Cyanogen chloride (CK)	2.3	
d. Choking agents		
Chlorine (CL)	2.3	
Phosgene (CG)	2.3	
e. Irritants		
CS	6.1	
CR	6.1	
CN	6.1	
OC	2.2 (subsequent risk 6.1)	
f. Biological agents and toxins		
Anthrax	6.2	
Mycotoxin	6.1 or 6.2	
Plague	6.2	
Tularemia	6.2	

Identifying Planning the Response: Emergency Decontamination Procedures

(new) 3-3.4.5
Describe the procedure listed in the local ERP or the organization's SOP for decontamination of a large number of people exposed to hazardous materials.

Implementing the Planned Response: Establishing and Enforcing Scene Control Procedures

(add to) 3-4.1.6
Identify the items to be considered in a safety briefing prior to allowing personnel to work at the following:
    a. Hazardous materials incident
    b. Hazardous materials incident with criminal or terrorist activities
*The following are some examples of items to be considered in a safety briefing for criminal or terrorist related incidents:
    a. Secondary events intended to incapacitate emergency responders
    b. Armed resistance
    c. Use of weapons
    d. Booby traps
    e. Secondary contamination from handling patients

Implementing the Planned Response: Performing Defensive Control Actions

(new) 3-4.4.6

Describe procedures, such as those listed in the local Emergency Response Plan or the organization's Standard Operating Procedures, to preserve evidence at hazardous materials incidents involving suspected criminal or terrorist acts.

Technician

Analyzing the Incident: Surveying the Hazardous Materials Incident

(new) 4.2.1.1.6

For each of the following, describe a method that can be used to detect them:
 a. Nerve agents
 b. Vesicants (blister agents)
 c. Biological agents and toxin
 d. Irritants (riot control agents)

Analyzing the Incident: Describing the Condition of the Container Involved in the Incident

(add to) 4-2.2.2

Describe the following terms and explain their significance in the risk assessment process: Acid, caustic; (b) Air reactivity; (c) Boiling point; (d) Catalyst; (e) Chemical interactions; (f) Chemical reactivity; (g) Compound, mixture; (h) Concentration; (i) Corrosivity (pH); (j) Critical temperatures and pressure; (k) Expansion ratio; (l) Flammable (explosive) range (LEL & UEL); (m) Fire point; (n) Flash point; (o) Halogenated hydrocarbon; (p) Ignition (autoignition) temperature; (q) Inhibitor; (r) Instability; (s) Ionic & covalent compounds; (t) Maximum safe storage temperature (MSST); (u) Melting point/freezing point; (v) Miscibility; (w) Organic and inorganic; (x) Oxidation potential; (y) pH; (z) Physical state (solid, liquid, gas); (aa) Polymerization; (bb) Radioactivity; (cc) Saturated, unsaturated, and aromatic hydrocarbons; (dd) Self-accelerating decomposition temperature (SADT); (ee) Solution, slurry; (ff) Specific gravity; (gg) Strength; (hh) Sublimation; (ii) Temperature of product; (jj) Toxic products of combustion; (kk) Vapor density; (ll) Vapor pressure; (mm) Viscosity; (nn) Volatility; (oo) Water reactivity; (pp) Water solubility; *(qq) Nerve agents; (rr) Vesicants (blister agents); (ss) Biological agents and toxins; and (tt) Irritants (riot control agents).*

(new) 4-2.3.1.6

Demonstrate a method for collecting samples of the following:
 a. liquid
 b. solid
 c. gas

Planning the Response: Developing a Plan of Action

(new) 4-3.5.6
Identify the procedures, equipment, and safety precautions for collecting legal
evidence at hazardous materials incidents.

Incident Commander

Analyzing the Incident: Estimating Potential Outcomes

(new) 5-2.2.4
Describe the health risks associated with the following:
   a. Nerve agents
   b. Vesicants (blister agents)
   c. Blood agents
   d. Choking agents
   e. Biological agents and toxins
   f. Irritants (riot control agents)
*Some examples are as follows (text added to NFPA 472 by HMEP author team):
   a. Nerve agents
   Liquids of low volatility that are rapidly absorbed through the eyes, lungs, or
   skin. They are highly toxic and have a NFPA 704 rating of 4/1/1 (tabun has
   a rating of 4/2/1). These materials inhibit acetylcholinesterase (the enzyme
   that removes the acetyl-choline after a nerve impulse has been transmitted)
   in tissue, and their effects are caused by the resulting excess acetylcholine
   (the chemical that carries nerve impulses from one nerve cell to another).
   Health effect are rapid on set with symptoms of organophosphate (pesti-
   cide) poisoning.

Common Name	NFPA 704	Military Abbreviation	PEL/TWA mg/m³	LD50 (mg min/m³)
Sarin	411	GB	0.0001	70
Soman	411	GD	0.00003	70
Tabun	421	GA	0.0001	133
V agent	411	VX	0.0001	10 (percutaneous) 30 (vapor)

   b. Vesicants (blister agents)
   Liquids of low volatility that are rapidly absorbed through tissue. They are
   highly toxic and have a NFPA 704 rating of 4/1/1. These materials cause severe burns to the skin, eyes, and tissue in the respiratory tract. Systemic poisoning can occur if significant exposure occurs,

Common Name	NFPA 704	Military Abbreviation	PEL/TWA mg/m³	LD50 (mg min/m³)
Mustard	411	H, HD	0.003	1500
Lewisite	411	L	0.003	1000–1500

c.   Blood Agents

Liquids under pressure that can interfere with the blood's ability to transfer oxygen to the cells. They are highly toxic and have a NFPA 704 rating of 4/4/2. Health effects are rapid onset of difficulty breathing, vomiting and headache.

d.   Choking Agents

Liquid under pressure that cause severe irritation to human tissue. Damage to respiratory tissue can result in pulmonary edema, congestive heart failure and death. These materials are highly toxic industrial chemicals.

e.   Biological agents and toxins

Biological Agents are generally divided into three groups:

Bacteria—single-celled organisms which cause a variety of diseases in animals, plants, and humans. They may also produce extremely potent toxins inside the body. Examples are Anthrax, Plague, Tularemia and Q fever. These agents show exposure symptoms in a period of 1–10 days and have an associated high fatality rate.

Viruses—much smaller than bacteria, and work inside individual cells. Examples are Smallpox, Venezuelan Equine Encephalitis and Viral Hemorrhagic Fever. Symptoms of exposure can be sudden to 1–3 days. Viral Hemorrhagic Fever is the most toxic of these viruses and is almost always fatal.

Toxins—potent poisons produced by a variety of living organisms including bacteria, plants, and animals. Examples are Botulinum Toxin, Staphylococcal Enterotoxin B, Ricin and Mycotoxins. Symptoms of exposure are 3 hours to 24 hours. These materials are the most toxic known substances, Botulinum Toxin is 100,000 times more toxic than Sarin, one of the well known opranophosphate nerve agents.

Common Name	Days/Latency	Fatal
Anthrax	1–5	Yes
Botulism	2–3	Yes
Cholera	2–5	Yes
Encephalitis	2–5	Yes
Plague	1–3	Yes
Tularemia	1–10	Yes

f.   Irritants (riot control agents)

solids with low vapor pressure and are dispersed in fine particles or in solution. Immediate onset of pain, burning, and irritation of exposed mucous membranes and skin. Exposure to these agents is almost never fatal to the healthy individual.

Planning the Response: Approving the Level of Personal Protective Equipment

(new) 5-3.3.5

Identify the limitations of military chemical/biological protective clothing.

Planning the Response: Developing a Plan of Action

(add to) 5-3.4.3

Given the local emergency response plan and/or the organization's standard operating procedures, identify which agency will perform the following:

- **(a)** Receive the initial notification
- **(b)** Provide secondary notification and activation of response agencies
- **(c)** Make ongoing assessments of the situation
- **(d)** Command on-scene personnel (incident management system)
- **(e)** Coordinate support and mutual aid
- **(f)** Provide law enforcement and on-scene security (crowd control)
- **(g)** Provide traffic control and rerouting
- **(h)** Provide resources for public safety protective action (evacuation or shelter in-place)
- **(i)** Provide fire suppression services when appropriate
- **(j)** Provide on-scene medical assistance (ambulance) and medical treatment (hospital)
- **(k)** Provide public notification (warning)
- **(l)** Provide public information (news media statements)
- **(m)** Provide on-scene communications support
- **(n)** Provide emergency on-scene decontamination when appropriate
- **(o)** Provide operational-level hazard control services
- **(p)** Provide technician-level hazard mitigation services
- **(q)** Provide environmental remedial action ("cleanup") services
- **(r)** Provide environmental monitoring
- **(s)** Implement on-site accountability
- **(t)** On-site responder identification
- **(u)** Command post security
- **(v)** Crime scene investigation
- **(w)** Evidence collection and sampling

Terminating the Incident: Reporting and Documenting the Hazardous Materials Incident

(new) 5-6.4.6

Identify the procedures required for legal documentation and chain of custody/ continuity described in the organization's standard operating procedure or the local emergency operating plan.

> NOTE: The domestic terrorism sub-committee for NFPA Standard 473, EMS Responder Levels 1 and 2, is preparing supplemental competencies for EMS Level 1 and Level 2 responders to terrorist hazardous materials incidents. These competencies have been drafted and will be undergoing extensive national review and comment prior to their implementation. These draft competencies are displayed below for reference but will not be added to the recommended portions of the response training guidelines, Guidelines for Public Sector Hazardous Materials Training, until the competencies have been finalized by NFPA.

EMS Level 1

Analyzing the Incident

(new) 2-2.3

Identify types of locations that may become targets for criminal or terrorist activity using hazardous materials.

*The following are some examples of locations:
  a. Public assembly
  b. Public buildings
  c. Mass transit systems
  d. Places with high economic impact
  e. Telecommunications facilities
  f. Places with historical or symbolic significance

(new) 2-2.4

Identify at least four indicators of possible criminal or terrorist activity involving hazardous materials.

*The following are some examples of indicators:
  a. Hazardous materials or lab equipment that is not relevant to the occupancy
  b. Intentional releases of hazardous materials
  c. Unexplained patterns of sudden onset of similar, non-traumatic illnesses or deaths. Pattern may be geographic, by employer, or other
  d. Unusual odors or tastes
  e. Unexplained signs of skin, eye or airway irritation
  f. Unusual security, locks, bars on windows, covered windows, and barbed wire

g.   Unexplained vapor clouds, mists, and plumes
h.   Patients twitching, tightness in chest, sweating, pin-point pupils (miosis) runny nose (rhinorrhea), and nausea/vomiting

Planning the Response

(new) 2-3.5
Identify the procedures, equipment, and safety precautions for collecting legal evidence at hazardous materials incidents.

Implementing the Planned Response

(add to) 2-4.2(b)
List the common signs and symptoms and describe the EMS treatment protocols for the following:
a.   Corrosives (e.g., acid, alkali)
b.   Pulmonary irritants (e.g., ammonia, chlorine)
c.   Pesticides (e.g., organophosphates, carbamates)
d.   Chemical asphyxiants (e.g., cyanide, carbon monoxide)
e.   Hydrocarbon solvents (e.g., xylene, methylene chloride)
f.   Nerve agents
g.   Vesicants (blister agents)
h.   Blood agents (cyanide)
i.   Choking agents (pulmonary agents)
j.   Irritants (riot control agents)
k.   Biological agents and toxins
l.   Incapacitating agents (BZ, LSD)

(new) 2-4.4
Identify the specific actions necessary when an incident is suspected to involve criminal or terrorist activity
*The following are some examples of action:
a.   Communicate the suspicion during the notification process
b.   Isolate potentially exposed people
c.   Document the initial observation
d.   Attempt to preserve evidence while performing operational duties

(new) 2-4.5
Given either a facility or transportation scenario of hazardous materials with or without criminal or terrorist activities, the Level 1 EMS/HM Responder shall identify the appropriate initial notifications to be made and how to make them, consistent with the local emergency response plan or the organization's standard operating procedures.

(new) 2-4.6

Given an incident involving the suspicion of a biological warfare agent, the EMT shall

    a.   Identify the correct body substance isolation procedures to be followed

    b.   Identify the proper decontamination procedures in accordance with their standard operating procedures or guidelines

    c.   Identify the necessary post-exposure reporting

*This is important to facilitate post-exposure prophylaxis when available

EMS Level 2

Analyzing the Incident

(new) 3-2.3

Given an emergency involving potential domestic terrorism, the Level II Responder shall determine the availability of basic tools for identification of the substance, detection devices appropriate to the substance, and where these detection devices are available locally.

Appendix L
Terrorism Incident Annex (Federal Response Plan)

Signatory Agencies: Department of Defense
Department of Energy
Department of Health and Human Services
Department of Justice
Federal Bureau of Investigation
Environmental Protection Agency
Federal Emergency Management Agency

I. Introduction

Presidential Decision Directive 39 (PDD-39), U.S. Policy on Counterterrorism, establishes policy to reduce the Nation's vulnerability to terrorism, deter and respond to terrorism, and strengthen capabilities to detect, prevent, defeat, and manage the consequences of terrorist use of weapons of mass destruction (WMD). PDD-39 states that the United States will have the ability to respond rapidly and decisively to terrorism directed against Americans wherever it occurs, arrest or defeat the perpetrators using all appropriate instruments against the sponsoring organizations and governments, and provide recovery relief to victims, as permitted by law.

Responding to terrorism involves instruments that provide crisis management and consequence management. "Crisis management" refers to measures to identify, acquire, and plan the use of resources needed to anticipate, prevent, and/or resolve a threat or act of terrorism. The Federal Government exercises primary authority to prevent, preempt, and terminate threats or acts of terrorism and to apprehend and prosecute the perpetrators; State and local governments provide assistance as required.

Crisis management is predominantly a law enforcement response. "Consequence management" refers to measures to protect public health and safety, restore essential government services, and provide emergency relief to governments, businesses, and individuals affected by the consequences of terrorism. State and local governments exercise primary authority to respond to the consequences of terrorism; the Federal Government provides assistance as required. Consequence management is generally a multifunction response coordinated by emergency management.

Based on the situation, a Federal crisis management response may be supported by technical operations, and by Federal consequence management, which may operate concurrently (see Figure L-1). "Technical operations" include actions to identify, assess, dismantle, transfer, dispose of, or decontaminate personnel and property exposed to explosive ordnance or WMD.

A. Purpose

The purpose of this annex is to ensure that the Federal Response Plan (FRP) is adequate to respond to the consequences of terrorism within the United States, including terrorism involving WMD. This annex:

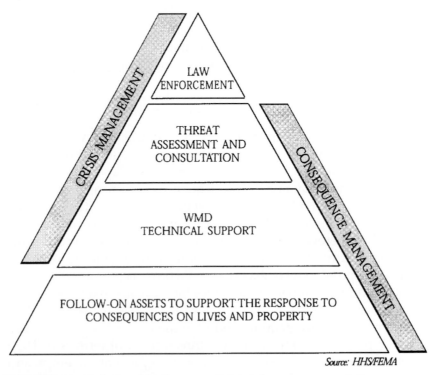

Source: HHS/FEMA

FIGURE L-1 Relationship Between Crisis Management and Consequence Management

1. Describes crisis management. Guidance is provided in other Federal emergency operation plans;
2. Defines the policies and structures to coordinate crisis management with consequence management; and
3. Defines consequence management, which uses the FRP process and structure, supplemented as necessary by resources normally activated through other Federal emergency operations plans.

B. Scope

This annex:

1. Applies to all threats or acts of terrorism within the United States that the White House determines require a response under the FRP;
2. Applies to all Federal departments and agencies that may be directed to respond to the consequences of a threat or act of terrorism within the United States; and
3. Builds upon the process and structure of the FRP by addressing *unique* policies, situations, operating concepts, responsibilities, and funding guidelines required for response to the consequences of terrorism.

II. Policies

A. PDD-39 validates and reaffirms existing lead agency responsibilities for all facets of the U.S. counterterrorism effort.
B. The Department of Justice is designated as the lead agency for threats or acts of terrorism within U.S. territory. The Department of Justice assigns lead responsibility for operational response to the Federal Bureau of Investigation (FBI). Within that role, the FBI operates as the on-scene manager for the Federal Government. It is FBI policy that crisis management will involve only those Federal agencies requested by the FBI to provide expert guidance and/or assistance, as described in the PDD-39 Domestic Deployment Guidelines (classified) and the FBI WMD Incident Contingency Plan.
C. The Federal Emergency Management Agency (FEMA) is designated as the lead agency for consequence management within U.S. territory. FEMA retains authority and responsibility to act as the lead agency for consequence management throughout the Federal response. It is FEMA policy to use FRP structures to coordinate all Federal assistance to State and local governments for consequence management.
D. To ensure that there is one overall Lead Federal Agency (LFA), PDD-39 directs FEMA to support the Department of Justice (as delegated to

the FBI) until the Attorney General transfers the overall LFA role to FEMA. FEMA supports the overall LFA as permitted by law.

III. Situation

A. Conditions

1. FBI assessment of a potential or credible threat of terrorism within the United States may cause the FBI to direct other members of the law enforcement community and to coordinate with other Federal agencies to implement a pre-release response.
   a. FBI requirements for assistance from other Federal agencies will be coordinated through the Attorney General and the President, with coordination of National Security Council (NSC) groups as warranted.
   b. FEMA will advise and assist the FBI and coordinate with the affected State and local emergency management authorities to identify potential consequence management requirements and with Federal consequence management agencies to increase readiness.
2. An act that occurs without warning and produces major consequences may cause FEMA to implement a post-release consequence management response under the FRP. FEMA will exercise its authorities and provide concurrent support to the FBI as appropriate to the specific incident.

B. Planning Assumptions

1. No single agency at the local, State, Federal, or private-sector level possesses the authority and expertise to act unilaterally on many difficult issues that may arise in response to a threat or act of terrorism, particularly if WMD are involved.
2. An act of terrorism, particularly an act directed against a large population center within the United States involving WMD, may produce major consequences that would overwhelm the capabilities of many local and State governments almost immediately.
3. Major consequences involving WMD may overwhelm existing Federal capabilities as well, particularly if multiple locations are affected.
4. Local, State, and Federal responders will define working perimeters that may overlap. Perimeters may be used to control access to the area, target public information messages, assign operational sectors among responding organizations, and assess potential effects on the population and the environment. Control of these perimeters may

be enforced by different authorities, which will impede the overall response if adequate coordination is not established.

5. If appropriate personal protective equipment is not available, entry into a contaminated area (i.e., a hot zone) may be delayed until the material dissipates to levels that are safe for emergency response personnel. Responders should be prepared for secondary devices.

6. Operations may involve geographic areas in a single State or multiple States, involving responsible FBI Field Offices and Regional Offices as appropriate. The FBI and FEMA will establish coordination relationships as appropriate, based on the geographic areas involved.

7. Operations may involve geographic areas that spread across U.S. boundaries. The Department of State is responsible for coordination with foreign governments.

IV. Concept of Operations

A. Crisis Management

Source: FBI, National Security Division, Domestic Terrorism/Counterterrorism Planning Section

1. PDD-39 reaffirms the FBI's Federal lead responsibility for crisis management response to threats or acts of terrorism that take place within U.S. territory or in international waters and that do not involve the flag vessel of a foreign country. The FBI provides a graduated, flexible response to a range of incidents, including:

   a. A credible threat, which may be presented in verbal, written, intelligence-based, or other form;

   b. An act of terrorism that exceeds the local FBI field division's capability to resolve;

   c. The confirmed presence of an explosive device or WMD capable of causing a significant destructive event, prior to actual injury or property loss;

   d. The detonation of an explosive device, utilization of a WMD, or other destructive event, with or without warning, that results in *limited* injury or death; and

   e. The detonation of an explosive device, utilization of a WMD, or other destructive event, with or without warning, that results in *substantial* injury or death.

2. The FBI notifies FEMA and other Federal agencies providing direct support to the FBI of a credible threat of terrorism. The FBI initiates a threat assessment process that involves close coordination with Federal agencies with technical expertise, in order to determine the

viability of the threat from a technical as well as tactical and behavioral standpoints.

3. The FBI provides initial notification to law enforcement authorities within the affected State of a threat or occurrence that the FBI confirms as an act of terrorism.

4. If warranted, the FBI implements an FBI response and simultaneously advises the Attorney General, who notifies the President and NSC groups as warranted, that a Federal crisis management response is required. If authorized, the FBI activates multiagency crisis management structures at FBI Headquarters, the responsible FBI Field Office, and the incident scene (see Figure L-2). Federal agencies requested by the FBI, including FEMA, will deploy a representative(s) to the FBI Headquarters Strategic Information and Operations Center (SIOC) and take other actions as necessary and appropriate to support crisis management. (The FBI provides guidance on the crisis management response in the FBI WMD Incident Contingency Plan.)

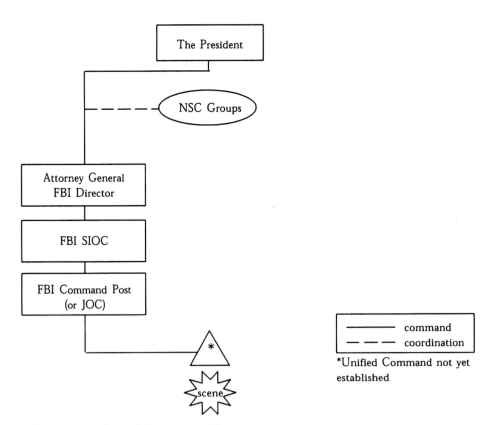

FIGURE L-2 Crisis Management Structures

5. If the threat involves WMD, the FBI Director may recommend to the Attorney General, who notifies the President and NSC groups as warranted, to deploy a Domestic Emergency Support Team (DEST). The mission of the DEST is to provide expert advice and assistance to the FBI On-Scene Commander (OSC) related to the capabilities of the DEST agencies and to coordinate follow-on response assets. When a Joint Operations Center (JOC) is formed, DEST components merge into the JOC structure as appropriate. (The FBI provides guidance on the DEST in PDD-39 Domestic Deployment Guidelines (classified).)

6. During crisis management, the FBI coordinates closely with local law enforcement authorities to provide a successful law enforcement resolution to the incident. The FBI also coordinates with other Federal authorities, including FEMA.

7. The FBI Field Office responsible for the incident site modifies its Command Post to function as a JOC and establishes a Joint Information Center (JIC). The JOC structure includes the following standard groups: Command, Operations, Support, and Consequence Management. Representation within the JOC includes some Federal, State, and local agencies (see Figure L-3).

8. The JOC Command Group plays an important role in ensuring coordination of Federal crisis management and consequence management actions. Issues arising from the response that affect multiple agency authorities and responsibilities will be addressed by the FBI OSC and the other members of the JOC Command Group, who are all working in consultation with other local, State, and Federal representatives. While the FBI OSC retains authority to make Federal crisis management decisions at all times, operational decisions are made cooperatively to the greatest extent possible. The FBI OSC and the Senior FEMA Official at the JOC will provide, or obtain from higher authority, an immediate resolution of conflicts in priorities for allocation of critical Federal resources (such as airlift or technical operations assets) between the crisis management and the consequence management response.

9. A FEMA representative coordinates the actions of the JOC Consequence Management Group, expedites activation of a Federal consequence management response should it become necessary, and works with an FBI representative who serves as the liaison between the Consequence Management Group and the FBI OSC. The JOC Consequence Management Group monitors the crisis management response in order to advise on decisions that may have implications for consequence management, and to provide continuity should a

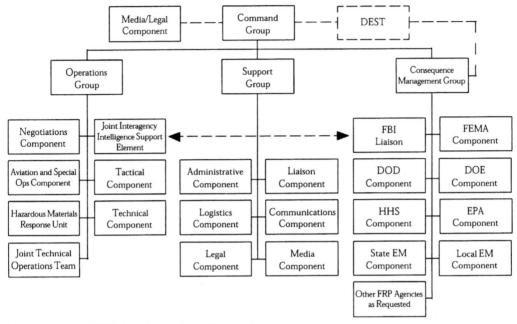

FIGURE L-3   FBI Joint Operations Center Structure

Federal consequence management response become necessary. Coordination will also be achieved through the exchange of operational reports on the incident. Because reports prepared by the FBI are "law enforcement sensitive," FEMA representatives with access to the reports will review them, according to standard procedure, in order to identify and forward information to Emergency Support Function (ESF) #5—Information and Planning that may affect operational priorities and action plans for consequence management.

B. Consequence Management

1. Pre-Release

a. FEMA receives initial notification from the FBI of a credible threat of terrorism. Based on the circumstances, FEMA Headquarters and the responsible FEMA region(s) may implement a standard procedure to alert involved FEMA officials and Federal agencies supporting consequence management.

b. FEMA deploys representatives with the DEST and deploys additional staff for the JOC, as required, in order to provide support to the FBI regarding consequence management. FEMA determines the appropriate agencies to staff the JOC Consequence

Management Group and advises the FBI. With FBI concurrence, FEMA notifies consequences management agencies to request that they deploy representatives to the JOC. Representatives may be requested for the JOC Command Group, the JOC Consequence Management Group, and the JIC.

c. When warranted, FEMA will consult immediately with the Governor's office and the White House in order to determine if Federal assistance is required and if FEMA is permitted to use authorities of the Robert T. Stafford Disaster Relief and Emergency Assistance Act to mission-assign Federal consequence management agencies to pre-deploy assets to lessen or avert the threat of a catastrophe. These actions will involve appropriate notification and coordination with the FBI, as the overall LFA.

d. FEMA Headquarters may activate an Emergency Support Team (EST) and may convene an executive-level meeting of the Catastrophic Disaster Response Group (CDRG). When FEMA activates the EST, FEMA will request FBI Headquarters to provide liaison. The responsible FEMA region(s) may activate a Regional Operations Center (ROC) and deploy a representative(s) to the affected State(s). When the responsible FEMA region(s) activates a ROC, the region(s) will notify the responsible FBI Field Office(s) to request a liaison.

2. **Post-Release**

a. If an incident involves a transition from joint (crisis/consequence) response to a threat of terrorism to joint response to an act of terrorism, then consequence management agencies providing advice and assistance at the JOC pre-release will reduce their presence at the JOC post-release as necessary to fulfill their consequence management responsibilities. The Senior FEMA Official and staff will remain at the JOC until the FBI and FEMA agree that liaison is no longer required.

b. If an incident occurs without warning that produces major consequences and appears to be caused by an act of terrorism, then FEMA and the FBI will initiate consequence management and crisis management actions concurrently. FEMA will consult immediately with the Governor's office and the White House to determine if Federal assistance is required and if FEMA is permitted to use the authorities of the Stafford Act to mission-assign Federal agencies to support a consequence management response. If the President directs FEMA to implement a Federal consequence management response, then FEMA will support the FBI as required and will lead a concurrent Federal consequence management response (see Figure L-4).

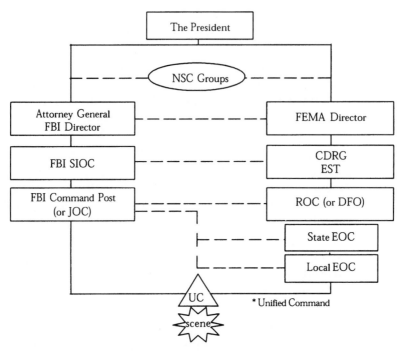

FIGURE L-4 Coordination Relationships

c. The overall LFA (either the FBI or FEMA when the Attorney General transfers the overall LFA role to FEMA) will establish a Joint Information Center in the field, under the operational control of the overall LFA's Public Information Officer, as the focal point for the coordination and provision of information to the public and media concerning the Federal response to the emergency. Throughout the response, agencies will continue to coordinate incident-related information through the JIC. FEMA and the FBI will ensure that appropriate spokespersons provide information concerning the crisis management and consequence management responses. Before a JIC is activated, public affairs offices of responding Federal agencies will coordinate the release of information through the FBI SIOC.

d. During the consequence management response, the FBI provides liaison to either the ROC Director of the Federal Coordinating Officer (FCO) in the field, and a liaison to the EST Director at FEMA Headquarters. While the ROC Director or FCO retains authority to make Federal consequence management decisions at all times, operational decisions are made cooperatively to the greatest extent possible.

e. As described previously, resolution of conflicts between the crisis management and consequence management responses will be provided by the Senior FEMA Official and the FBI OSC at the JOC or, as necessary, will be obtained from higher authority. Operational reports will continue to be exchanged. The FBI liaisons will remain at the EST and the ROC or DFO until FEMA and the FBI agree that a liaison is no longer required.

3. Disengagement

a. If an act of terrorism does not occur, the consequence management response disengages when the FEMA Director, in consultation with the FBI Director, directs FEMA Headquarters and the responsible region(s) to issue a cancellation notification by standard procedure to appropriate FEMA officials and FRP agencies. FRP agencies disengage according to standard procedure.

b. If an act of terrorism occurs that results in major consequences, each FRP component (the EST, CDRG, ROC, and DFO if necessary) disengages at the appropriate time according to standard procedure. Following FRP disengagement, operations by individual Federal agencies or by multiple Federal agencies under other Federal plans may continue, in order to support the affected State and local governments with long-term hazard monitoring, environmental decontamination, and site restoration (cleanup).

V. Responsibilities

A. Department of Justice

PDD-39 validates and reaffirms existing lead agency responsibilities for all facets of the U.S. counterterrorism effort. The Department of Justice is designated as the overall LFA for threats of acts of terrorism that take place within the United States until the Attorney General transfers the overall LFA role to FEMA. The Department of Justice delegates this overall LFA role to the FBI for the operational response. On behalf of the Department of Justice, the FBI will:

1. Consult with and advise the White House, through the Attorney General, on policy matters concerning the overall response;

2. Designate and establish a JOC in the field;

3. Appoint an FBI OSC to manage and coordinate the Federal operational response (crisis management and consequence management). As necessary, the FBI OSC will convene and chair meetings of operational decision makers representing lead State and local crisis

management agencies, FEMA, and lead State and local consequence management agencies in order to provide an initial assessment of the situation, develop an action plan, monitor and update operational priorities, and ensure that the overall response (crisis management and consequence management) is consistent with U.S. law and achieves the policy objectives outlined in PDD-39. The FBI and FEMA may involve supporting Federal agencies as necessary; and

4. Issue and track the status of actions assigned by the overall LFA.

B. Federal Bureau of Investigation

Under PDD-39, the FBI supports the overall LFA by operating as the lead agency for crisis management. The FBI will:

1. Determine when a threat of an act of terrorism warrants consultation with the White House, through the Attorney General;

2. Advise the White House, through the Attorney General, when the FBI requires assistance for a Federal crisis management response, in accordance with the PDD-39 Domestic Deployment Guidelines;

3. Work with FEMA to establish and operate a JIC in the field as the focal point for information to the public and the media concerning the Federal response to the emergency;

4. Establish the primary Federal operations centers for the crisis management response in the field and Washington, DC;

5. Appoint an FBI OSC (or subordinate official) to manage and coordinate the crisis management response. Within this role, the FBI OSC will convene meetings with operational decision makers representing Federal, State, and local law enforcement and technical support agencies, as appropriate, to formulate incident action plans, define priorities, review status, resolve conflicts, identify issues that require decisions from higher authorities, and evaluate the need for additional resources;

6. Issue and track the status of crisis management actions assigned by the FBI; and

7. Designate appropriate liaison and advisory personnel to support FEMA.

C. Federal Emergency Management Agency

Under PDD-39, FEMA supports the overall LFA by operating as the lead agency for consequence management until the overall LFA role is transferred to FEMA. FEMA will:

1. Determine when consequences are "imminent" for the purposes of the Stafford Act;
2. Consult with the Governor's office and the White House to determine if a Federal consequence management response is required and if FEMA is directed to use Stafford Act authorities. This process will involve appropriate notification and coordination with the FBI, as the overall LFA;
3. Work with the FBI to establish and operate a JIC in the field as the focal point for information to the public and the media concerning the Federal response to the emergency;
4. Establish the primary Federal operations centers for consequence management in the field and Washington, DC;
5. Appoint a ROC Director or FCO to manage and coordinate the Federal consequence management response in support of State and local governments. In coordination with the FBI, the ROC Director or FCO will convene meetings with decision makers of Federal, State, and local emergency management and technical support agencies, as appropriate, to formulate incident action plans, define priorities, review status, resolve conflicts, identify issues that require decisions from higher authorities, and evaluate the need for additional resources;
6. Issue and track the status of consequence management actions assigned by FEMA; and
7. Designate appropriate liaison and advisory personnel to support the FBI.

D. Federal Agencies Supporting Technical Operations

1. Department of Defense

As directed in PDD-39, the Department of Defense (DOD) will activate technical operations capabilities to support the Federal response to threats or acts of WMD terrorism. DOD will coordinate military operations within the United States with the appropriate civilian lead agency(ies) for technical operations.

2. Department of Energy

As directed in PDD-39, the Department of Energy (DOE) will activate technical operations capabilities to support the Federal response to threats or acts of WMD terrorism. In addition, the FBI has concluded formal agreements with potential LFAs of the Federal Radiological

Emergency Response Plan (FRERP) that provide for interface, coordination, and technical assistance in support of the FBI's mission. If the FRERP is implemented concurrently with the FRP:

   a. The Federal On-Scene Commander under the FRERP will coordinate the FRERP response with the FEMA official (either the ROC Director or the FCO), who is responsible under PDD-39 for coordination of all Federal support to State and local governments.

   b. The FRERP response may include on-site management, radiological monitoring and assessment, development of Federal protective action recommendations, and provision of information on the radiological response to the public, the White House, Members of Congress, and foreign governments. The LFA of the FRERP will serve as the primary Federal source of information regarding on-site radiological conditions and off-site radiological effects.

   c. The LFA of the FRERP will issue taskings that draw upon funding from the responding FRERP agencies.

3. Department of Health and Human Services

As directed in PDD-39, the Department of Health and Human Services (HHS) will activate technical operations capabilities to support the Federal response to threats or acts of WMD terrorism. HHS may coordinate with individual agencies identified in the HHS Health and Medical Services Support Plan for the Federal Response to Acts of Chemical/Biological (C/B) Terrorism, to use the structure, relationships, and capabilities described in the HHS plan to support response operations. If the HHS plan is implemented:

   a. The HHS on-scene representative will coordinate, through the ESF #8—Health and Medical Services Leader, the HHS plan response with the FEMA official (either the ROC Director or the FCO), who is responsible under PDD-39 for on-scene coordination of all Federal support to State and local governments.

   b. The HHS plan response may include threat assessment, consultation, agent identification, epidemiological investigation, hazard detection and reduction, decontamination, public health support, medical support, and pharmaceutical support operations.

   c. HHS will issue taskings that draw upon funding from the responding HHS plan agencies.

4. Environmental Protection Agency

As directed in PDD-39, the Environmental Protection Agency (EPA) will activate technical operations capabilities to support the Federal response to acts of WMD terrorism. EPA may coordinate with individual agencies identified in the National Oil and Hazardous Substances Pollution Contingency Plan (NCP) to use the structure, relationships, and capabilities of the National Response System as described in the NCP to support response operations. If the NCP is implemented:

    a. The Hazardous Materials On-Scene Coordinator under the NCP will coordinate, through the ESF #10—Hazardous Materials Chair, the NCP response with the FEMA official (either the ROC Director or the FCO), who is responsible under PDD-39 for on-scene coordination of all Federal support to State and local governments.

    b. The NCP response may include threat assessment, consultation, agent identification, hazard detection and reduction, environmental monitoring, decontamination, and long-term site restoration (environmental cleanup) operations.

VI. Funding Guidelines

    A. As stated in PDD-39, Federal agencies directed to participate in the resolution of terrorist incidents or conduct of counterterrorist operations bear the costs of their own participation, unless otherwise directed by the President. This responsibility is subject to specific statutory authorization to provide support without reimbursement. In the absence of such specific authority, the Economy Act applies, and reimbursement cannot be waived.

    B. FEMA can use limited pre-deployment authorities in advance of a Stafford Act declaration to "lessen or avert the threat of a catastrophe" only if the President expresses intention to go forward with a declaration. This authority is further interpreted by congressional intent, to the effect that the President must determine that assistance under existing Federal programs is inadequate to meet the crisis, before FEMA may directly intervene under the Stafford Act. The Stafford Act authorizes the President to issue "emergency" and "major disaster" declarations.

        1. Emergency declarations may be issued in response to a Governor's request, or in response to those rare emergencies, including some acts of terrorism, for which the Federal Government is assigned in the laws of the United States the exclusive or preeminent responsibility and authority to respond.

        2. Major disaster declarations may be issued in response to a Governor's request for any natural catastrophe or, regardless of cause, any fire, flood, or explosion that has caused damage of sufficient severity and magnitude, as determined by the President, to warrant major disaster assistance under the Act.

        3. If a Stafford Act declaration is provided, funding for consequence management may continue to be allocated from responding

agency operating budgets, the Disaster Relief Fund, and supplemental appropriations.

C. If the President directs FEMA to use Stafford Act authorities, FEMA will issue mission assignments through the FRP to support consequence management.

1. Mission assignments are reimbursable work orders, issued by FEMA to Federal agencies, directing completion of specific tasks. Although the Stafford Act states that "Federal agencies *may* [emphasis added] be reimbursed for expenditures under the Act" from the Disaster Relief Fund, it is FEMA policy to reimburse Federal agencies for eligible work performed under mission assignments.

2. Mission assignments issued to support consequence management will follow FEMA's Standard Operating Procedures for the Management of Mission Assignments or applicable superseding documentation.

D. FEMA provides the following funding guidance to the FRP agencies:

1. Commitments by individual agencies to take precautionary measures in anticipation of special events will not be reimbursed under the Stafford Act, unless mission-assigned by FEMA to support consequence management.

2. Stafford Act authorities do not pertain to law enforcement functions. Law enforcement or crisis management actions will not be mission-assigned for reimbursement under the Stafford Act.

VII. References

A. Presidential Decision Directive 39, U.S. Policy on Counterterrorism (classified). An unclassified extract may be obtained from FEMA.

B. PDD-39 Domestic Deployment Guidelines (classified).

C. PDD-62, Protection Against Unconventional Threats to the Homeland and Americans Overseas (classified).

D. FBI WMD Incident Contingency Plan.

E. HHS Health and Medical Services Support Plan for the Federal Response to Acts of Chemical/Biological Terrorism.

VIII. Terms and Definitions

A. Biological Agents

The FBI WMD Incident Contingency Plan defines biological agents as microorganisms or toxins from living organisms that have infectious or noninfectious properties that produce lethal or serious effects in plants and animals.

B. Chemical Agents

The FBI WMD Incident Contingency Plan defines chemical agents as solids, liquids, or gases that have chemical properties that produce lethal or serious effects in plants and animals.

C. Consequence Management

FEMA defines consequence management as measures to protect public health and safety, restore essential government services, and provide emergency relief to governments, businesses, and individuals affected by the consequences of terrorism.

D. Credible Threat

The FBI conducts an interagency threat assessment that indicates that the threat is credible and confirms the involvement of a WMD in the developing terrorist incident.

E. Crisis Management

The FBI defines crisis management as measures to identify, acquire, and plan the use of resources needed to anticipate, prevent, and/or resolve a threat or act of terrorism.

F. Domestic Emergency Support Team (DEST)

PDD-39 defines the DEST as a rapidly deployable interagency support team established to ensure that the full range of necessary expertise and capabilities are available to the on-scene coordinator. The FBI is responsible for the DEST in domestic incidents.

G. Lead Agency

The FBI defines lead agency, as used in PDD-39, as the Federal department or agency assigned lead responsibility to manage and coordinate a specific function—either crisis management or consequence management. Lead agencies are designated on the basis of their having the most authorities, resources, capabilities, or expertise relative to accomplishment of the specific function. Lead agencies support the overall Lead Federal Agency during all phases of the terrorism response.

H. Nuclear Weapons

The Effects of Nuclear Weapons (DOE, 1977) defines nuclear weapons as weapons that release nuclear energy in an explosive manner as the result of nuclear chain reactions involving fission and/or fusion of atomic nuclei.

I. Senior FEMA Official

The official appointed by the Director of FEMA or his representative to represent FEMA on the Command Group at the Joint Operations Center. The Senior FEMA Official is not the Federal Coordinating Officer.

J. Technical Operations

As used in this annex, technical operations include actions to identify, assess, dismantle, transfer, dispose of, or decontaminate personnel and property exposed to explosive ordnance or WMD.

K. Terrorist Incident

The FBI defines a terrorist incident as a violent act, or an act dangerous to human life, in violation of the criminal laws of the United States or of any State, to intimidate or coerce a government, the civilian population, or any segment thereof in furtherance of political or social objectives.

L. Weapon of Mass Destruction (WMD)

Title 18, U.S.C. 2332a, defines a weapon of mass destruction as (1) any destructive device as defined in section 921 of this title, [which reads] any explosive, incendiary, or poison gas, bomb, grenade, rocket having a propellant charge of more than four ounces, missile having an explosive or incendiary charge of more than one-quarter ounce, mine or device similar to the above; (2) poison gas; (3) any weapon involving a disease organism; or (4) any weapon that is designed to release radiation or radioactivity at a level dangerous to human life.

Appendix M
NBC Indicator Matrix
Defense Protective Service

A handout developed by the Defense Protective Service to provide first responders with a tool to assess general indicators that may be present at a Nuclear, Biological, or Chemical incident until such time as a more technical assessment can be made and a process for reporting that assessment to appropriate authorities.

Defense Protective Service
1155 Defense Pentagon
Washington, D.C. 20301-1155
Phone: (703) 695-4088

Instructions for Using NBC Indicator Matrix

The NBC Indicator Matrix process is an assessment of general indicators that may be present at a Nuclear, Biological, or Chemical Incident. Developed for use as part of the Defense Protective Service (DPS) NBC Response Plan, the use of this matrix should not be considered as a authoritative determination of the type of NCB Incident that is occurring, but is rather a general guide for First Responding personnel (police, fire, or medical) until a more technical assessment can be made.

The matrix itself is a combination of symptoms, observations, and other indicators (listed in the order that such indicators are most likely to be noticed) that may be present for each of the agents/materials listed at the top. Primarily designed for use in Communications Centers (911, Dispatch Centers, etc.) or Emergency Operations Centers (fixed or mobile) to record indications in which the indicator would be noticed by first responding personnel to the scene of a potential NBC emergency. However, should these operations centers also be affected by the emergency it is possible for field units to use the matrix as well. The intended result is to give some indication of what type of NBC materials may be involved in the emergency to the initial responding units until such time as a expert determination can be made.

How To Use

(The following steps assume responding units are reporting information to their Dispatchers)

1. Units arrive at scene where even at a distance it is apparent that multiple persons are affected.
2. Units should "***STOP, LOOK, and LISTEN***" and relay observations to their dispatchers (See note on Radio Transmit Code).
3. Dispatcher personnel will record each of the relayed observations on the matrix by placing a check mark in the "Indicator Present" column for each indicator so observed.
4. For every row in which the Indicator Present column is checked the dispatcher will place another check mark in all "*unshaded*" boxes (including those with words inside the box) on that row. (Note: Boxes with words are designed to help classify the indicator, for example, vomiting is listed for all NBC materials, but "bloody" vomiting is a sign of blister agent use.)
5. At the bottom of each page the total number of check marks for each column should be added and recorded as page totals and these totals transferred to Page 3 of the matrix. Total up all the page totals listed on Page 3, and the column with the highest number of indicators should be considered that agent/material most likely present.

RADIO TRANSMIT CODE NOTE: For those agencies requiring a means to transmit information over a radio net or by unsecured phone line using code a column marked "Radio Transmit Code" lists a numeric code for each indicator (i.e. 1-1 for Prostration) listed on the matrix. In addition each of the types of NBC materials has an alphabetic letter designation. The use of these matrix codes would require that both the sending unit and the receiving station have the NBC Indicator Matrix available. Examples on possible uses are as follows:

1. "NBC Matrix 1-1 and 2-2" would indicate the prostration and painless blisters indicators were observed.
2. "NBC Matrix 1-9 and 5-7 Bravo" would indicate vomiting and 'bloody' diarrhea indicators were observed.
3. "NBC Matrix Total Indicators Present is 12 with A (Alpha) = 6, B (Bravo) = 4, and G (George) = 3" would indicate that a total of 12 indicators were listed on page 3 with 6 indicators for nerve agents,

4 indicators for blister agent, and 3 indicators for biological agents. This example could be used to transmit indicator totals to another agency using the matrix (such as the USPHS). The rules for using transmit codes or to transmit in the clear is left up to the agency using the matrix.

Special Thanks to the Marietta Fire and Emergency Services and the Georgia Mutual Aid Group (GMAG) for the suggested format change that resulted in a 3 page matrix instead of the original 8 page matrix from earlier editions of 10-90 Gold.

(Place a check mark for each indicator noted at the incident in each unshaded box across the row for that indicator. At bottom of each page total up the number of check marks made for each column and at last page total up all page totals. Highest score is indicator for which NBC material is involved. *For Official Use Only by Police, Fire, and Medical Service Agencies.*)

Indicator Noticed by First Responder *(These indicators are listed in the order in which the indicator would be noticed by first responding personnel to the scene of an NBC Emergency)* **STOP, LOOK and LISTEN** *(Resist rushing in, approach incident from upwind, stay clear of all spills, vapors, fumes and smoke. Be extremely mindful of enclosed or confined spaces.)*	Indicator Present	Radio Transmit Code	Nerve Agent	Blister Agents	Blood Agents	Choking Agents	Irritating Agents	Incapacitating Agents	Biological Agents	Radiological Materials
APPEARANCE (at a distance, multiple persons affected)			A	B	C	D	E	F	G	H
Prostration		1-1								
Involuntary twitching and jerking		1-2								
Convulsions		1-3								
Coma		1-4								
Confusion		1-5								
Bleeding from orifices (nose, ears, mouth, rectum)		1-6								
Coughing		1-7								
Sneezing, violent and persistent		1-8								
Vomiting		1-9		Bloody						
SKIN										
Reddening of lips and skin		2-1								
Blisters, painless (ask victim)		2-2								
Blisters, painful (ask victim)		2-3								
Gray area of dead skin that does not blister		2-4								
Sunburn like appearance (erythema)		2-5								
Pain, stinging or deep aching (ask victim)		2-6								
Clammy skin		2-7								
Skin, lesions, multiple pinpoint		2-8								
Hair Loss, large quantities		2-9								
EYES										
Pinpointing of pupils		3-1								
Enlargement of pupils		3-2								
Lesions		3-3								
Involuntary closing		3-4								
PAGE ONE TOTALS										

400

Symptom	Code									
Tears or tearing	3-5									
Eyes, immediate burning sensation & gritty feeling	3-6									
Pain in and above eyes, aggravated by bright light	3-7									
Dimness of vision (ask victim)	3-8									
RESPIRATORY										
Coughing-up of frothy sputum	3-9								Bloody	
Severe and uncontrollable coughing	3-10									
Hoarseness, (may progress to loss of voice)	3-11									
Runny Nose - copious	4-1									
Breathing Rate decreased	4-2									
Breathing Rate increased	4-3									
Breathing Depth increased	4-4									
Breathing, difficult (Observed or ask victim)	4-5									
Dry Throat (ask victim)	4-6									
Tightness in chest (ask victim)	4-7									
EXAMINATION (with protection if significant indicators above are found)										
CARDIOVASCULAR										
Pulse slow	4-8									
Blood pressure low	4-9									
Blood pressure high	5-1									
Heart action rapid and feeble	5-2									
Heart beat, rapid	5-3									
Headache (ask victim)	5-4									
Headaches, Frontal (ask victim)	5-5									
Dizziness (ask victim)	5-6									
DIGESTIVE SYSTEM (GI, GU, Glands)										
Diarrhea	5-7	Bloody								
Involuntary defecation and urination	5-8									
Nausea	5-9									
Localized Sweating	5-10									
Excessive Sweating	5-11									
TEMPERATURE										
Fever	6-1									
Temperature, subnormal	6-2									
PAGE TWO TOTALS										

(continued)

HISTORY or ENVIRONMENTAL	Radio Transmit Code	Nerve Agent	Blister Agents	Blood Agents	Choking Agents	Irritating Agents	Incapacitating Agents	Biological Agents	Radiological Materials
Odor - Apple Blossom	6-3						□		
Odor - Pepper like	6-4						□		
Odor - Garlic	6-5		□						
Odor - Horseradish	6-6		□						
Odor - Bitter Almonds (faint)	6-7			□					
Odor - Sour Fruit	6-8						□		
Odor - Peach kernels (faint)	6-9			□					
Odor - New mown hay or freshly cut grass	6-10				□				
Odor - Fruity to geranium like	7-1	□							
Unscheduled and unusual spray being disseminated.	7-2								
Unusual Liquid Droplets, oily, no recent rain.	7-3								
Abandoned spray devices	7-4								
Dead Animals, Birds, Fish.	7-5								
Dead Weeds, Trees, Bushes, Lawns, etc.	7-6								
Illness associated with specific geographic area, i.e. victims have different treatment locations, but all work within same area.	7-7								
Immediate Fatalities, not associated with trauma	7-8								
Lack of Insect Life.	7-9								
Low Lying Clouds not explained by surroundings.	7-10								
Reports of colleagues, medical community, media, etc., with similar unexplained illness.	8-1								
TOTALS		A	B	C	D	E	F	G	H
TOTAL INDICATORS FROM PAGE 1									
TOTAL INDICATORS FROM PAGE 2									
TOTAL INDICATORS FROM PAGE 3									

Bottom legend columns: Indicator Present | Radio Transmit Code | Nerve Agent | Blister Agents | Blood Agents | Choking Agents | Irritating Agents | Incapacitating Agents | Biological Agents | Radiological Materials

1. Put on Respiratory Protection
2. Report all observations to Communications
3. Report wind conditions (speed & direction)
4. Calm Victims: "Help is on the Way!!!"
5. Direct walking wounded to a collection point
6. Touch nothing and no victim until in PPE
7. REMAIN CALM

TOTAL INDICATORS OF ALL PAGES

Appendix N
Advantages and Limitations of Selected NBC Equipment Used by the Federal Government

FEMA—Rapid Response Information System
http://www.fema.gov/rris

Advantages and Limitations of Selected NBC Equipment Used by the Federal Government

ITEM	DESCRIPTION	ADVANTAGES	LIMITATIONS	MANUFACTURER	COST
M256A1 Chemical Detector Kit	The M256A1 kit is a "wet chemistry" chemical agent detector kit on a card. Each sampler-detector contains a square test spot to detect blister agents, a circular test spot to detect blood agents, a star test spot to detect nerve agents, and a tablet with a rubbing tab to detect lewisites. The M256A1 is used right after a chemical attack to detect and classify dangerous concentrations of toxic agents (both liquid and vapor) by the color-changing chemical reaction. The M256A1 is also used to determine when the area is "all clear" (negative reaction of M256A1).	• Will detect nerve, blister, and blood agents • Most widely used U.S. Army detection kit • Detects below IDLH for nerve and blood agents; below incapacitating dose for mustard agents • Also contains M8 paper	• Will not detect choking agents • Takes up to 15 minutes to use • Requires hand manipulations of card component • Once used the cards are mixed hazardous waste	Anachemia Canada, Inc., PO Box 147, Lachine (Montreal), Quebec, Canada H8S 4A7 800-361-0209 http://www.anachemia.com/ Truetech, Inc. 680 Elton treet Riverhead, NY 11901 516-727-8600 FAX 516-727-7592	$140 per kit (10 test cards) (Based on small order only—may be less if larger order made)
M256A1 (Simulator)	M28/29 Trainer Kits for the M256A1 Chemical Detector Kit			Anachemia Canada, Inc., PO Box 147, Lachine (Montreal), Quebec, Canada H8S 4A7 800-361-0209 http://www.anachemia.com/	

M18A2 Chemical Agent Detector Kit	The M18A2 kit is a "wet chemistry" chemical agent detector kit that is used to confirm the results of the M256A1 Kit. It is a portable, expendable item capable of detecting dangerous concentrations of vapors, aerosols, and liquid droplets of chemical agents. The kit's capability provides for the sampling of unknown NBC agents. The presence of a chemical agent is detected by distinctive color changes.	• Detects and identifies choking, nerve, blood, and blister agents • Samples unknown agents • Approximately twice as sensitive as the M256A1 for mustard agents • Contains detector tubes, detector tickets, M8 paper and instruction cards	• About 20 times less sensitive than the M256A1 for nerve agents • Does not detect chlorine • Requires manipulation of tubes, detector tickets and reagent vials • Complete series of tests takes 24 minutes • Not currently in production (a production run of 125 items each is possible) • No training item developed	Truetech, Inc. 680 Elton Street Riverhead, NY 11901 516-727-8600 FAX 516-727-7592	$248 per kit (Based on small order only—may be less if larger order made)
ICAD	Electronic Chemical Agent Detector	• Simultaneous detection of nerve, blood, blister, & choking agents • Visual/audible warning • No radioactive source • Light-weight, small • Minimal training • No common interferents	• Gross level vapor detection for AC & CG • Detects L (Blister) below detection level of M256A1 • Alarms below LD_{50} for nerve but above IDLH	ETG Defense Systems Division, Environmental Technologies Group Inc., 1400 Taylor Ave., Baltimore, MD 21284 410-321-5200	$2,700

(Continued)

Advantages and Limitations of Selected NBC Equipment Used by the Federal Government (Continued)

ITEM	DESCRIPTION	ADVANTAGES	LIMITATIONS	MANUFACTURER	COST
Chemical Agent Monitor/Improved Chemical Agent Monitor (CAM/ICAM)	The CAM is an electronic hand-held, semi-quantitative point detection device. It reduces the burden and enhances the efficiency of the decontamination process. The CAM is employed in both monitoring and survey missions and is primarily used to sort contaminated versus clean vehicles, equipment, and personnel. The CAM is no longer in production; the *Improved* Chemical Agent Monitor is an improvement over the original CAM. The ICAM is less expensive to repair, requires less maintenance, and starts up faster after prolonged storage.	• Detect and differentiate between low levels of nerve (G, VX) and mustard (HD, HN3) on personnel and equipment • Light-weight, hand held • Minimal training • ICAM can run approximately 10 times longer (without maintenance) than the CAM	• The ICAM requires a license from the U.S. Nuclear Regulatory Commission (NRC) under Title 10 Code of Federal Regulations • CAM/ICAM must be purchased through the Government or with Government's approval • Detects Blister and Nerve agents only • The CAM may give false reading when used in enclosed spaces or when sampling near strong vapor sources (i.e. in dense smoke).	**CAM:** Graseby Ionics, Park Avenue, Bushey, Waterford, Hertfordshire WD2 2BW, UK +44 1923 23 8483 ETG Defense Systems Division, Environmental Technologies Group Inc., 1400 Taylor Ave., Baltimore, MD 21284 410-321-5200 **ICAM:** Intellitech Div., Technical Products Group, Inc., 2000 Brunswick Lane DeLand, FL 32724 904-736-1700 FAX 904-736-2250	Approximately $4,000 for the ICAM through the federal government; otherwise, approximately $6333.

Product	Description	Advantages	Disadvantages	Supplier	Price
M8 Paper	M8 Paper is Chemical Agent Detector Paper. It comes in a book of 25 tan sheets of chemically treated, dye-impregnated paper, perforated for easy removal. The M8 Paper willl turn yellow in the presence of liquid G-type nerve agents, red for liquid H blister (mustard) agents and green for liquid V-type nerve agents. A color comparison bar chart is printed on the inside of the front cover of the book.	• Qualitative detection of liquid nerve and blister agents with color change in 30 sec. • Little training required to use • Inexpensive • Ability to "wick" small droplets to make visible spots	• Some vapors known to give false readings are: aromatic vapors (perfumes, food flavorings, some aftershaves, peppermints, cough lozenges, and menthol cigarettes, cleaning compounds (disinfectants, methyl salicylate, menthol, etc.) smokes and fumes, and some wood preservative treatments • Detects vapors only • Bar warning only • No audible warning. • Gross liquid detection only (not vapor)—the paper must touch the liquid agent • Prone to false positives (mainly from organic solvents)	Truetech, Inc. 680 Elton Street Riverhead, NY 11901 516-727-8600 FAX 516-727-7592 Anachemia Canada, Inc., PO Box 147, Lachine (Montreal), Quebec, Canada H8S 4A7 800-361-0209	$4.50 per booklet (25 sheets per booklet)

(Continued)

Advantages and Limitations of Selected NBC Equipment Used by the Federal Government (Continued)

ITEM	DESCRIPTION	ADVANTAGES	LIMITATIONS	MANUFACTURER	COST
		• Different color changes for VX versus G versus H agents.	• Prone to false negatives when detecting chemical agents in water or aerosol agents in the air • Prone to false positives from some interferents such as certain cleaning solvents, insect repellent, DS2, and petroleum products		
M9 Paper	M9 Paper is Chemical Agent Detector Paper that can be torn into smaller lengths and attached to protective clothing, vehicles, equipment or supplies. The M9 paper contains an indicator dye that responds with marked, contrasting color change when exposed to liquid nerve and blister agents. The red or pink spots will appear on the green paper when it comes into contact with droplets of liquid toxic agent.	• Qualitative detection of liquid nerve and blister agents with color change in 20 sec. • Little training required to use • Inexpensive • Ability to "wick" small droplets to make visible spots • More sensitive and more rapid response than M8 Paper.	• Gross liquid detection only (not vapor) • Does not identify the specific agent. • Prone to false positives (mainly from organic solvents) • Single color change • Prone to false negatives when detecting agent in water • Prone to false positives from 'scuffing' • Protective gloves must be worn.	Truetech, Inc. 680 Elton Street Riverhead, NY 11901 516-727-8600 FAX 516-727-7592 Anachemia Chemicals Inc. 3 Lincoln Blvd. Rouses Point, NY 12979 514-489-5711	$6.77 per roll (both rolls and booklets available)

Item	Description	Features	Requirements	Source	Price
Mark I Nerve Agent Antidote Kit (NAAK)	The NAAK consists of one small auto-injector containing atropine and a second autoinjector containing pralidoxime chloride (PAM CI). The atropine autoinjector consist of a hard plastic tube containing 2 mg (0.7 milliliter (ml)) of atropine in solution. It has a pressure activated coiled spring mechanism which triggers the needle for injection of the antidote solution. The 2 PAM CI autoinjector is a hard plastic tube that dispenses 600 mg/2 ml of 2 PAM CI (300 mg/ml) solution when activated. It has a pressure activated coiled spring mechanism identical to that in the atropine autoinjector.	• Simple to use	• Possible legal restrictions on administration by nonmedically qualified personnel • Requires purchase order and prescription directly from city • Must protect the NAAK from freezing • Shelf-life of 5 years	Meridian Medical Technologies (MMT) 10240 Old Columbia Road Columbia, Maryland 21046 Phone: 800-638-8093 Fax: 410-309-1475 or 410-309-6830 E-mail: info@meridianmt.com http://www.meridianmeds.com/	$17.00 each *Special purchase order requirements apply to all Mark 1 orders.*
Mark I Nerve Agent Antidote Kit (NAAK) Training Device	Similar to the Mark I NAAK, but contains no medications or needles			Meridian Medical Technologies 10240 Old Columbia Road Columbia, Maryland 21046 Phone: 800-638-8093 Fax: 410-309-1475 or 410-309-6830 E-mail: info@meridianmt.com http://www.meridianmeds.com/	$11.00 each 25 per box

(Continued)

Advantages and Limitations of Selected NBC Equipment Used by the Federal Government (Continued)

ITEM	DESCRIPTION	ADVANTAGES	LIMITATIONS	MANUFACTURER	COST
CANA (Convulsant Antidote for Nerve Agent)	The CANA is similar to the Mark 1 (NAAK) autoinjector, but it holds 2 milliliter volume of diazepam. The CANA is a disposable device for intramuscular delivery of diazepam to a person who is incapacitated by nerve agent poisoning.	• Simple to use	• Legal restrictions on administration by nonmedically qualified personnel • Using more than one causes overdose • Shelf-life of 2 years.	Meridian Medical Technologies 10240 Old Columbia Road Columbia, Maryland 21046 Phone: 800-638-8093 Fax: 410-309-1475 or 410-309-6830 E-mail: info@meridianmt.com http://www.meridianmeds.com/	$13.35 each *Special purchase order requirements apply to all CANA orders.*
Jane's NBC Protection Equipment (Annually Published Catalogue)	*Jane's NBC Protection Equipment* is a reference source for NBC protective equipment. The catalogue lists equipment in use worldwide and lists the addresses for manufacturers of NBC protective equipment.			Jane's Information Group Inc., 1340 Braddock Place, Suite 300, Alexandria, VA 22314-1651 800-824-0768 http://www.janes.com/	$320.00[b]

Appendix O
Terrorism Internet Addresses

The following pages are a listing of some of the best Internet resources that we have found for research on this topic. Like with all other Internet addresses these links are subject to change or expiration, but as of this publication date these links are functional. Please contact us at *IMS4EMS@ hotmail.com* should you have any suggestions or discover that a link is no longer functioning.

INTERNET ADDRESSES

WEB SITE NAME	WEB SITE ADDRESS	DESCRIPTION
Organizational Strategic Solutions	http://www.ossgrp.com	Multiple links to references & services
Center for Disease control (CDC) NCEH (National Center of Environmental Health)	http://www.cdc.gov	Multiple links through the CDC on environmental health. Has a search option.
CHEMDEX	http://www.shef.ac.uk/~chem/chemdex	Chemistry resources on the net. Over 4,000. Updated 10-99
National Protection Center	http://npc.sbccom.army.mil	Reference materials
ATSDR Tox FAQs	http://www.atsdr.cdc.gov/toxfaq.html	Chemical profiles
Chemystery	http://library.thinkquest.org/3659	A virtual text book on high school chemistry.
CBIAC: Chemical & Biological Defense Information and Analysis Center	http://www.cbiac.apgea.army.mil	Collects, reviews, analyzes and summarizes CW/CBD information
The National Guard	http://www.ngb.dtic.mil.htm	National Guard topics
NBC Protect.com	http://www.nbcprotect.com/new/cwagents.htm	Chemical warfare chemical agents
Johns Hopkins University	http://www.hopkins-biodefense.org/index.html	Agents, events, library and current news stories
U.S. Army Center for Health Promotion & Preventative Medicine	http://www.chppm.com	Many links available, Updated 9-99
Terrorism with CB Weapons: Calibrating Risks and Responses	http://www.cbaci.org	Provides research, analysis, technical support, and education.
U.S. Army Medical Research Institute of Chemical Defense	http://chemdef.apgea.army.mil	Data links for literature on medical management of chemical casualties.
Military Unique MSDS	http://www.sbccom.army.mil	Chem agent MSDS
Department of Army Homeland Defense	http://hld.sbccom.army.mil	Information for first responders, technical, training, and press releases.
U.S. Environmental Protection Agency	http://www.epa.gov	
Chemical Emergency Preparedness & Prevention Office	http://www.epa.gov/swercepp	Many resource links and information
Terrorism Section of the above page	http://www.epa.gov/swercepp/cntr-ter.html	Terrorism related

Organization	URL	Description
EPA Envirofacts Database	*http://www.epa.gov/enviro/html/index_java.html*	Superfund, hazardous waste, toxic release inventory, chemical searches
RCRA, Superfund and EPCRA Hotline	*http://www.epa.gov/epaoswer/hotline*	Up-to-date info on several programs and regulations
USEPA Office of Pesticide Programs	*http://www.epa.gov/pesticides*	Pesticide page
EPA CEPPO page	*http://www.epa.gov/ceppo/lepclist.htm*	A listing, by state of all LEPCs and contact information.
American Chemistry Council	*http://www.cmahq.com*	Digest and links
Disaster Management Central Resource	*http://206.39.77.2/DMCR/dmrhome.html*	Lackland AFB site with information on civilian support resources, triage of MCI situations, medicine, injuries, etc
Disaster Resource Guide	*http://disaster-resource.com*	Source of information on commercial firms, which can Assist during emergencies.
National Institute of Justice	*http://www.nlectc.org/ccfp*	Homepage for the "Center for Civil Force Protection" Provides information on security to state and local governments as well as some private industry.
Henry L. Stimson Center	*http://www.stimson.org*	Searching for solutions to security challenges
ERRI, Emergence Response Research Institute	*http://www.emergency.com*	References and open source intel info
FEMA, Terrorism Incident	*http://www.fema.gov/r-n-r/frp/frpterr.htm*	Background Information
FEMA Preparedness	*http://www.fema.gov/fema/pre2.html*	Links to planning, training, exercises, information, community and family preparedness.
FEMA Preparedness, Training & Exercises Directorate	*http://www.fema.gov/pte*	
FEMA, terrorism site	*http://www.rris.fema.gov*	Rapid Response Information System Reference Guide, training aid, planning resource for NBC.
U.S. Fire Administration	*http://www.usfa.fema.gov/assist/fednews.htm*	Federal fire and emergency service current news and information.
FEMA	*http://www.fema.gov/library/terrorf.htm*	Fact sheet on terrorism.
Office of Justice Programs	*http://www.ojp.usdoj.gov/new.htm*	Current news.
U.S. Office of the Surgeon General	*http://www.nbc-med.org*	A distributed learning and reference source for medical NBC.
Terrorism Research Center	*http://www.terrorism.com/terrorism/index.html*	References, news stories, profiles, documents and links.

INTERNET ADDRESSES (Continued)

WEB SITE NAME	WEB SITE ADDRESS	DESCRIPTION
U.S. Department of Defense Terrorism Links	http://www.defenselink.mil/other_info/terrorism.html	Defense efforts to counter terrorism, and links.
National Fire Protection Association	http://www.nfpa.org	Homepage
U.S. Department of Health and Human Services	http://www.dhhs.gov	Counter terrorism program in the Office of Emergency Preparation
Federation of American Scientists, Chemical & Biological Weapons Verification Program.	http://www.fas.org/bwc/index.html	Reports of use of alleged use or release of BW and toxins.
Morbidity & Mortality Weekly Report (MMWR)	http://www.cdc.gov/mmwr/mmwr.html	Weekly MMWR Report
Emergency Net News	http://www.emergency.com/hzmtpage.html	Hazmat operations archives, news and lots of links.
National Park Service	http://www.nature.nps.gov/toxic	Environmental contaminants encyclopedia.
Emergency Planning for Chemical	http://www.chemicalspill.org	Lots of EPCRA, BOLDER, first responder, chemical cross reference in Spanish and other information.
National Institute for Occupational Safety & Health (NIOSH) database.	http://www.cdc.gov/niosh/homepage.html	Hazards to human health & safety. NIOSH pocket guide, access to registry of toxic effects of chemical substances
National Research Council	http://www.nas.edu/nrc	Homepage has search option.
Organization for the Prohibition of Chemical Weapons (OPCW)	http://www.opcw.nl	Hazards to human health & safety. NIOSH pocket guide, terms.
SIRI MSDS Collection	http://hazard.com/msds	Industrial based MSDS
MSDS Sheets on line	http://www.ilpi.com/msds/index/html	Many different sources for MSDS sheets.
Stockholm International Peace Research Institute (SIPRI)	http://www.sipri.se/projects/chembio.html	Extensive research in arms control, technology, CB weapons. Excellent external Web links.
Cal Poly, San Luis Obispo Chemical Biological Weapons Page.	http://www.calpoly.edu/~drjones/chemwarf.html	History, weapons, disarmament, Information on CBW
USEPA/OPP Chemical Ingredient Database query	http://www.cdpr.ca.gov/docs/epa/epachem.htm	Pesticide Info database
Toxicology Information Briefs	http://ace.ace.orst.edu/info/extoxnet/tibs/ghindex.html	Pesticide toxicology Info

Chemfinder Searching	http://chemfinder.camsoft.com	Chemical database
Interactive Weather Information Network	http://iwin.nws.noaa.gov/iwin/graphicsversion/main.html	Graphic weather information
The Weather Channel	http://www.weather.com/homepage.html	Weather information
University of Illinois Weather	http://www.atmos.uiuc.edu	Weather Info and links
Emergency Managers Weather	http://iwin.nws.noaa.gov/emwin	Weather homepage
Haz Mat safety	http://hazmat.dot.gov	DOT transportation haz mat info
National Council on Radiation Protection and Measurement	http://www.ncrp.com	Radiation health & safety
U.S. National Response Team	http://www.nrt.org	Homepage
DHHS Office of Emergency Preparedness	http://ndms.dhhs.gov/CT_Program/ct_program.html	Counterterrorism Initiatives
DOE Office of Emergency Response	http://www.dp.doe.gov/emergencyresponse	Defense
White House Homepage	http://www.whitehouse.gov	Directory of commonly requested services. Daily official news briefings
29 CFR Standards	http://www.osha-slc.gov/OshStd_toc/OSHA_Std_toc.html	OSHA standards
OSHA Publications	http://www.osha-slc.gov/OshDoc/Additional.html	
Mass Decontamination Information	http://www.2.sbccom.army.mil/hld/cwirp/cwirp_guidelines_mass_casualty_decon_download.com	Links to provide responder assistance with definition, technique and application of mass decontamination process
	http://wonder.cdc.gov/wonder/prevguid/p0000018/p0000018.asp	
	http://wonder.cdc.gov/wonder/prevguid/p0000019/p0000019.asp	

Appendix P
Material Safety
Data Sheets

Notice—The SBCCOM Web page contained a VX MSDS with erroneous information. If you have downloaded the VX MSDS or any of the MSDSs from our site prior to November 8, 2000, you should download and replace all existing MSDSs with the updated/corrected copies. If you require further information please feel free to contact us.

MSDS approved for release

- HD - Mustard
- GD - Soman
- GA - Tabun
- VX - Nerve Agent
- GB - Sarin
- Lewisite

Safety Office: (410) 436-4411

Disclaimer: While the Edgewood Chemical Biological Center, Department of the Army, believes that the data contained herein are factual and the opinions expressed are those of qualified experts regarding the results of the tests conducted, the data is not to be taken as warranty or representation for which the Department of the Army or Edgewood Chemical Biological Center assumes legal responsibility. They are offered solely for your consideration, investigation, and verification. Any use of this data and information must be determined by the user to be in accordance with applicable Federal, State, and local laws regulations.

Distilled Mustard (HD)

Date: 22 September 1988
Revised: 29 September 1999

In the event of an emergency:
Telephone the SBCCOM Operations
Center's 24-hour emergency
Number: 410-436-2148

Section I—General Information

Manufacturer's Address:
U.S. Army Soldier and Biological Chemical Command (SBCCOM)
Edgewood Chemical Biological Center (ECBC)
ATTN: AMSSB-RCB-RS
Aberdeen Proving Ground, MD 21010-5424

CAS Numbers: 505-60-2, 39472-40-7, 68157-62-0

Chemical Name: Bis-(2-chloroethyl)sulfide

Trade name and synonyms:
H; HD; HS
Mustard Gas
Sulfur mustard; Sulphur mustard gas
Sulfide, bis (2-chloroethyl)
Bis(beta-chloroethyl)sulfide
1,1'-thiobis(2-chloroethane)
1-chloro-2(beta-chloroethylthio)ethane
Beta, beta'-dichlorodiethyl sulfide
2,2'dichlorodiethyl sulfide
Di-2-chloroethyl sulfideBeta, beta'-dichloroethyl sulfide Iprit S-Lost;
S-yperite; Schewefel-lost
Senfgas
Yellow Cross Liquid
Yperite;Y
EA 1033

Chemical Family: Chlorinated sulfur compound

Formula/Chemical Structure:
C4 H8 C12 S

C1 CH2 CH2—S—CH2 CH2 C1

NFPA 704 Signal:

Health - 4
Flammability - 1
Reactivity - 1
Special - 0

Section II—Ingredients

Ingredients/Name: Sulfur Mustard

Percentage by Weight: 100%

Threshold Limit Value (TLV): 0.003mg/m³

Section III—Physical Data

Boiling Point °F (°C): Calculated 423.5 °F (217.5 °C) (decomposed)

Vapor Pressure (mm Hg):

0.069 @ 20 °C
0.11 @ 25 °C

Vapor Density (Air = 1): 5.5

Solubility (g/100g solvent): Negligible in water (0.92 @ 22 °C). Soluble in fats and oils, gasoline, kerosene, acetone, carbon tetrachloride, alcohol, tetrachloroethane, ethylbenzoate, and ether. Miscible with the organophosphorus nerve agents.

Specific Gravity (H₂0 = 1): 1.27 @ 25 °C

Freezing/Melting Point (°C): 14.45

Liquid Density (g/mL):

1.274 g/mL @ 20 °C
1.268 g/mL @ 25 °C

Volatility (mg/m³):

600 @ 20 °C
910 @ 25 °C

Viscosity (Centipoise): 5.175 @ 20 °C

Molecular Weight (g/mol): 159.08

Appearance and Odor: Normally amber to black colored liquid with garlic or horseradish odor. Water clear if pure. The odor threshold for HD is 0.6 mg/m^3 (0.0006 mg/L).

Section IV—Fire and Explosion Data

Flashpoint: 105 °C (Can be ignited by large explosive charges)

Flammability Limits (% by volume): Unknown

Extinguishing Media: Water, fog, foam, CO_2. Avoid use of extinguishing methods that will cause splashing or spreading of HD.

Special Fire Fighting Procedures: All persons not engaged in extinguishing the fire should be immediately evacuated from the area. Fires involving HD should be contained to prevent contamination to uncontrolled areas. When responding to a fire alarm in buildings or areas containing agents, firefighting personnel should wear full firefighter protective clothing (flame resistant) during chemical agent firefighting and fire rescue operations. Respiratory protection is required. Positive pressure, full facepiece, NIOSH-approved self-contained breathing apparatus (SCBA) will be worn where there is danger of oxygen deficiency and when directed by the fire chief or chemical accident/incident (CAI) operations officer. In cases where firefighters are responding to a chemical accident/incident for rescue/reconnaissance purposes, they will wear appropriate levels of protective clothing (See Section VIII). Do not breathe fumes. Skin contact with nerve agents must be avoided at all times. Although the fire may destroy most of the agent, care must still be taken to ensure the agent or contaminated liquids do not further contaminate other areas or sewers. Contact with the agent, liquid or vapor, can be fatal.

Section V—Health Hazard Data

Airborne Exposure Limit (AEL): The AEL for HD is 0.003 mg/m^3 as found in "DA Pam 40-173, Occupational Health Guidelines for the Evaluation and Control of Occupational Exposure to Mustard Agents H, HD, and HT." To date, the Occupational Safety and Health Administration (OSHA) has not promulgated a permissible exposure concentration for HD.

Effects of Overexposure: HD is a vesicant (causing blisters) and alkylating agent producing cytotoxic action on the hematopoietic (blood-forming) tissues which are especially sensitive. The rate of detoxification of HD in the body is very slow and repeated exposures produce a cumulative effect. HD has been found to be a human carcinogen by the International Agency for Research on Cancer (IARC).

Median doses of HD in man are:
LD50 (skin) = 100 mg/kg
ICt50 (skin) = 2000 mg-min/m^3 at 70–80 °F (humid environment)
= 1000 mg-min/m^3 at 90 °F (dry environment)
ICt50 (eyes) = 200 mg-min/m^3
ICt50 (inhalation) = 1500 mg-min/m^3
LD50 (oral) = 0.7 mg/kg
Maximum safe Ct for skin and eyes are 5 and 2 mg-min/m^3, respectively.

Acute Physiological Action of HD is Classified as Local and Systemic.

Local Actions: HD effects both the eyes and the skin. Skin damage occurs after percutaneous absorption. Being lipid soluble, HD can be absorbed into all organs. Skin penetration is rapid without skin irritation. Swelling (blisters) and reddening (erythema) of the skin occurs after a latency period of 4–24 hours following the exposure, depending on degree of exposure and individual sensitivity. The skin healing process is very slow. Tender skin, mucous membrane and perspiration-covered skin are more sensitive to the effects of HD. HD's effect on the skin, however, is less than on the eyes. Local action on the eyes produces severe necrotic damage and loss of eyesight. Exposure of eyes to HD vapor or aerosol produces lactimation, photophobia, and inflammation of the conjunctiva and cornea.

Systemic Actions: Occurs primarily through inhalation and ingestion. The HD vapor or aerosol is less toxic to the skin or eyes than the liquid form. When inhaled, the upper respiratory tract (nose, throat, trachea) is inflamed after a few hours latency period, accompanied by sneezing, coughing, and bronchitis, loss of appetite, diarrhea, fever, and apathy. Exposure to nearly lethal doses of HD can produce injury to bone marrow, lymph nodes, and spleen as shown by a drop in white blood cell count, thus resulting in increased susceptibility to local and systemic infections. Ingestion of HD will produce severe stomach pains, vomiting, and bloody stools after a 15–20 minute latency period.

Chronic Exposure: HD can cause sensitization, chronic lung impairment, (cough, shortness of breath, chest pain), cancer of the mouth, throat, respiratory tract and skin, and leukemia. It may also cause birth defects.

EMERGENCY AND FIRST AID PROCEDURES:

Inhalation: Hold breath until respiratory protective mask is donned. Remove from the source **Immediately.** If breathing is difficult, administer oxygen. If breathing has stopped, give artificial respiration. Mouth-to-mouth resuscitation should be used when approved mask-bag or oxygen delivery systems are not available. Do not use mouth-to-mouth resuscitation when facial contamination is present. Seek medical attention **Immediately.**

Eye Contact: Speed in decontaminating the eyes is absolutely essential. Remove the person from the liquid source, flush the eyes **Immediately** with water for at least 15 minutes by tilting the head to the side, pulling the eyelids apart with the fingers and pouring water slowly into the eyes. Do not cover eyes with bandages but, if necessary, protect eyes by means of dark or opaque goggles. Transfer the patient to a medical facility **Immediately.**

Skin Contact: Don respiratory protective mask. Remove the victim from agent sources **Immediately.** Immediately wash skin and clothes with 5% solution of sodium hypochlorite or liquid household bleach within one minute. Cut and remove contaminated clothing, flush contaminated skin area again with 5% sodium hypochlorite solution, then wash contaminated skin area with soap and water. Seek medical attention **Immediately.**

Ingestion: Do not induce vomiting. Give victim milk to drink. Seek medical attention **Immediately.**

Section VI—Reactivity Data

Stability: Stable at ambient temperatures. Decomposition temperature is 100–351 °F (149–117 °C). Mustard is a persistent agent depending on pH and moisture and has been known to remain active for up to three years in soil.

Incompatibility: Rapidly corrosive to brass @ 65 °C. Will corrode steel at a rate of 0.0001 in. of steel per month @ 65 °C.

Hazardous Decomposition: Mustard will hydrolyze to form HCl and thiodiglycol.

Hazardous Polymerization: Does not occur.

Section VII—Spill, Leak, and Disposal Procedures

Steps to Be Taken in Case Material Is Released or Spilled: Only personnel in full protective clothing (See Section VIII) will be allowed in an area where HD is spilled. See Section V for emergency and first aid instructions.

Recommended Field Procedures: The HD should be contained using vermiculite, diatomaceous earth, clay, or fine sand and neutralized as soon as possible using copious amounts of 5.25% sodium hypochlorite solution. Scoop up all material and place in an approved DOT container. Cover the contents with decontaminating solution as above. The exterior of the container will be decontaminated and labeled according to EPA and DOT regulations. All leaking containers will be over packed with sorbent (e.g. vermiculite) placed between the interior and exterior containers. Decontaminate and label according to EPA and DOT regulations. Dispose of the material in accordance with waste disposal methods provided below. Conduct general area monitoring with an approved monitor to confirm that the atmospheric concentrations do not exceed the airborne exposure limits (See Sections II and VIII). If 5.25% sodium hypochlorite solution is not available then the following decontaminants may be used instead and are listed in the order of preference: Calcium Hypochlorite, contamination Solution No. 2 (DS2), and Super Tropical Bleach Slurry (STB).

Warning: Pure, undiluted calcium hypochlorite (HTH) will burn on contact with liquid HD.

Recommended Laboratory Procedures:

Decontamination solution for each gram of HD. Allow 24 hours for decontamination to take place. Agitate solution at least one hour. Agitation is not necessary after the first hour. Testing for presence of active chlorine by use of acidic potassium iodide solution to give free iodine color. Adjust the resulting solution pH to between 10 and 11.

Place three milliliters (ml) of decontaminated solution in a test tube. Add several crystals of potassium iodine and swirl to dissolve. Add 3 ml of 50 wt.% sulfuric acid: water and swirl. IMMEDIATE iodine color shows the presence of active chlorine. If negative, add additional decontaminant to the decontamination solution, wait two hours, and test again for active chlorine. This works for either 5.5% sodium hypochlorite or 10% calcium hypochlorite decontamination solution.

Scoop up all materials and clothing and place in an approved DOT container. The exterior of the container will be decontaminated and labeled according to EPA and DOT regulations. All leaking containers will be over packed with sorbent (e.g. vermiculite) placed between the interior and exterior containers. Decontaminate and label according to EPA and DOT regulations. Dispose of the material in accordance with waste disposal methods provided below. Conduct general area monitoring with an approved monitor to confirm that the atmospheric concentrations do not exceed the airborne exposure limits (See Section VIII).

Note: Surfaces contaminated with HD, then rinsed and decontaminated may evolve sufficient HD vapor to produce a physiological response. HD on laboratory glassware may be oxidized by its vigorous reaction with concentrated nitric acid.

Waste Disposal Method: Open pit burning or burying of HD or items containing or contaminated with HD in any quantity is prohibited. Decontamination of waste or excess material shall be accomplished according to the procedures outlined above and can be destroyed by incineration in EPA approved incinerators according to appropriate provisions of Federal, State, and local Resource Conservation Recovery Act (RCRA) regulations.

Note: Some decontaminant solutions are hazardous waste according to RCRA regulations and must be disposed of according to those regulations.

Section VIII—Special Protection Information

RESPIRATORY PROTECTION:

Concentration

$<=0.003$ mg/m^3 as an 8-hr TWA

Respiratory Protective Equipment

Protective mask not required to be worn provided that:

(a) Monitoring will be conducted to confirm that engineering controls are properly maintaining concentrations $<=0.003$ mg/m^3 as an 8-hr TWA.

(b) M40-series mask is available for emergency escape purposes.

(c) Exposure has been limited to the extent practicable by engineering controls (remote operations, ventilation, and process isolation) and work practices.

If these conditions are not met, then follow the guidance for $>=0.003$ mg/m^3 as an 8-hr TWA.

Concentration

$>=0.003$ mg/m^3 as an 8-hr TWA

Respiratory Protective Equipment

NIOSH/MSHA approved, pressure demand full face piece SCBA suitable for use in high agent concentrations with protective ensemble. (See DA Pam 396-61 for examples).

VENTILATION:
Local Exhaust: Mandatory. Must be filtered or scrubbed. Air emissions shall meet local, state, and federal regulations.

Special: Chemical laboratory hoods will have an average inward face velocity of 100 linear feet per minute (lfpm) ±20% with the velocity at any point not deviating from the average face velocity by more than 20%. Existing laboratory hoods will have an inward face velocity of 150 lfpm ±20%. Laboratory hoods will be located such that cross drafts do not exceed 20% of the inward face velocity. A visual performance test using smoke producing devices will be performed in assessing the ability of the hood to contain agent HD.

Other: Recirculation of exhaust air from agent areas is prohibited. No connection between agent area and other areas through the ventilation system is permitted. Emergency backup power is necessary. Hoods should be tested semiannually or after modification or maintenance operations. Operations should be performed 20 centimeters inside hoods.

Protective Gloves: Butyl Rubber gloves M3 and M4 Norton, Chemical Protective Glove Set

Eye Protection: As a minimum, chemical goggles will be worn. For splash hazards use goggles and face shield.

Other Protective Equipment: For laboratory operations, wear lab coats, gloves, and have mask readily accessible. In addition, daily clean smocks, foot covers, and head covers will be required when handling contaminated lab animals.

Monitoring: Available monitoring equipment for agent HD is the M8/M9 detector paper, blue band tube, M256/M256A1 kits, bubbler, Depot Area Air Monitoring System (DAAMS), Automated Continuous Air Monitoring System (ACAMS), CAM-M1, Hydrogen Flame Photometric Emission Detector (HYFED), the Miniature Chemical Agent Monitor (MINICAM), and Real Time Analytical Platform (RTAP). Real-time, low-level monitors (with alarm) are required for HD operations. In their absence, an Immediately Dangerous to Life and Health (IDLH) atmosphere must be presumed. Laboratory operations conducted in appropriately maintained and alarmed engineering controls require only periodic low-level monitoring.

Section IX—Special Precautions

Precautions to Be Taken in Handling and Storing: When handling agents, the buddy system will be incorporated. No smoking, eating, or drinking in areas containing agents is permitted. Containers should be periodically inspected for leaks, (either visually or using a detector kit). Stringent control over all personnel practices must be exercised. Decontaminating equipment will be conveniently located. Exits must be designed to permit rapid evacuation. Chemical showers, eyewash stations, and personal cleanliness facilities must be provided. Wash hands before meals and shower thoroughly with special attention given to hair, face, neck, and hands using plenty of soap and water before leaving at the end of the work day.

Other Precautions: HD should be stored in containers made of glass for Research, Development, Test and Evaluation (RDTE) quantities or one-ton steel containers for large quantities. Agent containers will be stored in a single containment system within a laboratory hood or in a double-containment system.

For additional information see "AR 385-61, The Army Toxic Chemical Agent Safety Program," "DA Pam 385-61, Toxic Chemical Agent Safety Standards," and "DA Pam 40-173, Occupational Health Guidelines for the Evaluation and Control of Occupational Exposure to HD Agents H, HD, and HT."

Section X—Transportation Data

Note: Forbidden for transport other than via military (Technical Escort Unit) transport according to 49 CFR 172

Proper Shipping Name: Toxic liquids, n.o.s.

DOT Hazard Class: 6.1, Packing Group I, Hazard Zone B

DOT Label: Poison

DOT Marking: Toxic liquids, n.o.s. Bis-(2-chloroethyl) sulfide UN 2810, Inhalation Hazard

DOT Placard: Poison

Emergency Accident Precautions and Procedures: See Sections IV, VII, and VIII.

Precautions to Be Taken in Transportation: Motor vehicles will be placarded regardless of quantity. Drivers will be given full information regarding shipment and conditions in case of an emergency. AR 50-6 deals specifically with the shipment of chemical agents. Shipment of agents will be escorted in accordance with AR 740-32.

The Edgewood Chemical Biological Center (ECBC), Department of the Army, believes that the data contained herein are actual and are the results of the tests conducted by ECBC experts. The data are not to be taken as a warranty or representation for which the Department of the Army or ECBC assumes legal responsibility. They are offered solely for consideration. Any use of this data and information contained in this MSDS must be determined by the user to be in accordance with applicable Federal, State, and local laws and regulations.

Addendum A

Additional Information for Thickened HD

Trade Name and Synonyms: Thickened HD, THD

Trade Name and Synonyms for Thickener:
Acrylic acid butyl ester
Polymer with styrene
Butyl acrylate-styrene polymer
Butyl acrylate-styrene copolymer
N-Butyl acrylate-styrene polymer
Polymer with styrene acrylic acid butyl ester
2-Propenoic acid
Butyl ester
Polymer with ethenylbenzene Acronal 4D
Acronal 290D
Acronal 295D Mowilith DM60
Sokrate LX 75
OSH22097

Hazardous Ingredients: Styrene-butyl acrylate copolymer is used to thicken HD and is not known to be hazardous except in a finely-divided, powder form.

Physical Data: Essentially the same as HD.

Fire and Explosion Data: Same as HD and slight fire hazard when exposed to heat or flame.

Health Hazard Data: Same as HD except for skin contact. For skin contact, don respiratory protective mask and remove contaminated clothing

IMMEDIATELY. IMMEDIATELY scrape the HD from the skin surface, then wash the contaminated surface with acetone. Seek medical attention IMMEDIATELY.

Spill, Leak, and Disposal Procedures: If spills or leaks of HD occur, follow the same procedures as those for HD, but dissolve THD in acetone before introducing any decontaminating solution. Containment of THD is generally not necessary. Spilled THD can be carefully scraped off the contaminated surface and placed in a fully removable head drum with a high density, polyethylene lining. THD can then be decontaminated, after it has been dissolved in acetone, using the same procedures used for HD. Contaminated surfaces should be treated with acetone, then decontaminated using the same procedures as those used for HD.

Note: Surfaces contaminated with THD and then rinse-decontaminated may evolve sufficient HD vapor to produce a physiological response.

Special Protection Information: Same as HD.

Special Precautions: Same as HD with the following addition. Handling the THD requires careful observation of the "stringers" (elastic, thread like attachments) formed when the agents are transferred or dispensed. These stringers must be broken cleanly before moving the contaminating device or dispensing device to another location, or unwanted contamination of a working surface will result. Avoid contact with strong oxidizers, excessive heat, sparks, or open flame.

Transportation Data: Same as HD.

Soman (GD)

Date: 14 September 1988
Revised: 29 September 1999

In the event of an emergency
Telephone the SBCCOM Operations
Center's 24-hour emergency
Number: 410-436-2148

Section I—General Information

Manufacturer's Address:
U.S. Army Soldier and Biological Chemical Command (SBCCOM)
Edgewood Chemical Biological Center (ECBC)
ATTN: AMSSB-RCB-RS
Aberdeen Proving Ground, MD 21010-5424

CAS Registry Numbers:
96-64-0, 50642-24-5

Chemical Name: Pinacolyl methyl phosphonofluoridate

Alternate Chemical Names:
Phosphonofluoridic acid, methyl-,1,2,2-trimethylpropyl ester O-Pinalcolyl methylphosphonofluoridate

Trade Name and Synonyms:
3,3 dimethyl-n-but-2-yl methylphosphonofluridate
1,2,2-Trimethylpropyl methylphosphonofluridate
Methylpinacolyloxyfluorophosphine oxide
Pinacolyloxymethylphosphonyl fluoride
Pinacolyl methanefluorophosphonate
Methylfluoropinacolylphosphonate
Fluoromethylpinacolyloxyphosphine oxide
Methylpinacolyloxyphosphonyl fluoride
Pinacolyl methylfluorophosphonate
1,2,2-Trimethylpropoxyfluoromethylphosphine oxide
GD
EA 1210
Soman
Zoman
PFMP

Chemical Family: Fluorinated organophosphorus compound

Formula/Chemical Structure:

$$C_7 H_{16} F O_2 P$$

NFPA 704 Signal:

Health - 4
Flammability - 1
Reactivity - 1
Special - 0

Section II—Ingredients

Ingredients/Name: GD

Percentage by Weight: 100%

Threshold Limit Value (TLV): 0.00003 mg/m^3

Section III—Physical Data

Boiling Point @ 760 mm Hg: 388 °F (198 °C)

Vapor Pressure (mm Hg): 0.40 @ 25 °C

Vapor Density (Air = 1 STP): 6.29 @ 25 °C

Solubility (g/100g solvent): Slightly soluble in water; 3.4 @ 0 °C; 2.1 @ 20 °C. Soluble in sulfur mustard, gasoline, alcohols, fats, and oils.

Specific Gravity (H$_2$O = 1 g/mL @ 25 °C): 1.0252

Freezing/Melting Point (°C): −42 °C

Liquid Density (g/cc): 1.0222 @ 25 °C

Volatility (mg/m3): 3900 @ 25 °C

Viscosity (CENTISTOKES): 3.098 @ 25 °C

Appearance and Odor: When pure, colorless liquid with a fruity odor. With impurities, amber or dark brown with oil of camphor odor.

Section IV—Fire and Explosion Data

Flashpoint: 121 °C (Open Cup Method)

Flammability Limits (% By Volume): Not Available

Lower Explosive Limit: Not Available

Upper Explosive Limit: Not Available

Extinguishing Media: Water mist, fog, and foam, CO_2. Avoid using extinguishing methods that will cause splashing or spreading of the GD.

Special Fire Fighting Procedures: GD will react with steam or water to produce toxic and corrosive vapors. All persons not engaged in extinguishing the fire should be immediately evacuated from the area. Fires involving GD should be contained to prevent contamination to uncontrolled areas. When responding to a fire alarm in buildings or areas containing GD, fire-fighting personnel should wear full fire-fighter protective clothing during chemical agent fire-fighting and fire rescue operations. Respiratory protection is required. Positive pressure, full face piece, NIOSH-approved self-contained breathing apparatus (SCBA) will be worn where there is danger of oxygen deficiency and when directed by the fire chief or chemical accident/incident (CAI) operations officer. In cases where fire-fighters are responding to a chemical accident/incident for rescue/reconnaissance purposes they will wear appropriate levels of protective clothing (See Section VIII).

Do not breathe fumes. Skin contact with nerve agents must be avoided at all times. Although the fire may destroy most of the agent, care must still be taken to assure the agent or contaminated liquids do not further contaminate other areas or sewers. Contact with liquid GD or vapors can be fatal.

Unusual Fire and Explosion Hazards: Hydrogen produced by the corrosive vapors reacting with metals, concrete, etc., may be present.

Section V—Health Hazard Data

Airborne Exposure Limits (AEL): The permissible airborne exposure concentration for VX for an 8-hour workday of a 40-hour workweek is an 8-hour time weighted average (TWA) of 0.00003 mg/m^3. This value can be found in "DA Pam 40-8, Occupational Health Guidelines for the Evaluation and Control of Occupational Exposure to Nerve Agents GA, GB, GD, and VX." To date, however, the Occupational Safety and Health Administration (OSHA) has not promulgated a permissible exposure concentration for GD.

GD is not listed by the International Agency for Research on Cancer (IARC), American Conference of Governmental Industrial Hygienists (ACGIH), Occupational Safety and Health Administration (OSHA), or National Toxicology Program (NTP) as a carcinogen.

Effects of Overexposure: GD is a lethal cholinesterase inhibitor. Doses that are potentially life threatening may be only slightly larger than those producing least effects.

Route	Form	Effect	Type	Dosage
Ocular	vapor	miosis	ECt50	<2 mg-min/m^3
Inhalation	vapor	runny nose	ECt50	<2 mg-min/m^3
Inhalation (15 1/min)	vapor	severe incapacitation	ICt50	<35 mg-min/m^3
Inhalation (15 1/min)	vapor	death	LCt50	<70 mg-min/m^3
Percutaneous	liquid	death	LD50	<350 mg/70 kg man

Effective dosages for vapor are estimated for exposure durations of 2–10 minutes.

Symptoms of overexposure may occur within minutes or hours, depending upon dose, They include: miosis (constriction of pupils) and visual effects, headaches and pressure sensation, runny nose and nasal congestion, salivation, tightness in the chest, nausea, vomiting, giddiness, anxiety, difficulty in thinking and sleeping, nightmares, muscle twitches, tremors, weakness, abdominal cramps, diarrhea, involuntary urination and defecation. With severe exposure symptoms progress to convulsions and respiratory failure.

EMERGENCY AND FIRST AID PROCEDURES:

Inhalation: Hold breath until respiratory protective mask is donned. If severe signs of agent exposure appear (chest tightens, pupil constriction, in coordination, etc.), immediately administer, in rapid succession, all three Nerve Agent Antidote Kit(s), Mark I injectors (or atropine if directed by a physician). Injections using the Mark I kit injectors may be repeated at 5 to 20 minute intervals if signs and symptoms are progressing until three series of injections have been administered. No more injections will be given unless directed by medical personnel. In addition, a record will be maintained of all injections given. If breathing has stopped, give artificial respiration. Mouth-to-mouth resuscitation should be used when mask-bag or oxygen delivery systems are not available. Do not use mouth-to-mouth resuscitation when facial contamination exists. If breathing is difficult, administer oxygen. Seek medical attention **Immediately.**

Eye Contact: Immediately flush eyes with water for 10–15 minutes, then don respiratory protective mask. Although miosis (pinpointing of the pupils) may be an early sign of agent exposure, an injection will not be administered when miosis is the only sign present. Instead, the individual will be taken **Immediately** to a medical treatment facility for observation.

Skin Contact: Don respiratory protective mask and remove contaminated clothing. **Immediately** wash contaminated skin with copious amounts of soap and water, 10% sodium carbonate solution, or 5% liquid household bleach. Rinse well with water to remove excess decontaminant. Administer nerve agent antidote kit, Mark I, only if local sweating and muscular twitching symptoms are observed. Seek medical attention **Immediately.**

Ingestion: Do not induce vomiting. First symptoms are likely to be gastrointestinal. **Immediately** administer Nerve Agent Antidote Kit, Mark I. Seek medical attention **Immediately.**

Section VI—Reactivity Data

Stability: Stable in steel for 3 months at 65 °C. **Incompatibility:** GD corrodes steel at the rate of 1×10^{-5} inch/month.

Hazardous Decomposition Products: GD will hydrolyze to form HF and

$$
\begin{array}{ccc}
 & H & O \\
 & | & || \\
(CH_3)_3C & - C - O - & P - OH \\
 & | & | \\
 & CH_3 & CH_3
\end{array}
$$

Hazardous Polymerization: Does not occur.

Section VII—Spill, Leak, and Disposal Procedures

Steps To Be Taken In Case Material Is Released Or Spilled: If leaks or spills of GD occur, only personnel in full protective clothing will remain in the area (See Section VIII). In case of personnel contamination, see Section V for emergency and first aid instructions.

Recommended Field Procedures: Spills must be contained by covering with vermiculite, diatomaceous earth, clay, fine sand, sponges, and paper or cloth towels. Decontaminate with copious amounts of aqueous sodium hydroxide solution (a minimum 10 wt.%). Scoop up all material and place in a DOT approved container. Cover the contents with decontaminating solution as above. After sealing, decontaminate the exterior and labeled according to EPA and DOT regulations. All leaking containers will be over packed with sorbent (e.g. vermiculite) placed between the interior and exterior containers. Decontaminate and label according to EPA and DOT regulations. Dispose of material according to federal, state, and local laws. Conduct general area monitoring to confirm that the atmospheric concentrations do not exceed the airborne exposure limits (See Sections II and VIII).

If 10 wt.% aqueous sodium hydroxide is not available then the following decontaminants may be used instead and are listed in the order of preference: Decontaminating Agent (DS2), Sodium Carbonate, and Supertropical Bleach Slurry (STB).

Recommended Laboratory Procedures: A minimum of 55 grams of decon solution is required per gram of GD. Decontaminant/agent solution is allowed to agitate for a minimum of one hour. Agitation is not necessary following the first hour provided a single phase is obtained. At the end of the first hour, the pH should be checked and adjusted up to 11.5 with additional NaOH as required. An alternate solution for the decontamination of GD is

10% sodium carbonate in place of the 10% Sodium Hydroxide solution above. Continue with 55 grams of decon per gram of GD. Agitate for one hour and allow to react for three hours. At the end of the third hour, adjust the pH to above 10. It is also permitted to substitute 5.25% sodium hypochlorite for the 10% sodium hydroxide solution above. Continue with 55 grams of decon per gram of GD. Agitate for one hour and allow to react for three hours then adjust the pH to above 10. Scoop up all material and clothing. Place all material in a DOT approved container. Cover the contents with decontaminating solution as above. After sealing, decontaminate the exterior of the container and label according to EPA and DOT regulations. All leaking containers will be over packed with sorbent placed between the interior and exterior containers. Decontaminate and label according to EPA and DOT regulations. Dispose of contents and decontaminate according to Federal, State, and local laws. Conduct general area monitoring to confirm that the atmospheric concentrations do not exceed the airborne exposure limits (See Sections II and VIII).

Waste Disposal Method: Open pit burning or burying of GD or items containing or contaminated with GD in any quantity is prohibited. The detoxified GD (using procedures above) can be thermally destroyed by incineration in EPA approved incinerators according to appropriate provisions of Federal, state, and local Resource Conservation and Recovery Act (RCRA) Regulations.

Note: Some decontaminate solutions are hazardous wastes according to RCRA regulations and must be disposed of according to those regulations.

Section VIII—Special Protection Information

Concentration	Respiratory Protective Equipment
<0.00003 mg/m³	A full face piece, chemical canister, air-purifying protective mask will be on hand for escape. M40-series masks are acceptable for this purpose. Other masks certified as equivalent may be used.
>0.00003 or = 0.06 mg/m³	A NIOSH/MSHA approved pressure demand full face piece SCBA or supplied air respirators with escape air

| | cylinder may be used. Alternatively, a full face piece, chemical canister air-purifying protective mask is acceptable for this purpose (See DA Pam 385-61 for determination of appropriate level). |
| >0.06 mg/m³ or unknown | NIOSH/MSHA approved pressure demand full face piece SCBA suitable for use in high agent concentrations with protective ensemble (See DA Pam 385-61 for examples). |

VENTILATION:

Local exhaust: Mandatory. Must be filtered or scrubbed to limit exit concentrations to <0.00003 mg/m³. Air emissions will meet local, state, and federal regulations.

Special: Chemical laboratory hoods will have an average inward face velocity of 100 linear feet per minute (lfpm)±20% with the velocity at any point not deviating from the average face velocity by more than 20%. Existing laboratory hoods will have an inward face velocity of 150 lfpm ±20%. Laboratory hoods will be located such that cross-drafts do not exceed 20% of the inward face velocity. A visual performance test using smoke-producing devices will be performed in assessing the ability of the hood to contain agent GD.

Other: Recirculation or exhaust air from chemical areas is prohibited. No connection between chemical areas and other areas through ventilation system is permitted. Emergency backup power is necessary. Hoods should be tested at least semiannually or after modification or maintenance operations. Operations should be performed 20 centimeters inside hood face.

Protective Gloves: Butyl Rubber Glove M3 and M4 Norton, Chemical Protective Glove Set

Eye Protection: At a minimum chemical goggles will be worn. For splash hazards use goggles and face shield.

Other Protective Equipment: For laboratory operations, wear lab coats, gloves and have mask readily accessible. In addition, daily clean smocks, foot covers, and head covers will be required when handling contaminated lab animals.

Monitoring: Available monitoring equipment for agent GD is the M8/M9 detector paper, detector ticket, M256/M256A1 kits, bubbler, Depot Area Air Monitoring System (DAAMS), Automated Continuous Air Monitoring System (ACAMS), Real-Time Monitor (RTM), Demilitarization Chemical Agent Concentrator (DCAC), M8/M43, M8A1/M43A1, CAM-M1, Hydrogen Flame Photometric Emission Detector (HYFED), the Miniature Chemical Agent Monitor (MINICAM), and the Real Time Analytical Platform (RTAP).

Real-time, low-level monitors (with alarm) are required for GD operations. In their absence, an Immediately Dangerous to Life and Health (IDLH) atmosphere must be presumed. Laboratory operations conducted in appropriately maintained and alarmed engineering controls require only periodic low-level monitoring.

Section IX—Special Precautions

Precautions To Be Taken In Handling and Storing: When handling agents, the buddy system will be incorporated. No smoking, eating, or drinking in areas containing agents is permitted. Containers should be periodically inspected for leaks, (either visually or using a detector kit). Stringent control over all personal practices must be exercised. Decontaminating equipment will be convenietly located. Exits must be designated to permit rapid evacuation. Chemical showers, eyewash stations, and personal cleanliness facilities must be provided. Wash hands before meals and shower thoroughly with special attention given to hair, face, neck, and hands using plenty of soap and water before leaving at the end of the workday.

Other Precautions: Agent containers will be stored in a single containment system within a laboratory hood or in a double containment system.

For additional information see "AR 385-61, The Army Toxic Chemical Agent Safety Program," "DA Pam 385-61, Toxic Chemical Agent Safety Standards," and "DA Pam 40-173, Occupational Health Guidelines for the Evaluation and Control of Occupational Exposure to Nerve Agents GA, GB, GD, and VX."

Section X—Transportation Data

Note: Forbidden for transport other than via military (Technical Escort Unit) transport according to 49 CFR 172

Proper Shipping Name: Toxic liquids, organic, n.o.s.

DOT Hazard Class: 6.1, Packing Group I, Hazard Zone B

DOT Label: Poison

DOT Marking: Toxic liquids, organic, n.o.s. (Pinacolyl methyl phosphono-fluoridate) uN 2810, Inhalation Hazard

DOT Placard: Poison

Emergency Accident Precautions And Procedures: See Sections IV, VII, and VIII.

Precautions To Be Taken In Transportation: Motor vehicles will be placarded regardless of quantity. Drivers will be given full information regarding shipment and conditions in case of an emergency. AR 50-6 deals specifically with the shipment of chemical agents. Shipment of agents will be escorted in accordance with AR 740-32.

The Edgewood Chemical Biological Center (ECBC), Department of the Army believes that the data contained herein are actual and are the results of the tests conducted by ECBC experts. The data are not to be taken as a warranty or representation for which the Department of the Army or ECBC assumes legal responsibility. They are offered solely for consideration. Any use of this data and information contained in this MSDS must be determined by the user to be in accordance with applicable Federal, State, and local laws and regulations.

Addendum A

Additional Information for Thickened GD

Trade Name And Synonyms: Thickened GD, TGD

Trade Name And Synonyms for Thickener:
Acrylic acid butyl ester
Polymer with styrene
Butyl acrylate-styrene polymer
Butyl acrylate-styrene copolymer
N-Butyl acrylate-styrene polymer
Polymer with styrene acrylic acid butyl ester
2-Propenoic acid
Butyl ester
Polymer with ethenylbenzene
Styrene-butyl acrylate polymer
Acronal 4D
Acronal 290D

Acronal 295D
Acronal 320D
Mowilith DM60
Sokrate LX 75
OSH22097

Hazardous Ingredients: Styrene-butyl acrylate copolymer is used to thicken GD and is not known to be hazardous except in a finely-divided, powder form.

Physical Data: Essentially the same as GD.

Fire And Explosion Data: Same as GD. Thickener poses a slight fire hazard when exposed to heat or flame.

Health Hazard Data: Same as GD except for skin contact. For skin contact, don respiratory protective mask and remove contaminated clothing **Immediately. Immediately** scrape the TGD from the skin surface, then wash the contaminated surface with acetone. Administer Nerve Agent Antidote Mark I Kit, only if local sweating and muscular twitching symptoms are observed. Seek medical attention **Immediately.**

Spill, Leak, And Disposal Procedures: If spills or leaks of GD occur, follow the same procedures as those for GD, but dissolve TGD in acetone before introducing any decontaminating solution. Containment of TGD is generally not necessary. Spilled TGD can be carefully scraped off the contaminated surface and placed in a fully removable head drum with a high density, polyethylene lining. TGD can then be decontaminated, after it has been dissolved in acetone, using the same procedures used for GD. Contaminated surfaces should be treated with acetone, then decontaminated using the same procedures as those used for GD.

Special Protection Information: Same as GD.

Special Precautions: Same as GD with the following addition. Handling the TGD requires careful observation of the "stringers" (elastic, thread like attachments) formed when the agents are transferred or dispensed. These stringers must be broken cleanly before moving the contaminating device or dispensing device to another location, or unwanted contamination of a working surface will result. Avoid contact with strong oxidizers, excessive heat, sparks, or open flame.

Transportation Data: Same as GD.

Tabun (GA)

Date: 22 September 1988
Revised: 28 September 1999

In the event of an emergency:
Telephone the SBCCOM Operations
Center's 24-hour emergency
Number: 410-436-2148

Section I—General Information

Manufacturer's Address:
U.S. Army Soldier and Biological Chemical Command (SBCCOM)
Edgewood Chemical Biological Center (ECBC)
ATTN: AMSSB-RCB-RS
Aberdeen Proving Ground, MD 21010-5424

CAS Registry Numbers: 77-81-6

Chemical Name: Ethyl N,N-dimethylphosphoramidocyanidate

Trade Name and Synonyms:
Ethyl dimethylphosphoramidocyanidate
Dimethylaminoethoxy-cyanophosphine oxide
Dimethylamidoethoxyphosphoryl cyanide
Ethyldimethylaminocyanophosphonate
Ethyl ester of dimethylphosphoroamidocyanidic acid
Ethyl phosphorodimethylamidocyanidate
GA
EA1205
Tabun

Chemical Family: Organophosphorus compound

Formula/Chemical Structure:

$$C_5 \, H_{11} \, N_2 \, O_2 \, P$$

```
                        O        CH₃
                        ||      /
    CH₃ CH₂ – O – P – – N
                        |      \
                        CN      CH₃
```

NFPA 704 Signal:

Health - 4
Flammability - 1
Reactivity - 1
Special - 0

Section II—Ingredients

Ingredients/Name: GA

Percentage by Weight: 100%

Threshold Limit Value (TLV): 0.0001 mg/m^3

Section III—Physical Data

Boiling Point @ 760 mm Hg: 478 °F (248 °C)

Vapor Pressure: 0.057 mm Hg @ 25 °C

Vapor Density (Air = 1 STP): 5.59 @25 °C

Solubility (g/100g solvent): Slightly soluble in water, 9.8 @ 0 °C; 7.2 @ 20 °C. Readily soluble in organic solvents.

Specific Gravity (H$_2$0 = 1 g/mL@25 °C): 1.076

Freezing/Melting Point (°C): −50 °C

Liquid Density (g/cc): 1.073 @ 25 °C

Volatility (mg/m³): 490 @ 25 °C

Viscosity (CENTISTOKES): 2.18 @ 25 °C

Appearance and Odor: Colorless to brown liquid, faintly fruity odor. Odorless in pure foam.

Section IV—Fire And Explosion Data

Flashpoint: 78 °C (Closed Cup Method)

Flammability Limits (% By Volume): Not Available

Lower Explosive Limit: Not Available

Upper Explosive Limit: Not Available

Extinguishing Media: Water mist, fog, and foam, CO_2. Avoid using extinguishing methods that will cause splashing or spreading of the GA.

Special Fire Fighting Procedures: GA will react with steam or water to produce toxic and corrosive vapors. All persons not engaged in extinguishing the fire should be immediately evacuated from the area. Fires involving GA should be contained to prevent contamination to uncontrolled areas. When responding to a fire alarm in buildings or areas containing GA, fire-fighting personnel should wear full fire-fighter protective clothing during chemical agent fire-fighting and fire rescue operations. Respiratory protection is required. Positive pressure, full face piece, NIOSH-approved self-contained breathing apparatus (SCBA) will be worn where there is danger of oxygen deficiency and when directed by the fire chief or chemical accident/incident (CAI) operations officer. In cases where fire-fighters are responding to a chemical accident/incident for rescue/reconnaissance purposes they will wear apropriate levels of protective clothing (See Section VIII). Do not breathe fumes. Skin contact with nerve agents must be avoided at all times. Although the fire may destroy most of the agent, care must still be taken to assure the agent or contaminated liquids do not further contaminate other areas or sewers. Contact with liquid GA or vapors can be fatal.

Unusual Fire and Explosion Hazards: Fires involving this chemical may result in the formation of hydrogen cyanide.

Section V—Health Hazard Data

Airborne Exposure Limits (AEL): The permissible airborne exposure concentration for GA for an 8-hour workday of a 40-hour workweek is an 8-hour time weighted average (TWA) of 0.0001 mg/m^3. This value can be found in "DA Pam 40-8, Occupational Health Guidelines for the Evaluation and Control of Occupational Exposure to Nerve Agents GA, GB, GD, and VX." To date, however, the Occupational Safety and Health Administration (OSHA) has not promulgated a permissible exposure concentration for GA.

GA is not listed by the International Agency for Research on Cancer (IARC), American Conference of Governmental Industrial Hygienists (ACGIH), Occupational Safety and Health Administration (OSHA), or National Toxicology Program (NTP) as a carcinogen.

Effects of Overexposure: GA is a lethal cholinesterase inhibitor similar in action to GB. Although only about half as toxic as GB by inhalation, GA in low concentrations is more irritating to the eyes than GB. The number and severity of symptoms that appear are dependent on the quantity and rate of entry of the nerve agent introduced into the body. (Very small skin dosages sometimes cause local sweating and tremors with few other effects). Individuals poisoned by GA display approximately the same sequence of symptoms' despite the route by which the poison enters the body (whether by inhalation, absorption, or ingestion). These symptoms, in normal order of appearance, are: runny nose; tightness of the chest; dimness of vision and pin pointing of the eye pupils; difficulty in breathing; drooling and excessive sweating; nausea; vomiting, cramps, and involuntary defecation and urination; twitching, jerking, and staggering; and headache, confusion, drowsiness, coma, and convulsions. These symptoms are followed by cessation of breathing and death.

Onset Time of Symptoms: Symptoms appear much more slowly from a skin dosage than from a respiratory dosage. Although skin absorption great enough to cause death may occur in 1 to 2 minutes, death may be delayed for 1 to 2 hours. Respiratory lethal dosages kill in 1 to 10 minutes, and liquid in the eye kills almost as rapidly.

Median Lethal Dosage, Animals:

LD50 (monkey, percutaneous) = 9.3 mg/kg (shaved skin)

LCt50 (monkey, inhalation) = 187 mg-min/m(t = 10)

Median Lethal Dosage, Man:

LCt50 (man, inhalation) = 135 mg-min/m[(t = 0.5–2 min) at RMV (Respiratory Minute Volume) of 15 l/min];

200 mg-min/m[T at RMV of 10 l/min]

Emergency and First Aid Procedures:

Inhalation: Hold breath until respiratory protective mask is donned. If severe signs of agent exposure appear (chest tightens, pupil constriction, in coordination, etc.), immediately administer, in rapid succession, all three Nerve Agent Antidote Kit(s), Mark I injectors (or atropine if directed by a physician). Injections using the Mark I kit injectors may be repeated at 5 to 20 minute intervals if signs and symptoms are progressing until three series of injections have been administered. No more injections will be given unless directed by medical personnel. In addition, a record will be maintained of all injections given. If breathing has stopped, give artificial respiration. Mouth-to-mouth resuscitation should be used when mask-bag or oxygen delivery systems are not available. Do not use mouth-to-mouth resuscitation when facial contamination exists. If breathing is difficult, administer oxygen. Seek medical attention **Immediately.**

Eye Contact: Immediately flush eyes with water for 10–15 minutes, then don respiratory protective mask. Although miosis (pinpointing) of the pupils) may be an early sign of agent exposure, an injection will not be administered when miosis is the only sign present. Instead, the individual will be taken **Immediately** to medical treatment facility for observation.

Skin Contact: Don respiratory protective mask and remove contaminated clothing. **Immediately** wash contaminated skin with copious amounts of soap and water, 10% sodium carbonate solution, or 5% liquid household bleach. Rinse well with water to remove excess decontaminant. Administer nerve agent antidote kit, Mark I, only if local sweating and muscular twitching symptoms are observed. Seek medical attention **Immediately.**

Ingestion: Do not induce vomiting. First symptoms are likely to be gastrointestinal. **Immediately** administer Nerve Agent Antidote Kit, Mark I. Seek medical attention **Immediately.**

Section VI—Reactivity Data

Stability: Compound is stable in steel for several years.

Incompatibility: Not Available.

Hazardous Decomposition Products: Decomposes within 3 months at 65 °C. Complete decomposition in 195 minutes at 150 °C. May produce hydrogen cyanide (HCN), oxides of nitrogen, oxides of phosphorus, and carbon monoxide.

Hazardous Polymerization: Not available.

Section VII—Spill, Leak, And Disposal Procedures

Steps To Be Taken In Case Material Is Released Or Spilled: If leaks or spills of BA occur, only personnel in full protective clothing will remain in the area (See Section VIII). In case of personnel contamination, see Section V for emergency and first aid instructions. **Recommended Field Procedures:** Spills must be contained by covering with vermiculite, diatomaceous earth, clay, fine sand, sponges, and paper or cloth towels. Decontaminate with copious amounts of aqueous sodium hydroxide solution (a minimum 10 wt. %). Scoop up all material and place in a DOT approved container. The decontaminant solution must be treated with excess bleach to destroy the HCN formed during the hydrolysis. Cover the contents with decontaminating solution as above. After sealing, decontaminate the exterior and labeled according to EPA and DOT regulations. All leaking containers will be over packed with sorbent (e.g. vermiculite) placed between the interior and exterior containers. Decontaminate and label according to EPA and DOT regulations. Dispose of material according to Federal, state, and local laws. Conduct general area monitoring to confirm that the atmospheric concentrations do not exceed the airborne exposure limits (See Sections II and VIII).

If 10 wt.% aqueous sodium hydroxide is not available, then the following decontaminants may be used instead and are listed in the order of preference: Decontaminating Agent (DS2), Sodium Carbonate, and Supertropical Bleach Slurry (STB).

Recommended Laboratory Procedures:

A minimum of 56 grams of decon solution is required for each gram of GA. The decontamination solution is agitated while GA is added and the agitation is maintained for at least one hour. The resulting solution is allowed to react for 24 hours. At the end of 24 hours, the solution must be titrated to a pH between 10 and 12. After completion of the 24-hour period, the decontamination solution must be treated with excess bleach (2.5 mole OCl/mole GA) to destroy the CN formed during the hydrolysis.

Scoop up all material and clothing. Place all material in a DOT approved container. Cover the contents with decontaminating solution as above. After sealing, decontaminate the exterior of the container and label according to EPA and DOT regulations. All leaking containers will be over packed with sorbent placed between the interior and exterior containers. Decontaminate and label according to EPA and DOT regulations. Dispose of contents and decontaminant according to Federal, State, and local laws. Conduct general area monitoring to confirm that the atmospheric concentrations do not exceed the airborne exposure limits (See Sections II and VIII).

Waste Disposal Method: Open pit burning or burying of GA or items containing or contaminated with GA in any quantity is prohibited. The detoxified GA (using procedures above) can be thermally destroyed by incineration in EPA approved incinerators according to appropriate provisions of Federal, state, and local Resource Conservation and Recovery Act (RCRA) Regulations.

Note: Some decontaminant solutions are hazardous wastes according to RCRA regulations and must be disposed of according to those regulations.

Section VIII—Special Protection Information

RESPIRATORY PROTECTION:

Concentration
<0.0001 mg/m^3

Respiratory Protective Equipment

A full face piece, chemical canister air-purifying protective mask will be on hand for escape. M40-series masks are acceptable for this purpose. Other masks certified as equivalent may be used.

Concentration
>0.0001 or $=0.02$ mg/m^3

Respiratory Protective Equipment

A NIOSH/MSHA approved pressure demand full face piece SCBA or supplied air respirators with escape air cylinder may be used. Alternatively, a full face piece, chemical canister air-purifying protective mask is acceptable for this purpose (See DA Pam 385-61 for determination of appropriate level)

Concentration
>0.02 mg/m³ or unknown

Respiratory Protective Equipment

NIOSH/MSHA approved pressure demand full face piece SCBA suitable for use in high agent concentrations with protective ensemble. (See DA Pam 385-61 for examples)

VENTILATION:

Local exhaust: Mandatory. Must be filtered or scrubbed to limit exit concentrations to >0.0001 mg/m³. Air emissions will meet local, state and federal regulations.

Special: Chemical laboratory hoods will have an average inward face velocity of 100 linear feet per minute (lfpm)±20% with the velocity at any point not deviating from the average face velocity by more than 20%. Existing laboratory hoods will have an inward face velocity of 150 lfpm±20%. Laboratory hoods will be located such that cross-drafts do not exceed 20% of the inward face velocity. A visual performance test using smoke-producing devices will be performed in assessing the ability of the hood to contain agent GA.

Other: Recirculation or exhaust air from chemical areas is prohibited. No connection between chemical areas and other areas through ventilation system is permitted. Emergency backup power is necessary. Hoods should be tested at least semiannually or after modification or maintenance operations. Operations should be performed 20 centimeters inside hood face.

Protective Gloves: Butyl Rubber Glove M3 and M4 Norton, Chemical Protective Glove Set

Eye Protection: At a minimum chemical goggles will be worn. For splash hazards use goggles and face shield.

Other Protective Equipment: For laboratory operations, wear lab coats and gloves and have mask readily accessible. In addition, daily clean smocks, foot covers, and head covers will be required when handling contaminated lab animals.

Monitoring: Available monitoring equipment for agent GA is the M8/M9 detector paper, detector ticket, M256/M256A1 kits, bubbler, Depot Area Air Monitoring System (DAAMS), Automated Continuous Air Monitoring System (ACAMS), Real-Time Monitor (RTM), Demilitarization Chemical Agent Concentrator (DCAC), M8/M43, M8A1/M43A1, CAM-M1, Hydrogen Flame Photometric Emission Detector (HYFED), the Miniature Chemical Agent Monitor (MINICAM), and the Real Time Analytical Platform (RTAP).

Real-time, low-level monitors (with alarm) are required for GA operations. In their absence, an Immediately Dangerous to Life and Health (IDLH) atmosphere must be presumed. Laboratory operations conducted in appropriately maintained and alarmed engineering controls require only periodic low-level monitoring.

Section IX—Special Precautions

Precautions To Be Taken In Handling and Storing: When handling agents, the buddy system will be incorporated. No smoking, eating, or drinking in areas containing agents is permitted. Containers should be periodically inspected for leaks, (either visually or using a detector kit). Stringent control over all personnel practices must be exercised. Decontaminating equipment will be conveniently located. Exits must be designated to permit rapid evacuation. Chemical showers, eyewash stations, and personal cleanliness facilities must be provided. Wash hands before meals and shower thoroughly with special attention given to hair, face, neck, and hands using plenty of soap and water before leaving at the end of the workday.

Other Precautions: Agent containers will be stored in a single containment system within a laboratory hood or in a double containment system.

For additional information see "AR 385-61, The Army Toxic Chemical Agent Safety Program," "DA Pam 385-61, Toxic Chemical Agent Safety Standards," and "DA Pam 40-173, Occupational Health Guidelines for the Evaluation and Control of Occupational Exposure to Nerve Agents GA, GB, GD, and VX."

Section X—Transportation Data

Note: Forbidden for transport other than via military (Technical Escort Unit) transport according to 49 CFR 172

Proper Shipping Name: Toxic liquids, organic, n.o.s.

Dot Hazard Class: 6.1, Packing Group I, Hazard Zone B

DOT Label: Poison

DOT Marketing: Toxic liquids, organic, n.o.s. (Ethyl dimethylphosphoramidocyanidate) UN 2810, Inhalation Hazard

DOT Placard: Poison

Emergency Accident Precautions And Procedures: See Sections IV, VII, and VIII.

Precautions To Be Taken In Transportation: Motor vehicles will be placarded regardless of quantity. Drivers will be given full information regarding shipment and conditions in case of an emergency. AR 50-6 deals specifically with the shipment of chemical agents. Shipment of agents will be escorted in accordance with AR 740-32.

> The Edgewood Chemical Biological Center (ECBC), Department of the Army believes that the data contained herein are actual and are the results of the tests conducted by ECBC experts. The data are not to be taken as a warranty or representation for which the Department of the Army or ECBC assumes legal responsibility. They are offered solely for consideration. Any use of this data and information contained in this MSDS must be determined by the user to be in accordance with applicable Federal, State, and local laws and regulations.

Lethal Nerve Agent (VX)

Date: 14 September 1988
Revised: 29 September 1999

In the event of an emergency:
Telephone the SBCCOM Operations
Center's 24-hour emergency
Number: 410-436-2148

Section I—General Information

Manufacturer's Address:
U.S. Army Soldier and Biological Chemical Command (SBCCOM)
Edgewood Chemical Biological Center (ECBC)
ATTN: AMSSB-RCB-RS
Aberdeen Proving Ground, MD 21010-5424

CAS Registry Numbers:
50782-69-9, 51848-47-6, 53800-40-1, 70938-84-0

Chemical Name:
O-ethyl-S-(2-diisopropylaminoethyl) methylphosphonothiolate

Trade Name And Synonyms:
Phosphonothioic acid, methyl-, S-(bis(1-methylethylamino)ethyl) 0-ethyl
ester O-ethyl S-(2-diisopropylaminoethyl)
methylphosphonothiolate
S-2-Diisopropylaminoethyl O-ethyl methylphosphonothioate
S-2((2-Diisopropylamino)ethyl) O-ethyl methylphosphonothiolate
O-ethyl S-(2-diisopropylaminoethyl) methylphosphonothioate
O-ethyl S-(2-diisopropylaminoethyl) methylthiolphosphonoate
S-(2-diisopropylaminoethyl) o-ethyl methyl phosphonothiolate
Ethyl-S-dimethylaminoethyl methylphosphonothiolate VX
EA 1701
TX60

Chemical Family: Sulfonated organophosphorous compound

Formula/Chemical Structure:

$$C_{11} H_{26} N O_2 P S$$

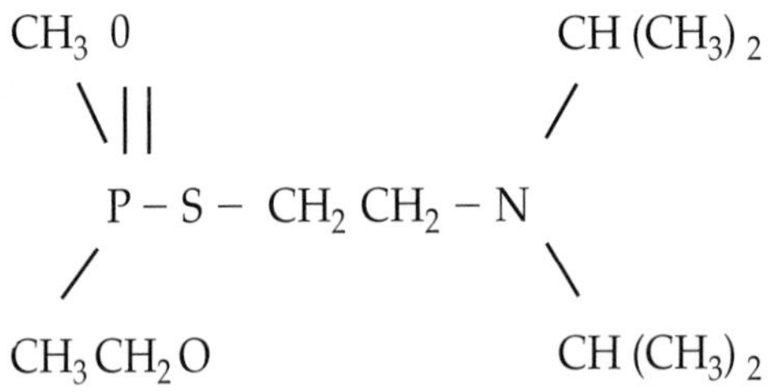

```
  CH₃  0                          CH (CH₃)₂
    \ ||                            /
     P – S –  CH₂ CH₂ – N
    /                               \
  CH₃ CH₂O                        CH (CH₃)₂
```

NFPA 704 Signal:
Health - 4
Flammability - 1
Reactivity - 1
Special - 0

Section II—Ingredients

Ingredients/Name: VX

Percentage by Weight: 100%

Threshold Limit Value (TLV): 0.00001 mg/m^3

Section III—Physical Data
Boiling Point @ 760 mm Hg: 568 °F (298 °C)

Vapor Pressure: 0.00063 mm Hg @ 25 °C

Vapor Density (Air = 1 STP): 9.2 @ 25 °C

Solubility (g/100g solvent): 5.0 @ 21.5 °C and 3.0 @ 25 °C in water. Soluble in organic solvents.

Specific Gravity (H$_2$O = 1g/mL@25°C): 1.0113

Freezing/Melting Point (°C): −50 °C

Liquid Density: 1.0083 g/mL@25 °C

Volatility: 8.9 mg/m^3 @ 25 °C

Viscosity (CENTISTOKES): 9.958 @ 25 °C

Appearance and Odor: Colorless to straw colored liquid and oderless, similar in appearance to motor oil.

Section IV—Fire and Explosion Data
Flashpoint: 159 °C (McCutchan-Young)

Flammability Limits (% By Volume): Not Available

Lower Explosive Limit: Not Applicable.

Upper Explosive Limit: Not Applicable.

Extinguishing Media: Water mist, fog, foam, CO$_2$. Avoid using extinguishing methods that will cause splashing or spreading of the VX.

Special Fire Fighting Procedures: All persons not engaged in extinguishing the fire should be immediately evacuated from the area. Fires involving VX should be contained to prevent contamination to uncontrolled areas. When responding to a fire alarm in buildings or areas containing VX, fire fighting personnel should wear full firefighter protective clothing during chemical agent firefighting and fire rescue operations. Respiratory protection is required. Positive pressure, full face piece, NIOSH-approved self-contained breathing apparatus (SCBA) will be worn where there is danger of oxygen deficiency and when directed by the fire chief or chemical accident/incident (CAI) operations officer. In cases where firefighters are responding to a chemical accident/incident for rescue/reconnaissance purposes they will wear appropriate levels of protective clothing (See Section VIII). Do not breathe fumes. Skin contact with nerve agents must be avoided at all times. Although the fire may destroy most of the agent, care must still be taken to assure the agent or contaminated liquids do not further contaminate other areas or sewers. Contact with liquid VX or vapors can be fatal.

Unusual Fire And Explosion Hazards: None known.

Section V—Health Hazard Data

Airborne Exposure Limits (AEL): The permissible airborne exposure concentration for VX for an 8-hour workday of a 40-hour work week is an 8-hour time weighted average (TWA) of 0.00001 mg/m^3. This value can be found in "DA Pam 40-8, Occupational Health Guidelines for the Evaluation and Control of Occupational Exposure to Nerve Agents GA, GB, GD, and VX." To date, however, the Occupational Safety and Health Administration (OSHA) has not promulgated a permissible exposure concentration for VX.

VX is not listed by the International Agency for Research on Cancer (IARC), American Conference of Governmental Industrial Hygienists (ACGIH), Occupational Safety and Health Administration (OSHA), or National Toxicology Program (NTP) as a carcinogen.

Effects of Overexposure: VX is a lethal cholinesterase inhibitor. Doses which are potentially life-threatening may be only slightly larger than those producing least effects. Death usually occurs within 15 minutes after absorption of a fatal dosage.

Route	Form	Effect	Type	Dosage
ocular	vapor	miosis	Ect50	<0.09 mg-min/m^3
Inhalation	vapor	runny nose	Ect50	<0.09 mg-min/m^3
Inhalation (15 l/min)	vapor	severe incapacitation	Ict50	<25 mg-min/m^3
Inhalation (15 l/min)	vapor	death	Ict50	<30 mg-min/m^3
Percutaneous	liquid	death	Lct50	<10 mg/70 kg man minutes

Effective dosages for vapor are estimated for exposure durations of 2–10 minutes.

Symptoms of overexposure may occur within minutes or hours, depending upon the dose. They include: miosis (constriction of pupils) and visual effects, headaches and pressure sensation, runny nose and nasal congestion, salivation, tightness in the chest, nausea, vomiting, giddiness, anxiety, difficulty in thinking, difficulty sleeping, nightmares, muscle twitches, tremors, weakness, abdominal cramps, diarrhea, involuntary urination and defecation. With severe exposure symptoms progress to convulsions and respiratory failure.

Emergency and First Aid Procedures:

Inhalation: Hold breath until respiratory protective mask is donned. If severe signs of agent exposure appear (chest tightens, pupil constriction, incoordination, etc.), immediately administer, in rapid succession, all three Nerve Agent Antidote Kit(s), Mark I injectors (or atropine if directed by a physician). Injections using the Mark I kit injectors may be repeated at 5 to 20 minute intervals if signs and symptoms are progressing until three series of injections have been administered. No more injections will be given unless directed by medical personnel. In addition, a record will be maintained of all injections given. If breathing has stopped, give artificial respiration. Mouth-to-mouth resuscitation should be used when mask-bag or oxygen delivery systems are not available. Do not use mouth-to-mouth resuscitation when facial contamination exists. If breathing is difficult, administer oxygen. Seek medical attention **Immediately.**

Eye Contact: Immediately flush eyes with water for 10–15 minutes, then don respiratory protective mask. Although miosis (pinpointing of the pupils) may be an early sign of agent exposure, an injection will not be administered when miosis is the only sign present. Instead, the individual will be taken **Immediately** to a medical treatment facility for observation.

Skin Contact: Don respiratory protective mask and remove contaminated clothing. Immediately wash contaminated skin with copious amounts of soap and water, 10% sodium carbonate solution, or 5% liquid household bleach. Rinse well with water to remove excess decontaminant. Administer nerve agent antidote kit, Mark I, only if local sweating and muscular twitching symptoms are observed. Seek medical attention **Immediately.**

Ingestion: Do not induce vomiting. First symptoms are likely to be gastrointestinal. **Immediately** administer Nerve Agent Antidote Kit, Mark I. Seek medical attention **Immediately.**

Section VI—Reactivity Data

Stability: Relatively stable at room temperature. Unstabilized VX of 95% purity decomposes at a rate of 5% a month at 71 °C.

Incompatibility: Negligible on brass, steel, and aluminum.

Hazardous Decomposition Products: During a basic hydrolysis of VX up to 10% of the agent is converted to diisopropylaminoethyl methylphosphonothioic acid (EA2192). Based on the concentration of EA2192 expected to be formed during hydrolysis and its toxicity (1.4 mg/kg dermal in rabbit at 24 hours in a 10/90 wt.% ethanol/water solution), a Class B poison would result. The large scale decon procedure, which uses both HTH and NaOH, destroys VX by oxidation and hydrolysis. Typically the large scale product contains 0.2–0.4 wt.% EA2192 at 24 hours. At pH 12, the EA2192 in the large scale product has a half-life of about 14 days. Thus, the 90-day holding period at pH 12 results in about a 64-fold reduction of EA2192 (six half-lives). This holding period is sufficient to reduce the toxicity of the product below that of a Class B poison. Other less toxic products are ethyl methylphosphonic acid, methylphosphinic acid, diisopropyaminoethyl mercaptan, diethyl methylphosphonate, and ethanol. The small scale decontamination procedure uses sufficient HTH to oxidize all VX thus no EA2192 is formed.

Hazardous Polymerization: Does not occur.

Section VII—Spill, Leak, And Disposal Procedures

Steps To Be Taken In Case Material Is Released Or Spilled:

If leaks or spills of VX occur, only personnel in full protective clothing (See Section VIII) will remain in the area. In case of personnel contamination see Section V for emergency and first aid instructions.

Recommended Field Procedures (For Quantities Greater Than 50 Grams):

Note: These procedures can only be used with the approval of the Risk Manager or qualified safety professionals. Spills must be contained by covering with vermiculite, diatomaceous earth, clay or fine sand. An alcoholic HTH mixture is prepared by adding 100 milliliters of denatured ethanol to a 900-milliliter slurry of 10% HTH in water. This mixture should be made just before use since the HTH can react with the ethanol. Fourteen grams of alcoholic HTH solution are used for each gram of VX. Agitate the decontamination mixture as the VX is added. Continue the agitation for a minimum of one hour. This reaction is reasonably exothermic and evolves substantial off gassing. The evolved reaction gases should be routed through a decontaminate filled scrubber before release through filtration systems. After completion of the one hour minimum agitation, 10% sodium hydroxide is added in a quantity equal to that necessary to assure that a pH of 12.5 is maintained for a period not less than 24 hours. Hold the material at a pH between 10 and 12 for a period not less than 90 days to ensure that a hazardous intermediate material is not formed (See Section VI). Scoop up all material and place in a DOT approved container. Cover the contents of the with decontaminating solution as above. After sealing the exterior, decontaminate and label according to EPA and DOT regulations. All leaking containers will be over packed with sorbent (e.g., vermiculite) placed between the interior and exterior containers. Decontaminate and label according to EPA and DOT regulations. Dispose of decontaminant according to Federal, state, and local laws. Conduct general area monitoring to confirm that the atmospheric concentrations do not exceed the airborne exposure limits (See Sections II and VIII).

If the alcoholic HTH mixture is not available, then the following decontaminants may be used instead and are listed in the order of preference: Decontaminating Agent (DS2), Supertropical Bleach Slurry (STB), and Sodium Hypochlorite.

Recommended Laboratory Procedures (For Quantities Less Than 50 Grams):

If the active chlorine of the Calcium Hypochlorite (HTH) is at least 55%, then 80 grams of a 10% slurry are required for each gram of VX. Proportionally more HTH is required if the chlorine activity of the HTH is lower than 55%. The mixture is agitated as the VX is added and the agitation is maintained for a minimum of one hour. If phasing of the VX/decon solution continues after 5 minutes, an amount of denatured ethanol equal to a 10 wt.% of the total agent/decon will be added to help miscibility. Place all material in a DOT approved container. Cover the contents with decontaminating solution as above. After sealing, decontaminate the exterior of the container and label according to EPA and DOT regulations. All leaking containers will be over packed with sorbent placed between the interior and exterior containers. Decontaminate and label according to EPA and DOT regulations. Dispose of according to Federal, State, and local laws. Conduct general area monitoring to confirm that the atmospheric concentrations do not exceed the airborne exposure limits (See Sections II and VIII).

Note: Ethanol should be reduced to prevent the formation of a hazardous waste.

Upon completion of the one hour agitation the decon mixture will be adjusted to a pH between 10 and 11. Conduct general area monitoring to confirm that the atmospheric concentrations do not exceed the airborne exposure limits (See Sections II and VIII).

Waste Disposal Method: Open pit burning or burying of VX or items containing or contaminated with VX in any quantity is prohibited. The detoxified VX (using procedures above) can be thermally destroyed by in a EPA approved incinerator according to appropriate provisions of Federal, State, or local Resource Conservation and Recovery Act (RCRA) regulations.

Note: Some decontaminant solutions are hazardous waste according to RCRA regulations and must be disposed of according to those regulations.

Section VIII—Special Protection Information

Concentration	Respiratory Protection Respiratory Protective Equipment
<0.00001 mg/m^3	A full face piece, chemical canister air-purifying protective mask will be on hand for escape. M40-series masks are acceptable for this purpose. Other masks certified as equivalent may be used.
>0.00001 or = 0.02 mg/m^3	A NIOSH/MSHA approved pressure demand full face piece SCBA or supplied air respirators with escape air cylinder may be used. Alternatively, a full face piece, chemical canister air-purifying protective mask is acceptable for this purpose (See DA Pam 385-61 for determination of appropriate level.
>0.02 or unknown	NIOSH/MSHA approved pressure demand full face piece SCBA suitable for use in high agent concentrations with protective ensemble. (See DA Pam 385-61 for examples)

VENTILATION:

Local exhaust: Mandatory. Must be filtered or scrubbed to limit exit concentrations to <0.00001 mg/m^3. Air emissions will meet local, state, and federal regulations.

Special: Chemical laboratory hoods will have an average inward face velocity of 100 linear feet per minute (lfpm) ±20% with the velocity at any point not deviating from the average face velocity by more than 20%. Existing laboratory hoods will have an inward face velocity of 150 lfpm ±20%. Laboratory hoods will be located such that cross-drafts do not exceed 20% of the inward face velocity. A visual performance test using smoke-producing devices will be performed in assessing the ability of the hood to contain agent VX.

Other: Recirculation or exhaust air from chemical areas is prohibited. No connection between chemical areas and other areas through ventilation system is permitted. Emergency backup power is necessary. Hoods should be tested at least semiannually or after modification or maintenance operations. Operations should be performed 20 centimeters inside hood face.

Protective Gloves: Butyl Rubber Glove M3 and M4 Norton, Chemical Protective Glove Set

Eye Protection: At a minimum chemical goggles will be worn. For splash hazards use goggles and face shield.

Other Protective Equipment: For laboratory operations, wear lab coats, gloves and have mask readily accessible. In addition, daily clean smocks, foot covers, and head covers will be required when handling contaminated lab animals.

Monitoring: Available monitoring equipment for agent VX is the M8/M9 detector paper, detector ticket, M256/M256A1 kits, bubbler, Depot Area Air Monitoring System (DAAMS), Automated Continuous Air Monitoring System (ACAMS), Real-Time Monitor (RTM), Demilitarization Chemical Agent Concentrator (DCAC), M8/M43, M8A1/M43A1, CAM-M1, Hydrogen Flame Photometric Emission Detector (HYFED), the Miniature Chemical Agent Monitor (MINICAM), and the Real Time Analytical Platform (RTAP). Real-time, low-level monitors (with alarm) are required for VX operations. In their absence, an Immediately Dangerous to Life and Health

(IDLH) atmosphere must be presumed. Laboratory operations conducted in appropriately maintained and alarmed engineering controls require only periodic low-level monitoring.

Section IX—Special Precautions

Precautions To Be Taken In Handling And Storing: When handling agents, the buddy system will be incorporated. No smoking, eating, or drinking in areas containing agents is permitted. Containers should be periodically inspected for leaks, (either visually or using a detector kit). Stringent control over all personnel practices must be exercised. Decontaminating equipment will be conveniently located.

Exits must be designed to permit rapid evacuation. Chemical showers, eyewash stations, and personal cleanliness facilities must be provided. Wash hands before meals and shower thoroughly with special attention given to hair, face, neck, and hands using plenty of soap and water before leaving at the end of the work day.

Other Precautions: Agent containers will be stored in a single containment system within a laboratory hood or in double containment system.

For additional information see "AR 385-61, The Army Toxic Chemical Agent Safety Program," "DA Pam 385-61, Toxic Chemical Agent Safety Standards," and "DA Pam 40-173, Occupational Health Guidelines for the Evaluation and Control of Occupational Exposure to Nerve Agents GA, GB, GD, and VX."

Section X—Transportation Data

Note: Forbidden for transport other than via military (Technical Escort Unit) transport according to 49 CFR 172.

Proper Shipping Name: Toxic liquids, organic, n.o.s.

DOT Hazard Class: 6.1, Packing Group I, Hazard Zone A.

DOT Label: Poison.

DOT Marketing: Toxic liquids, organic, n.o.s. (O-ethyl S-(2-diisopropylaminoethyl)methylphosphonothiolate) UN 2810, Inhalation Hazard.

DOT Placard: Poison.

Emergency Accident Precautions And Procedures: See Sections IV, VII, and VIII.

Precautions to be taken in transportation: Motor vehicles will be placarded regardless of quantity. Drivers will be given full information regarding shipment and conditions in case of an emergency. AR 50-6 deals specifically with the shipment of chemical agents. Shipment of agents will be escorted in accordance with AR 740-32.

The Edgewood Chemical Biological Center (ECBC), Department of the Army believes that the data contained herein are actual and are the results of the tests conducted by ECBC experts. The data are not to be taken as a warranty or representation for which the Department of the Army or ECBC assumes legal responsibility. They are offered solely for consideration. Any use of this data and information contained in this MSDS must be determined by the user to be in accordance with applicable Federal, State, and local laws and regulations.

Lethal Nerve Agent (GB)

Date: 22 September 1988
Revised: 29 September 1999

In the event of an emergency
Telephone the SBCCOM Operations
Center's 24-hour emergency
Number: 410-436-2148

Section I—General Information

Manufacturer's Address:
U.S. Army Soldier and Biological Chemical Command (SBCCOM)
Edgewood Chemical Biological Center (ECBC)
ATTN: AMSSB-RCB-RS
Aberdeen Proving Ground, MD 21010-5424

CAS Registry Numbers:

107-44-8, 50642-23-4

Chemical Name:
Isopropyl methylphosphonofluoridate

Alternate Chemical Names:
O-Isopropyl Methylphosphonofluoridate
Phosphonofluoridic acid, methyl-, isopropyl ester
Phosphonofluoridic acid, methyl-, 1-methylethyl ester

Trade Name And Synonyms:
Isopropyl ester of methylphosphonofluoridic acid
Methylisopropoxyfluorophosphine oxide
Isopropyl Methylfluorophosphonate
O-Isopropyl Methylisopropoxfluorophosphine oxide
Methylfluorophosphonic acid, isopropyl ester
Isopropoxymethylphosphonyl fluoride
Isopropyl methylfluorophosphate
Isopropoxmethylphosphoryl fluoride
GB
Sarin
Zarin

Chemical Family:
Fluorinated organophosphorous compound

Formula/Chemical Structure:

$$C_4 H_{10} F O_2 P$$

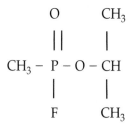

NFPA 704 Signal:
Health - 4
Flammability - 1
Reactivity - 1
Special - 0

Section II—Ingredients

Ingredients/Name: GB

Percentage by Weight: 100%

Threshold Limit Value (TLV): 0.0001 mg/m³

Section III—Physical Data

Boiling Point @ 760 mm Hg: 316 °F (158 °C)

Vapor Pressure (mm Hg): 2.9 @ 25 °C

Vapor Density (Air = 1 STP): 4.83 @ 25 °C

Solubility: Miscible with water. Soluble in all organic solvents.

Specific Gravity (H$_2$O − 1 g/mL): 1.0919 @ 25 °C

Freezing/Melting Point (°C): −56 °C

Liquid Density (g/cc): 1.0887 @ 25 °C
               1.102 @ 20 °C

Volatility (m/mg^3): 22,000 @ 25 °C

Viscosity (CENTISTOKES): 1.283 @ 25 °C

Appearance and Odor: Colorless liquid. Odorless in pure form.

Section IV—Fire and Explosion Data

Flashpoint: Did not flash to 280 °F (McCutchan-Young)

Flammability Limits (% By Volume): Not Applicable

Lower Explosive Limit: Not Applicable

Upper Explosive Limit: Not Applicable

Estinguishing Media: Water mist, fog, foam, CO$_2$. Avoid using extinguishing methods that will cause splashing or spreading of the GB.

Special Fire Fighting Procedures: GB will react with steam or water to produce toxic and corrosive vapors. All persons not engaged in extinguishing the fire should be immediately evacuated from the area. Fires involving GB should be contained to prevent contamination to uncontrolled areas. When responding to a fire alarm in buildings or areas containing GB, fire fighting personnel should wear full firefighter protective clothing during chemical agent firefighting and fire rescue operations. Respiratory protection is required. Positive pressure, full face piece, NIOSH-approved self-contained breathing apparatus (SCBA) will be worn where there is danger of oxygen deficiency and when directed by the fire chief or chemical accident/incident (CAI) operations officer. In cases where firefighters are responding to a chemical accident/incident for rescue/reconnaissance purposes they will wear appropriate levels of protective clothing (See Section VIII).

Do not breathe fumes. Skin contact with nerve agents must be avoided at all times. Although the fire may destroy most of the agent, care must still be taken to assure the agent or contaminated liquids do not further contaminate other areas or sewers. Contact with liquid GB or vapors can be fatal.

Unusual Fire And Explosion Hazards: Hydrogen may be present.

Section V—Health Hazard Data

Airborne Exposure Limits (AEL): The permissible airborne exposure concentration for GB for an 8-hour workday of a 40-hour work week is an 8-hour time weighted average (TWA) of 0.0001 mg/m^3. This value can be found in "DA Pam 40-8, Occupational Health Guidelines for the Evaluation and Control of Occupational Exposure to Nerve Agents GA, GB, GD, and VX." To date, however, the Occupational Safety and Health Administration (OSHA) has not promulgated a permissible exposure concentration for GB.

GB is not listed by the International Agency for Research on Cancer (IARC), American Conference of Governmental Industrial Hygienists (ACGIH), Occupational Safety and Health Administration (OSHA), or National Toxicology Program (NTP) as a carcinogen.

Effects Of Overexposure: GB is a lethal cholinesterase inhibitor. Doses which are potentially life-threatening may be only slightly larger than those producing least effects.

Route	Form	Effect	Type	Dosage
ocular	vapor	miosis	ECt50	<2 mg-min/m^3
Inhalation	vapor	runny nose	ECt50	<2 mg-min/m^3
Inhalation (15 1/min)	vapor	severe incapacitaion	ICt50	35 mg-min/m^3
Inhalation (15 1/min)	vapor	death	LCt50	70 mg-min/m^3
Percutaneous	liquid	death	LD50	1700 mg/70 kg man

Effective dosages for vapor are estimated for exposure durations of 2–10 minutes.

Symptoms of overexposure may occur within minutes or hours, depending upon the dose. They include: miosis (constriction of pupils) and visual effects, headaches and pressure sensation, runny nose and nasal congestion, salivation, tightness in the chest, nausea, vomiting, giddiness, anxiety, difficulty in thinking, difficulty sleeping, nightmares, muscle twitches, tremors, weakness, abdominal cramps, diarrhea, involuntary urination and defecation. With severe exposure symptoms progress to convulsions and respiratory failure.

Emergency And First Aid Procedures:

Inhalation: Hold breath until respiratory protective mask is donned. If severe signs of agent exposure appear (chest tightens, pupil constriction, in coordination, etc.), immediately administer, in rapid succession, all three Nerve Agent Antidote Kit(s), Mark I injectors (or atropine if directed by a physician). Injections using the Mark I kit injectors may be repeated at 5 to 20 minute intervals if signs and symptoms are progressing until three series of injections have been administered. No more injections will be given unless directed by medical personnel. In addition, a record will be maintained of all injections given. If breathing has stopped, give artificial respiration. Mouth-to-mouth resuscitation should be used when mask-bag or oxygen delivery systems are not available. Do not use mouth-to-mouth resuscitation when facial contamination exists. If breathing is difficult, administer oxygen. Seek medical attention **Immediately.**

Eye Contact: Immediately flush eyes with water for 10–15 minutes, then don respiratory protective mask. Although miosis (pinpointing of the pupils) may be an early sign of agent exposure, an injection will not be administered when miosis is the only sign present. Instead, the individual will be taken **Immediately** to a medical treatment facility for observation.

Skin Contact: Don respiratory protective mask and remove contaminated clothing. **Immediately** wash contaminated skin with copious amounts of soap and water, 10% sodium carbonate solution, or 5% liquid household bleach. Rinse well with water to remove excess decontaminant. Administer nerve agent antidote kit, Mark I, only if local sweating and muscular twitching symptoms are observed.

Ingestion: Do not induce vomiting. First symptoms are likely to be gastrointestinal. Immediately administer Nerve Agent Antidote Kit, Mark I. Seek medical attention **Immediately.**

Section VI—Reactivity Data

Stability: Stable when pure. Plant grade material stabilized with tri-n-buty-lamine can be stored in steel containers for long periods of time at temperatures up to 70 °C, but unstablized material tends to build-up pressure within a few weeks.

Incompatibility: Attacks tin, magnesium, cadmium plated steel, and some aluminum. Slightly attacks copper, brass, and lead; practically no attack on 1020 steels, Inconel and K-monel.

Hazardous Decomposition Products: Hydrolyzes to form HF under acid conditions and isopropyl alcohol and polymers under basic conditions.

Hazardous Polymerizations: Does not occur.

Section VII—Spill, Leak, and Disposal Procedures

Steps To Be Taken In Case Material Is Released Or Spilled: If leaks or spills of GB occur, only personnel in full protective clothing will remain in the area (See Section VIII). In case of personnel contamination see Section V for emergency and first aid instructions.

Recommended Field Procedures: Spills must be contained by covering with vermiculite, diatomaceous earth, clay, fine sand, sponges, and paper, or cloth towels. Decontaminate with copious amounts of aqueous sodium hydroxide solution (a minimum 10 wt.%). Scoop up all material and place in a DOT approved container. Cover the contents with decontaminating solution as above. After sealing, the exterior will be decontaminated and labeled according to EPA and DOT regulations. All leaking containers will be over packed with sorbent (e.g., vermiculite) placed between the interior and exterior containers. Decontaminate and label according to EPA and DOT regulations. Dispose of decontaminate according to Federal, state, and local laws. Conduct general area monitoring to confirm that the atmospheric concentrations do not exceed the airborne exposure limits (See Sections II and VIII).

If 10 wt.% aqueous sodium hydroxide is not available then the following decontaminants may be used instead and are listed in the order of preference: Decontaminating Agent (DS2), Sodium Carbonate, and Supertropical Bleach Slurry (STB).

Recommended Laboratory Procedures: A minimum of 56 grams of decon solution is required for each gram of GB. Decontaminant and agent solution is allowed to agitate for a minimum of one hour. Agitation is not necessary following the first hour. At the end of one hour, the resulting solution should be adjusted to a pH greater than 11.5. If the pH is below 11.5, NaOH should be added until a pH above 11.5 can be maintained for 60 minutes. An alternate solution for the decontamination of GB is 10 wt.% sodium carbonate in place of the 10% sodium hydroxide solution above. Continue with 56 grams of decon for each gram of agent. Agitate for one hour but allow three hours for the reaction. The final pH should be adjusted to above zero. It is also permitted to substitute 5.25.% sodium hypochlorite or 25 wt.% Monoethylamine (MEA) for the 10% sodium hydroxide solution above. MEA must be completely dissolved in water before addition of the agent. Continue with 56 grams of decon for each gram of GB and provide agitation for one hour. Continue with same ratios and time stipulations. Scoop up all material and clothing. Place all material in a DOT approved container. Cover the contents with decontaminating solution as above. After sealing, decontaminate the exterior of the container and label according to EPA and DOT regulations. All leaking containers will be over packed with sorbent placed between the interior and exterior containers. Decontaminate and label according to EPA and DOT regulations. Dispose of decontaminate according to Federal, State, and local laws. Conduct general area monitoring to confirm that the atmospheric concentrations do not exceed the airborne exposure limits (See Sections II and VIII).

Waste Disposal Method: Open pit burning or burying of GB or items containing or contaminated with GB in any quantity is prohibited. The detoxified GB (using procedures above) can be thermally destroyed by incineration in EPA approved incinerators according to appropriate provisions of Federal, state and local Resource Conservation and Recovery Act (RCRA) Regulations.

Note: Some decontaminate solutions are hazardous waste according to RCRA regulations and must be disposed of according to those regulations.

Section VIII—Special Protection Information

Respiratory Protection:

Concentration	Respiratory Protective Equipment
<0.0001 mg/m^3	A full face piece, chemical canister, air-purifying protective mask will be on hand for

	escape. M40-series masks are acceptable for this purpose. Other masks certified as equivalent may be used.
>0.0001 or = 0.2 mg/m³	A NIOSH/MSHA approved pressure demand full face piece SCBA or supplied air respirators with escape air cylinder may be used. Alternatively, a full face piece, chemical canister air-purifying protective mask is acceptable for this purpose (See DA Pam 385-61 for determination of appropriate level)
>0.2 mg/m³ or unknown	NIOSH/MSHA approved pressure demand full face piece SCBA suitable for use in high agent concentrations with protective ensemble (See DA Pam 385-61 for examples).

VENTILATION:

Local exhaust: Mandatory. Must be filtered or scrubbed to limit exit concentrations to <0.0001 mg/m³. Air emissions will meet local, state, and federal regulations.

Special: Chemical laboratory hoods will have an average inward face velocity of 100 linear feet per minute (lfpm) ±20% with the velocity at any point not deviating from the average face velocity by more than 20%. Existing laboratory hoods will have an inward face velocity of 150 lfpm ±20%. Laboratory hoods will be located such that cross-drafts do not exceed 20% of the inward face velocity. A visual performance test using smoke-producing devices will be performed in assessing the ability of the hood to contain agent GB.

Other: Recirculation or exhaust air from chemical areas is prohibited. No connection between chemical areas and other areas through ventilation system is permitted. Emergency backup power is necessary. Hoods should be tested at least semiannually or after modification or maintenance operations. Operations should be performed 20 centimeters inside hood face.

Protection Gloves: Butyl Rubber Glove M3 and M4 Norton, Chemical Protective Glove Set

Eye Protection: At a minimum chemical goggles will be worn. For splash hazards use goggles and face shield.

Other Protective Equipment: For laboratory operations, wear lab coats, gloves, and have mask readily accessible. In addition, daily clean smocks, foot covers, and head covers will be required when handling contaminated lab animals.

Monitoring: Available monitoring equipment for agent GB is the M8/M9 detector paper, detector ticket, M256/M256A1 kits, bubbler, Depot Area Air Monitoring System (DAAMS), Automated Continuous Air Monitoring System (ACAMS), Real-Time Monitor (RTM), Demilitarization Chemical Agent Concentrator (DCAC), M8/M43, M8A1/M43A1, CAM-M1, Hydrogen Flame Photometric Emission Detector (HYFED), the Miniature Chemical Agent Monitor (MINICAM), and the Real Time Analytical Platform (RTAP).

Real-time, low-level monitors (with alarm) are required for GB operations. In their absence, an Immediately Dangerous to Life and Health (IDLH) atmosphere must be presumed. Laboratory operations conducted in appropriately maintained and alarmed engineering controls require only periodic low-level monitoring.

Section IX—Special Precautions

Precautions To Be Taken In Handling And Storing: When handling agents, the buddy system will be incorporated. No smoking, eating, or drinking in areas containing agents is permitted. Containers should be periodically inspected for leaks (either visually or using a detector kit). Stringent control over all personnel practices must be exercised. Decontaminating equipment will be conveniently located. Exits must be designed to permit rapid evacuation. Chemical showers, eyewash stations, and personal cleanliness facilities must be provided. Wash hands before meals and shower thoroughly with special attention given to hair, face, neck, and hands using plenty of soap and water before leaving at the end of the work day.

Other Precautions: Agent containers will be stored in a single containment system within a laboratory hood or in double containment system.

For additional information see "AR 385-61, The Army Toxic Chemical Agent Safety Program," "DA Pam 385-61, Toxic Chemical Agent Safety Standards," and "DA Pam 40-173, Occupational Health Guidelines for the Evaluation and Control of Occupational Exposure to Nerve Agents GA, GB, GD, and VX."

Section X—Transportation Data

Note: Forbidden for transport other than via military (Technical Escort Unit) transport according to 49 CFR 172

Proper Shipping Name: Toxic liquids, organic, n.o.s.

DOT Hazard Class: 6.1, Packing Group I, Hazard Zone A

DOT Label: Poison

DOT Marking: Toxic liquids, organic, n.o.s. (Isopropyl methylphosphono-fluoridate) UN 2810, Inhalation Hazard

DOT Placard: Poison

Emergency Accident Precautions And Procedures: See Sections IV, VII, and VIII.

Precautions To Be Taken In Transportation: Motor vehicles will be placarded regardless of quantity. Drivers will be given full information regarding shipment and conditions in case of an emergency. AR 50-6 deals specifically with the shipment of chemical agents. Shipment of agents will be escorted in accordance with AR 740-32.

> The Edgewood Chemical Biological Center (ECBC), Department of the Army believes that the data contained herein are actual and are the results of the tests conducted by ECBC experts. The data are not to be taken as a warranty or representation for which the Department of the Army or ECBC assumes legal responsibility. They are offered solely for consideration. Any use of this data and information contained in this MSDS must be determined by the user to be in accordance with applicable Federal, State, and local laws and regulations.

Lewisite

Date: 16 April 1988
Revised: 4 October 1999

In the event of an emergency
Telephone the SBCCOM Operations
Center's 24-hour emergency
Number: 410-436-2148

Section I—General Information

Manufacturer's Address:
U.S. Army Soldier and Biological Chemical Command (SBCCOM)
Edgewood Chemical Biological Center (ECBC)
ATTN: AMSSB-RCB-RS
Aberdeen Proving Ground, MD 21010-5424

CAS Registry Number: 541-25-3

Chemical Name: Dichloro- (2-chlorovinyl) arsine

Trade name and synonyms:
Arsine, (2-chlorovinyl) dichloro-
Arsonous dichloride, (2-chloroethenyl)
Chlorovinylarsine dichloride
2-Chlorovinyldichloroarsine
Beta/Chlorovinyldichloarsine
Lewisite
L
EA 1034

Chemical Family: Arsenical (vesicant)

Formula/Chemical Structure:

$$C_2 \, H_2 \, As \, C_{13}$$

$$Cl \, CH = CH - As \begin{array}{l} Cl \\ \diagup \\ \diagdown \\ Cl \end{array}$$

NFPA 704 Signal:
Health - 4
Flammability - 1
Reactivity - 1
Special - 0

Section II—Ingredients

Ingredients/Name: Lewisite

Percentage by Weight: 100%

Threshold Limit Value (TLV): 0.003 mg/m³ (This is a ceiling value)

Section III—Physical Data

Boiling Point °F (°C): Calculated 374 °F (190 °C)

Vapor Pressure (mm Hg): 0.22 @ 20 °C 0.35 @ 25 °C

Vapor Density (Air = 1): 7.1

Solubility (g/100g solvent): Insoluble in water and dilute mineral acids. Soluble in organic solvents, oils and alcohol.

Specific Gravity (H₂0 = 1): 1.891 @ 20 °C

Freezing/Melting Point (°C): −18.2 to 0.1 (Depending on purity)

Liquid Density (g/mL): 1.888 @ 20 °C

Volatility (mg/m³): 2,500 @ 20 °C

Viscosity (Centipoise): 2.257 @ 20 °C

Molecular Weight (g/mol): 207.32

Appearance And Odor: Pure Lewisite is a colorless oily liquid. "War gas" is amber to dark brown liquid. A characteristic odor is usually geranium-like; very little odor when pure.

Section IV—Fire and Explosion Data

Flashpoint: Does not flash

Flammability Limits (% by volume): Not Applicable

Extinguishing Media: Water, fog, foam, CO_2. Avoid use of extinguishing methods that will cause splashing or spreading of L.

Special Fire Fighting Procedures: All persons not engaged in extinguishing the fire should be immediately evacuated from the area. Fires involving L should be contained to prevent contamination to uncontrolled areas. When responding to a fire alarm in buildings or areas containing agents, fire-fighting personnel should wear full firefighter protective clothing (flame resistant) during chemical agent fire-fighting and fire rescue operations. Respiratory protection is required. Positive pressure, full facepiece, NIOSH-approved self-contained breathing apparatus (SCBA) will be worn where there is danger of oxygen deficiency and when directed by the fire chief or chemical accident/incident (CAI) operations officer. In cases where firefighters are responding to a chemical accident/incident for rescue/reconnaissance purposes they will wear appropriate levels of protective clothing (See Section VIII).

Do not breathe fumes. Skin contact with nerve agents must be avoided at all times. Although the fire may destroy most of the agent, care must still be taken to assure the agent or contaminated liquids do not further contaminate other areas or sewers. Contact with the agent liquid or vapor can be fatal.

Unusual Fire and Explosion Hazards: None known

Section V—Health Hazard Data

Airborne Exposure Limit (AEL): The permissible airborne exposure concentration of L for an 8-hour workday or a 40-hour workweek is an 8-hour time weighted average (TWA) of 0.003 mg/m^3 as a ceiling value. A ceiling value may not be exceeded at any time. The ceiling value for Lewisite is based upon the present technologically feasible detection limits of 0.003 mg/m^3. This value can be found in "DA Pam 40-173, Occupational Health Guidelines for the Evaluation and Control of Occupational Exposure to Mustard H, HD, and HT." To date, however, the Occupational Safety and Health Administration (OSHA) has not promulgated permissible exposure concentration for L.

Effects Of Overexposure: L is a vesicant (blister agent), also, it acts as a systemic poison, causing pulmonary edema, diarrhea, restlessness, weakness, subnormal temperature, and low blood pressure. In order of severity and appearance of symptoms, it is a blister agent, a toxic lung irritant, absorbed in tissues, and a systemic poison. When inhaled in high concentrations, L may be fatal in as short a time as 10 minutes. L is not detoxified by the body. Common routes of entry into the body are ocular, percutaneous, and inhalation.

Lewisite is generally considered a suspect carcinogen because of its arsenic content.

Toxicological Data:

Man:
LCt50 (inhalation, man) = 1200 − 1500 mg min/m³
LCt50 (skin vapor exposure, man) = 100,000 mg min/m³
LDLO (skin, human) = 20 mg/kg
LCt50 (skin, man): >1500 mg/min³. L irritates eyes and skin and gives warning of its presence. Minimum effective dose (ED min) = 200 mg/m³ (30 min).
ICt50 (eyes, man): <300 mg min/m³.

Animal:
LD50 (oral, rat) = 50 mg/kg
LD50 (subcutaneous, rat) = 1 mg/kg
LCtLO (inhalation, mouse) = 150 mg/m³ 10m
LD50 (skin, dog = 15 mg/kg)
LD50 (skin, rabbit) = 6 mg/kg
LD50 (subcutaneous, rabbit) = 2 mg/kg
LD50 (intravenous, rabbit) = 2 mg/kg
LD50 (skin, guinea pig) = 12 mg/kg
LD50 (subcutaneous, guinea pig) = 1 mg/kg
LCt50 (inhalation, rat) = 1500 mg min/m³ (9 min)
LD50 (vapor skin, rat) = 20,000 mg min m 25 min)
LD50 (skin, rat) = 15 − 24 mg/kg
LD50 (ip, dog) = 2 mg/kg

Acute Exposure:

Eyes: Severe damage. Instant pain, conjunctivitis and blepharospasm leading to closure of eyelids, followed by corneal scarring and iritis. Mild exposure produces reversible eye damage if decontaminated instantly. More permanent injury or blindness is possible within one minute of exposure.

Skin: Immediate stinging pain increasing in severity with time. Erythema (skin reddening) appears within 30 minutes after exposure accompanied by pain with itching and irritation for 24 hours. Blisters appear within 12 hours after exposure with more pain that diminishes after 2–3 days. Skin burns are much deeper than with HD. Tender skin, mucous membrane, and perspiration-covered skin are more sensitive to the effects of L. This, however, is counteracted by L's hydrolysis by moisture, producing less vesicant and higher vapor pressure product.

Respiratory Tract: Irritating to nasal passages and produces a burning sensation followed by profuse nasal secretions and violent sneezing. Prolonged exposure causes coughing and production of large quantities of froth mucus. In experimental animals, injury to respiratory tracts, due to vapor exposure is similar to mustards; however, edema of the lung is more marked and frequently accompanied by pleural fluid.

Systemic Effects: L on the skin, and inhaled vapor, cause systemic poisoning. A manifestation of this is a change in capillary permeability, which permits loss of sufficient fluid from the bloodstream to cause hemoconcentration, shock and death. In nonfatal cases, hemolysis of erythrocytes has occurred with a resultant hemolytic anemia. The excretion of oxidized products into the bile by the liver produces focal necrosis of that organ, necrosis of the mucosa of the biliary passages with periobiliary hemorrhages, and some injury to the intestinal mucosa. Acute systematic poisoning from large skin burns causes pulmonary edema, diarrhea restlessness, weakness, subnormal temperature, and low blood pressure in animals.

Chronic Exposure: Lewisite can cause sensitization and chronic lung impairment. Also, by comparison to agent mustard and arsenical compounds, it can be considered as a suspected human carcinogen.

Emergency And First Aid Procedures:

Inhalation: Hold breath until respiratory protective mask is donned. Remove from the source **Immediately.** If breathing is difficult, administer oxygen. If breathing has stopped, give artificial respiration. Mouth-to-mouth resuscitation should be used when approved mask-bag or oxygen delivery systems are not available. Do not use mouth-to-mouth resuscitation when facial contamination is present. Seek medical attention **Immediately.**

Eye Contact: Speed in decontaminating the eyes is absolutely essential. Remove the person from the liquid source, flush the eyes **Immediately** with water for at least 15 minutes by tilting the head to the side, pulling the eyelids apart with the fingers and pouring water slowly into the eyes. Do not cover the eyes with bandages but, if necessary, protect eyes by means of dark or opaque goggles. Transfer patient to a medical facility **Immediately.**

Skin Contact: Don respiratory protective mask. Remove the victim from agent sources immediately. **Immediately** wash skin and clothes with 5% solution of sodium hypochlorite or liquid household bleach within one minute. Cut and remove contaminated clothing, flush contaminated skin area again with 5% sodium hypochlorite solution, then wash contaminated skin area with soap and water. Seek medical attention **Immediately.**

Ingestion: Do not induce vomiting. Give victim milk to drink. Seek medical attention **Immediately.**

Section VI—Reactivity Data

Stability: Stable in steel or glass containers at temperatures below 50 °C

Incompatibility: Corrosive to steel at a rate of 1×10^{-5} to 5×10^{-5} in/month at 65 °C

Hazardous Decomposition Products: Reasonably stable; however, in presence of moisture, it hydrolyses rapidly, losing its vesicant property. It also hydrolyses in acidic medium to form HCl and non-volatile (solid) chlorovinylarsenious oxide, which is less vesicant than Lewisite. Hydrolysis in alkaline medium, as in decontamination with alcoholic caustic or carbonate solution or DS2, produces acetylene and trisodium arsenate ($Na_3 As O_4$). Therefore, decontaminated solution would contain toxic arsenic.

Hazardous Polymerization: Does not occur.

Section VII—Spill, Leak, And Disposal Procedures

Steps To Be Taken In Case Material Is Released Or Spilled: If leaks or spills of L occur only personnel in full protective clothing will be allowed in the area (See Section VIII). See Section V for emergency and first aid instructions.

Recommended Field Procedures: Lewisite should be contained using vermiculite, diatomaceous earth, clay, or fine sand and neutralized as soon as possible using copious amounts of alcoholic caustic, carbonate, or Decontaminating Agent (DS2). Caution must be exercised when using these decontaminates since acetylene will be given off. Household bleach can also be used if accompanied by stirring to allow contact. Scoop up all material and place in a DOT approved container. Cover the contents with decontaminating solution as above. After sealing, the exterior decontaminated and labeled according to EPA and DOT regulations. All leaking containers will be over packed with sorbent (e.g. vermiculite) placed between the interior and exterior containers. Decontaminate and label according to EPA and DOT regulations. Dispose of decontaminate according to Federal, state, and local laws. Conduct general area monitoring to confirm that the atmospheric concentrations do not exceed the airborne exposure limits (See Sections II and VIII).

Recommended Laboratory Procedures: A 10 wt.% alcoholic sodium hydroxide solution is prepared by adding 100 grams of denatured ethanol to 900 grams of 10 wt.% NaOH in water. A minimum of 200 grams of decon is required for each gram of L. The decon and agent solution is agitated for a minimum of one hour. At the end of the hour the resulting pH should be checked and adjusted to above 11.5 using additional NaOH, if required. It is permitted to substitute 10 wt.% alcoholic sodium carbonate made and used in the same ratio as the NaOH listed above. Reaction time should be increased to 3 hours with agitation for the first hour. Final pH should be adjusted to above 10. Scoop up all material and place in an approved DOT container. Cover the contents with decontaminating solution as above. The exterior of the container will be decontaminated and labeled according to the EPA and DOT regulations. All leaking containers will be over packed with sorbent (e.g., vermiculite) placed between the interior and exterior containers. Decontaminate and label according to EPA and DOT regulations. Dispose of the material in accordance with waste disposal methods provided below. Conduct general area monitoring with an approved monitor to confirm that the atmospheric concentrations do not exceed the airborne exposure limits (See Sections II and VIII).

It is permitted to substitute 5.25% sodium hypochlorite for the 10% alcoholic sodium hydroxide solution above. Allow one hour with agitation for the reaction. Adjustment of the pH is not required. Conduct general area monitoring to confirm that the atmospheric concentrations do not exceed the airborne exposure limit (See Section VIII).

Waste Disposal Method: All neutralized material should be collected and contained for disposal according to land ban RCRA regulations or thermally decomposed in an EPA permitted incinerator equipped with a scrubber that will scrub out the chlorides and equipped with an electrostatic precipitator or other filter device and containerize and label according to DOT and EPA regulations. The arsenic will be disposed of according to land ban RCRA regulations. Any contaminated materials or protective clothing should be decontaminated using alcoholic caustic, carbonates, or bleach analyzed to assure it is free of detectable contamination (3X) level. The clothing should then be sealed in plastic bags inside properly labeled drums and held for shipment back to the DA issue point.

Note: Some decontaminate solutions are hazardous waste according to RCRA regulations and must be disposed of IAW those regulations.

Section VIII—Special Protection Information

Concentration	Respiratory Protective Equipment
<0.003 mg/m^3	A full face piece, chemical canister, air-purifying protective mask will be on hand for escape. M40-series masks are acceptable for this purpose. Other masks certified as equivalent may be used.
>or = 0.003 mg/m^3 or unknown	NIOSH/MSHA approved pressure demand full face piece SCBA suitable for use in high Lewisite concentrations with protective ensemble (See DA Pam 385-61 for examples).

VENTILATION:

Local Exhaust: Mandatory. Must be filtered or scrubbed. Air emissions shall meet local, state and federal regulations.

Special: Chemical laboratory hoods will have an average inward face velocity of 100 linear feet per minute (lfpm) ±20% with the velocity at any point not deviating from the average face velocity by more than 20%. Existing laboratory hoods will have an inward face velocity of 150 lfpm ±20%. Laboratory hoods will be located such that cross drafts do not exceed 20% of the inward face velocity. A visual performance test using smoke producing devices will be performed in assessing the ability of the hood to contain Lewisite.

Other: Recirculation of exhaust air from agent areas is prohibited. No connection between agent area and other areas through the ventilation system is permitted. Emergency backup power is necessary. Hoods should be tested semiannually or after modification or maintenance operations. Operations should be performed 20 centimeters inside hoods.

Protective Gloves: Butyl Rubber gloves M3 and M4 Norton, Chemical Protective Glove Set

Eye Protection: As a minimum, chemical goggles will be worn. For splash hazards use goggles and face shield.

Other Protective Equipment: For laboratory operations, wear lab coats, gloves and have mask readily accessible. In addition, daily clean smocks, foot covers, and head covers will be required when handling contaminated lab animals.

Monitoring: Available monitoring equipment for agent Lewisite is the M18A2 (yellow band), bubblers (arsenic and GC method), and M256 and A1 kits.

Real-time, low-level monitors (with alarm) are required for Lewisite operations. In their absence, an Immediately Dangerous to Life and Health (IDLH) atmosphere must be presumed. Laboratory operations conducted in appropriately maintained and alarmed engineering controls require only periodic low-level monitoring.

Section IX—Special Precautions

Precautions To Be Taken In Handling And Storing: When handling agents, the buddy system will be incorporated. No smoking, eating, or drinking in areas containing agents is permitted. Containers should be periodically inspected for leaks (either visually or using a detector kit). Stringent control over all personnel practices must be exercised. Decontaminating equipment will be conveniently located. Exits must be designated to permit rapid evacuation. Chemical showers, eyewash stations, and personal cleanliness facilities must be provided. Wash hands before meals and shower thoroughly with special attention given to hair, face, neck, and hands using plenty of soap and water before leaving at the end of the workday.

Other Precautions: L should be stored in containers made of glass for Research, Development, Test and Evaluation (RDTE) quantities or one-ton steel containers for large quantities. Agent will be stored in a single containment system within a laboratory hood or in a double containment system.

For additional information see "AR 385-61, The Army Toxic Chemical Agent Safety Program," "DA Pam 385-61, Toxic Chemical Agent Safety Standards," and "DA Pam 40-173, Occupational Health Guidelines for the Evaluation and Control of Occupational Exposure to Mustard Agents H, HD, and HT."

Section X—Transportation Data

Note: Forbidden for transport other than via military (Technical Escort Unit) transport according to 49 CFR 172

Proper Shipping Name: Toxic liquids, n.o.s.

Dot Hazard Class: 6.1, Packing Group I

Dot Label: Poison

Dot Marking: Toxic liquids, n.o.s.
Dichloro-(2-chlorovinyl)arsine UN 2810

Dot Placard: Poison

Emergency Accident Precautions And Procedures: See Sections IV, VII, and VIII.

Precautions To Be Taken In Transportations: Motor vehicles will be placarded regardless of quantity. Drivers will be given full information regarding shipment and conditions in case of an emergency. AR 50-6 deals specifically with the shipment of chemical agents. Shipment of agents will be escorted in accordance with AR 740-32.

The Edgewood Chemical Biological Center (ECBC), Department of the Army believes that the data contained herein are actual and are the results of the tests conducted by ECBC experts. The data are not to be taken as a warranty or representation for which the Department of the Army or ECBC assumes legal responsibility. They are offered solely for consideration. Any use of this data and information contained in this MSDS must be determined by the user to be in accordance with applicable Federal, State, and local laws and regulations.

Appendix Q
United States Government Interagency Domestic Terrorism Concept of Operations Plan (January 2001)

I. Introduction and Background

A. Introduction

The ability of the United States Government to prevent, deter, defeat and respond decisively to terrorist attacks against our citizens, whether these attacks occur domestically, in international waters or airspace, or on foreign soil, is one of the most challenging priorities facing our nation today. The United States regards all such terrorism as a potential threat to national security, as well as a violent criminal act, and will apply all appropriate means to combat this danger. In doing so, the United States vigorously pursues efforts to deter and preempt these crimes and to apprehend and prosecute directly, or assist other governments in prosecuting, individuals who perpetrate or plan such terrorist attacks.

In 1995, President Clinton signed Presidential Decision Directive 39 (PDD-39), the United States Policy on Counterterrorism. This Presidential Directive built upon previous directives for combating terrorism and further elaborated a strategy and an interagency coordination mechanism and management structure to be undertaken by the Federal government to combat both domestic and international terrorism in all its forms. This authority includes implementing measures to reduce our vulnerabilities, deterring terrorism through a clear public position, responding rapidly and effectively to threats or actual terrorist acts, and giving the highest priority to developing sufficient capabilitiees to combat and manage the consequences of terrorist incidents involving weapons of mass destruction (WMD).

To ensure this policy is implemented in a coordinated manner, the Concept of Operations Plan, hereafter referred to as the CONPLAN, is designed to provide overall guidance to Federal, State and local agencies concerning how the Federal government would respond to a potential or actual terrorist threat or incident that occurs in the United States, particularly one involving WMD. The CONPLAN outlines an organized and unified capability for a timely, coordinated response by Federal agencies to a terrorist threat or act. It establishes conceptual guidance for assessing and monitoring a developing threat, notifying appropriate Federal, State, and local agencies of the nature of the threat, and deploying the requisite advisory and technical resources to assist the Lead Federal Agency (LFA) in facilitating interagency/interdepartmental coordination of a crisis and consequence management response. Lastly, it defines the relationships between structures under which the Federal government will marshal crisis and consequence management resources to respond to a threatened or actual terrorist incident.

B. Purpose

The purpose of this plan is to facilitate an effective Federal response to all threats or acts of terrorism within the United States that are determined to be of sufficient magnitude to warrant implementation of this plan and the associated policy guidelines established in PDD-39 and PDD-62. To accomplish this, the CONPLAN:

- Establishes a structure for a systematic, coordinated and effective national response to threats or acts of terrorism in the United States;
- Defines procedures for the use of Federal resources to augment and support local and State governments; and
- Encompasses both crisis and consequence management responsibilities, and articulates the coordination relationships between these missions.

C. Scope

The CONPLAN is a strategic document that:

- Applies to all threats or acts of terrorism within the United States;
- Provides planning guidance and outlines operational concepts for the Federal crisis and consequence management response to a threatened or actual terrorist incident within the United States;
- Serves as the foundation for further development of detailed national, regional, State, and local operations plans and procedures;

- Includes guidelines for notification, coordination and leadership of response activities, supporting operations, and coordination of emergency public information across all levels of government;
- Acknowledges the unique nature of each incident, the capabilities of the local jurisdiction, and the activities necessary to prevent or mitigate a specific threat or incident; and
- Illustrates ways in which Federal, State and local agencies can most effectively unify and synchronize their response actions.

D. Primary Federal Agencies

The response to a terrorist threat or incident within the U.S. will entail a highly coordinated, multi-agency local, State, and Federal response. In support of this mission, the following primary Federal agencies will provide the core Federal response:

- Department of Justice (DOJ)/Fedeal Bureau of Investigation (FBI)*
- Federal Emergency Management Agency (FEMA)**
- Department of Defense (DOD)
- Department of Energy (DOE)
- Environmental Protection Agency (EPA)
- Department of Health and Human Services (DHHS)

*Lead Agency for Crisis Management
**Lead Agency for Consequence Management

Although not formally designated under the CONPLAN, other Federal departments and agencies may have authorities, resources, capabilities, or expertise required to support response operations. Agencies may be requested to participate in Federal planning and response operations, and may be asked to designate staff to function as liaison officers and provide other support to the LFA.

E. Primary Agency Responsibilities

1. **Department of Justice (DOJ)/Federal Bureau of Investigation (FBI)**
   The Attorney General is responsible for ensuring the development and implementation of policies directed at preventing terrorist attacks domestically, and will undertake the criminal prosecution of these acts of terrorism that violate U.S. law. DOJ has charged the FBI with execution of its LFA responsibilities for the management of a Federal response to terrorist threats or incidents that take place within U.S. territory or those occurring in international waters that do not involve the flag vessel of a foreign country. As the lead agency for crisis management, the FBI will implement a Federal

crisis management response. As LFA, the FBI will designate a Federal on-scene commander to ensure appropriate coordination of the overall United States Government response with Federal, State and local authorities until such time as the Attorney General transfers the overall LFA role to FEMA. The FBI, with appropriate approval, will form and coordinate the deployment of a Domestic Emergency Support Team (DEST) with other agencies, when appropriate, and seek appropriate Federal support based on the nature of the situation.

2. **Federal Emergency Management Agency (FEMA)**

As the lead agency for consequence management, FEMA will manage and coordinate any Federal consequence management response in support of State and local governments in accordance with its statutory authorities. Additionally, FEMA will designate appropriate liaison and advisory personnel for the FBI's Strategic Information and Operations Center (SIOC) and deployment with the DEST, the Joint Operations Center (JOC), and the Joint Information Center (JIC).

3. **Department of Defense (DOD)**

DOD serves as a support agency to the FBI for crisis management functions, including technical operations, and a support agency to FEMA for consequence management. In accordance with DOD Directives 3025.15 and 2000.12 and the Chairman Joint Chiefs of Staff CONPLAN 0300-97, and upon approval by the Secretary of Defense, DOD will provide assistance to the LFA and/or the CONPLAN primary agencies, as appropriate, during all aspects of a terrorist incident, including both crisis and consequence management. DOD assistance includes threat assessment; DEST participation and transportation; technical advice; operational support; tactical support; support for civil disturbances; custody, transportation and disposal of a WMD device; and other capabilities including mitigation of the consequences of a release.

DOD has many unique capabilities for dealing with a WMD and combating terrorism, such as the US Army Medical Research Institute for Infectious Diseases, Technical Escort Unit, and US Marine Corps Chemical Biological Incident Response Force. These and other DOD assets may be used in responding to a terrorist incident if requested by the LFA and approved by the Secretary of Defense.

4. **Department of Energy (DOE)**

DOE serves as a support agency to the FBI for technical operations and a support agency to FEMA for consequence management. DOE provides scientific-technical personnel and equipment in support of the LFA during all aspects of a nuclear/radiological WMD terrorist incident. DOE assistance can support both crisis and consequence management activities with capabilities such as threat assessment, DEST deployment, LFA advisory requirements, technical advice, forecasted modeling predictions, and operational support to include direct support of tactical operations. Deployable DOE scientific technical assistance and support includes capabilities such as search operations; access operations; diagnostic and device assessment; radiological assessment and monitoring; identificaton of material; development of Federal protective action recommendations; provision of information on the radiological response; render safe operations; hazards assessment; containment, relocation and storage of special nuclear material evidence; post-incident clean-up; and on-site management and radiological assessment to the public, the White House, and members of Congress and foreign governments. All DOE support to a Federal response will be coordinated through a Senior Energy Official.

5. **Environmental Protection Agency (EPA)**

EPA serves as a support agency to the FBI for technical operations and a support agency to FEMA for consequence management. EPA provides technical personnel and supporting equipment to the LFA during all aspects of a WMD terrorist incident. EPA assistance may include threat assessment, DEST and regional emergency response team deployment, LFA advisory requirements, technical advice and operational support for chemical, biological, and radiological releases. EPA assistance and advice includes threat assessment, consultation, agent identification, hazard detection and reduction, environmental monitoring; sample and forensic evidence collection/analysis; identification of contaminants; feasibility assessment and clean-up; and on-site safety, protection, prevention, decontamination, and restoration activities. EPA and the United States Coast Guard (USCG) share responsibilities for response to oil discharges into navigable waters and releases of hazardous substances, pollutants, and contaminants into the environment under the National Oil and Hazardous Substance Pollution Contingency Plan (NCP). EPA provides the predesignated Federal On-Scene Coordinator for inland areas and the USCG for coastal areas to coordinate containment, removal, and disposal efforts and resources during an oil, hazardous substance, or WMD incident.

6. **Department of Health and Human Services (HHS)**

HHS serves as a support agency to the FBI for technical operations and a support agency to FEMA for consequence management. HHS provides technical personnel and supporting equipment to the LFA during all aspects of a terrorist incident. HHS can also provide regulatory follow-up when an incident involves a product regulated by the Food and Drug Administration. HHS assistance supports threat assessment, DEST deployment, epidemiological investigation, LFA advisory requirements, and technical advice. Technical assistance to the FBI may include identification of agents, sample collection and analysis, on-site safety and protection activities, and medical management planning. Operational support to FEMA may include mass immunization, mass prophylaxis, mass fatality management, pharmaceutical support operations (National Pharmaceutical Stockpile), contingency medical records, patient tracking, and patient evacuation and definitive medical care provided through the National Disaster Medical System.

II. Policies

A. Authorities

The following authorities are the basis for the development of the CONPLAN:

- Presidential Decision Directive 39, including the Domestic Guidelines
- Presidential Decision Directive 62
- Robert T. Stafford Disaster Relief and Emergency Assistance Act

B. Other Plans and Directives

- Federal Response Plan, including the Terrorism Incident Annex
- Federal Radiological Emergency Response Plan
- National Oil and Hazardous Substances Pollution Contingency Plan
- HHS Health and Medical Services Support Plan for the Federal Response to Assets of Chemical/Biological Terrorism
- Chairman of the Joint Chiefs of Staff CONPLAN 0300/0400
- DODD 3025.15 Military Assistance to Civil Authorities
- Other Department of Defense Directives

C. Federal Agency Authorities

The CONPLAN does not supersede existing plans or authorities that were developed for response to incidents under department and agency statutory authorities. Rather, it is intended to be a coordinating plan between crisis and consequence management to provide an effective Federal response to terrorism. The CONPLAN is a Federal signatory plan among the six principal departments and agencies named in PDD-39. It may be updated and amended, as necessary, by consensus among these agencies.

D. Federal Response to a Terrorism Incident

The Federal response to a terrorist threat or incident provides a tailored, time-phased deployment of specialized Federal assets. The response is executed under two broad responsibilities:

1. **Crisis Management**

    Crisis management is predominantly a law enforcement function and includes measures to identify, acquire, and plan the use of resources needed to anticipate, prevent, and/or resolve a threat or act of terrorism. In a terrorist incident, a crisis management response may include traditional law enforcement missions, such as intelligence, surveillance, tactical operations, negotiation, forensics, and investigations, as well as technical support missions, such as agent identification, search, render safe procedures, transfer and disposal, and limited decontamination. In addition to the traditional law enforcement missions, crisis management also includes assurance of public health and safety.

    The laws of the United States assign primary authority to the Federal government to prevent and respond to acts of terrorism or potential acts of terrorism. Based on the situation, a Federal crisis management response may be supported by technical operations, and by consequence management activities, which should operate concurrently.

2. **Consequence Management**

    Consequence management is predominantly an emergency management function and includes measures to protect public health and safety, restore essential government services, and provide emergency relief to governments, businesses, and individuals affected by the consequences of terrorism. In an actual or potential terrorist incident, a consequence management response will be managed by FEMA using structures and resources of the Federal Response Plan (FRP). These efforts will include support missions as described in other Federal operations plans, such as predictive modeling, protective action recommendations, and mass decontamination.

    The laws of the United States assign primary authority to the State and local governments to respond to the consequences of terrorism; the Federal government provides assistance, as required.

E. Lead Federal Agency Designation

As mandated by the authorities references above, the operational response to a terrorist threat will employ a coordinated, interagency process organized through a LFA concept. PDD-39 reaffirms and elaborates on the U.S.

Government's policy on counterterrorism and expands the roles, responsibilities and management structure for combating terrorism. LFA responsibility is assigned to the Department of Justice, and is delegated to the FBI, for threats or acts of terrorism that take place in the United States or in international waters that do not involve the flag vessel of a foreign country. Within this role, the FBI Federal on-scene commander (OSC) will function as the on-scene manager for the U.S. Government. All Federal agencies and departments, as needed, will support the Federal OSC. Threats or acts of terrorism that take place outside of the United States or its trust territories, or in international waters and involve the flag vessel of a foreign country are outside the scope of the CONPLAN.

In addition, these authorities reaffirm that FEMA is the lead agency for consequence management within U.S. territory. FEMA retains authority and responsibility to act as the lead agency for consequence management throughout the Federal response. FEMA will use the FRP structure to coordinate all Federal assistance to State and local governments for consequence management. To ensure that there is one overall LFA, PDD-39 directs FEMA to support the Department of Justice (as delegated to the FBI) until the Attorney General transfers the LFA role to FEMA. At such time, the responsibility to function as the on-scene manager for the U.S. Government transfers from the FBI Federal OSC to the Federal Coordinating Officer (FCO).

F. Requests for Federal Assistance

Requests for Federal assistance by State and local governments, as well as those from owners and operators of critical infrastructure facilities, are coordinated with the lead agency (crisis or consequence) responsible under U.S. law for that function. In response to a terrorist threat or incident, multiple or competing requests will be managed based on priorities and objectives established by the JOC Command Group.

State and local governments will submit requests for Federal crisis management assistance through the FBI. State and local governments will submit requests for Federal consequence management assistance through standard channels under the Federal Response Plan. FEMA liaisons assigned to the DEST or JOC coordinate requests with the LFA to ensure consequence management plans and actions are consistent with overall priorities. All other requests for consequence management assistance submitted outside normal channels to the DEST or JOC will be forwarded to the Regional Operations Center (ROC) Director or the Federal Coordinating Officer (FCO) for action.

G. Funding

As mandated by PDD-39, Federal agencies directed to participate in counterterrorist operations or the resolution of terrorist incidents bear the costs of their own participation, unless otherwise directed by the President. This responsibility is subject to specific statutory authorization to provide support without reimbursement. In the absence of such specific authority, the Economy Act applies, and reimbursement cannot be waived.

H. Deployment/Employment Priorities

The multi-agency JOC Command Group, managed by the Federal OSC, ensures that conflicts are resolved, overall incident objectives are established, and strategies are selected for the use of critical resources. These strategies will be based on the following priorities:

1. Preserving life or minimizing risk to health. This constitutes the first priority of operations.
2. Preventing a threatened act from being carried out or an existing terrorist act from being expanded or aggravated.
3. Locating, accessing, rendering safe, controlling, containing, recovering, and disposing of a WMD that has not yet functioned.
4. Rescuing, decontaminating, transporting and treating victims. Preventing secondary casualties as a result of contamination or collateral threats.
5. Releasing emergency public information that ensures adequate and accurate communications with the public from all involved response agencies.
6. Restoring essential services and mitigating suffering.
7. Apprehending and successfully prosecuting perpetrators.
8. Conducting site restoration.

I. Planning Assumptions and Considerations

1. The CONPLAN assumes that no single private or government agency at the local, State, or Federal level possesses the authority and the expertise to act unilaterally on the difficult issues that may arise in response to threats or acts of terrorism, particularly if nuclear, radiological, biological, or chemical materials are involved.
2. The CONPLAN is based on the premise that a terrorist incident may occur at any time of day with little or no warning, may involve single or multiple geographic areas, and result in mass casualties.
3. The CONPLAN also assumes an act of terrorism, particularly an act directed against a large population center within the United States involving nuclear, radiological, biological, or chemical materials, will have major consequences that can overwhelm the capabilities of many local and State governments to respond and may seriously challenge existing Federal response capabilities, as well.
4. Federal participating agencies may need to respond on short notice to provide effective and timely assistance to State and local governments.

5. Federal departments and agencies would be expected to provide an initial response when warranted under their own authorities and funding. Decisions to mobilize Federal assets will be coordinated with the FBI and FEMA.
6. In the case of biological WMD attack, the effect may be temporally and geographically dispersed, with no determined or defined "incident site." Response operations may be conducted over a multi-jurisdictional, multi-State region.
7. A biological WMD attack employing a contagious agent may require quarantine by State and local health officials to contain the disease outbreak.
8. Local, State, and Federal responders will define working perimeters that overlap. Perimeters may be used by responders to control access to an affected area, to assign operational sectors among responding organizations, and to assess potential effects on the population and the environment. Control of these perimeters and response actions may be managed by different authorities, which will impede the effectiveness of the overall response if adequate coordination is not established.
9. If appropriate personal protective equipment and capabilities are not available and the area is contaminated with WMD materials, it is possible that response actions into a contaminated area may be delayed until the material has dissipated to a level that is safe for emergency response personnel to operate.

J. Training and Exercises

Federal agencies, in conjunction with State and local governments, will periodically exercise their roles and responsibilities designated under the CON-PLAN. Federal agencies should coordinate their exercises with the Exercise Subgroup of the Interagency Working Group on Counterterrorism and other response agencies to avoid duplication, and, more importantly, to provide a forum to exercise coordination mechanisms among responding agencies.

Federal agencies will assist State and local governments design and improve their response capabilities to a terrorist threat or incident. Each agency should coordinate its training programs with other response agencies to avoid duplication and to make its training available to other agencies.

III. Situation

A. Introduction

The complexity, scope, and potential consequences of a terrorist threat or incident require that there be a rapid and decisive capability to resolve the situation. The resolution to an act of terrorism demands an extraordinary level of coordination of crisis and consequence management functions and technical expertise across all levels of government. No single Federal, State, or local governmental agency has the capability or requisite authority to respond independently and mitigate the consequences of such a threat to national security. The incident may affect a single location or multiple locations, each of which may be a disaster scene, a hazardous scene and/or a crime scene simultaneously.

B. Differences Between WMD Incidents and Other Incidents

As in all incidents, WMD incidents may involve mass casualties and damage to buildings or other types of property. However, there are several factors surrounding WMD incidents that are unlike any other type of incidents that must be taken into consideration when planning a response. First responders' ability to identify aspects of the incident (e.g., signs and symptoms exhibited by victims) and report them accurately will be key to maximizing the use of critical local resources and for triggering a Federal response.

1. The situation may not be recognizable until there are multiple casualties. Most chemical and biological agents are not detectable by methods used for explosives and firearms. Most agents can be carried in containers that look like ordinary items.

2. There may be multiple events (e.g., one event in an attempt to influence another event's outcome).

3. Responders are placed at a higher risk of becoming casualties. Because agents are not readily identifiable, responders may become contaminated before recognizing the agent involved. First responders may, in addition, be targets for secondary releases or explosions.

4. The location of the incident will be treated as a crime scene. As such, preservation and collection of evidence is critical. Therefore, it is important to ensure that actions on-scene are coordinated between response organizations to minimize any conflicts between law enforcement authorities, who view the incident as a crime scene, and other responders, who view it as a hazardous materials or disaster scene.

5. Contamination of critical facilities and large geographic areas may result. Victims may carry an agent unknowingly to public transportation facilities, businesses, residences, doctors' offices, walk-in medical clinics, or emergency rooms because they don't realize that they are contaminated. First responders may carry the agent to fire or precinct houses, hospitals, or to the locations of subsequent calls.

6. The scope of the incident may expand geometrically and may affect mutual aid jurisdictions. Airborne agents flow with the air current and may disseminate via ventilation systems, carrying the agents far from the initial source.

7. There will be a stronger reaction from the public than with other types of incidents. The thought of exposure to a chemical or biological agent or radiation evokes terror in most people. The fear of the unknown also makes the public's response more severe.
8. Time is working against responding elements. The incident can expand geometrically and very quickly. In addition, the effects of some chemicals and biological agents worsen over time.
9. Support facilities, such as utility stations and 911 centers along with critical infrastructures, are at risk as targets.
10. Specialized State and local response capabilities may be overwhelmed.

C. Threat Levels

The CONPLAN establishes a range of threat levels determined by the FBI that serve to frame the nature and scope of the Federal response. Each threat level provides for an escalating range of actions that will be implemented concurrently for crisis and consequence management. The Federal government will take specific actions which are synchronized to each threat level, ensuring that all Federal agencies are operating with jointly and consistently executed plans. The Federal government will notify and coordinate with State and local governments, as necessary. The threat levels are described below:

1. **Level #4—Minimal Threat:**
   Received threats do not warrant actions beyond normal liaison notifications or placing assets or resources on a heightened alert (agencies are operating under normal day-to-day conditions).
2. **Level #3—Potential Threat:**
   Intelligence or an articulated threat indicates a potential for a terrorist incident. However, this threat has not yet been assessed as credible.
3. **Level #2—Credible Threat:**
   A threat assessment indicates that the potential threat is credible, and confirms the involvement of WMD in the developing terrorist incident. Intelligence will vary with each threat, and will impact the level of the Federal response. At this threat level, the situation requires the tailoring of response actions to use Federal resources needed to anticipate, prevent, and/or resolve the crisis. The Federal crisis management response will focus on law enforcement actions taken in the interest of public safety and welfare, and is predominantly concerned with preventing and resolving the threat. The Federal consequence management response will focus on contingency planning and pre-positioning of tailored resources, as re-

quired. The threat increases in significance when the presence of an explosive devise or WMD capable of causing a significant destructive event, prior to actual injury or loss, is confirmed or when intelligence and circumstances indicate a high probability that a device exists. In this case, the threat has developed into a WMD terrorist situation requiring an immediate process to identify, acquire, and plan the use of Federal resources to augment State and local authorities in lessening or averting the potential consequence of terrorist use or employment of WMD.

4. **Level #1—WMD Incident:**
   A WMD terrorism incident has occurred which requires an immediate process to identify, acquire, and plan the use of Federal resources to augment State and local authorities in response to limited or major consequences of a terrorist use or employment of WMD. This incident has resulted in mass casualties. The Federal response is primarily directed toward public safety and welfare and the preservation of human life.

D. Lead Federal Agency Responsibilities

The LFA, in coordination with the appropriate Federal, State and local agencies, is responsible for formulating the Federal strategy and a coordinated Federal response. To accomplish that goal, the LFA must establish multi-agency coordination structures, as appropriate, at the incident scene, area, and national level. These structures are needed to perform oversight responsibilities in operations involving multiple agencies with direct statutory authority to respond to aspects of a single major incident or multiple incidents. Oversight responsibilities include:

- Coordination. Coordinate the determination of operational objectives, strategies, and priorities for the use of critical resources that have been allocated to the situation, and communicate multi-agency decisions back to individual agencies and incidents.
- Situation Assessment. Evaluate emerging threats, prioritize incidents, and project future needs.
- Public Information. As the spokesperson for the Federal response, the LFA is responsible for coordinating information dissemination to the White House, Congress, and other Federal, State and local government officials. In fulfilling this responsibility, the LFA ensures that the release of public information is coordinated between crisis and consequence management response entities. The Joint Information Center (JIC) is established by the LFA, under the operational control of the LFA's Public Information Officer, as a focal point for the coordination

and provision of information to the public and media concerning the Federal response to the emergency. The JIC may be established in the same location as the FBI Joint Operations Center (JOC) or may be located at an on-scene location in coordination with State and local agencies. The following elements should be represented at the JIC: (1) FBI Public Information Officer and staff, (2) FEMA Public Information Officer and staff, (3) other Federal agency Public Information Officers, as needed, and (4) State and local Public Information Officers.

IV. Concept of Operations

A. Mission

The overall Lead Federal Agency, in conjunction with the lead agencies for crisis and consequence management response, and State and local authorities where appropriate, will notify, activate, deploy and employ Federal resources in response to a threat or act of terrorism. Operations will be conducted in accordance with statutory authorities and applicable plans and procedures, as modified by the policy guidelines established in PDD-39 and PDD-62. The overall LFA will continue operations until the crisis is resolved. Operations under the CONPLAN will then stand down, while operations under other Federal plans may continue to assist State and local governments with recovery.

B. Command and Control

Command and control of a terrorist threat or incident is a critical function that demands a unified framework for the preparation and execution of plans and orders. Emergency response organizations at all levels of government may manage command and control activities somewhat differently depending on the organization's history, the complexity of the crisis, and their capabilities and resources. Management of Federal, State and local response actions must, therefore, reflect an inherent flexibility in order to effectively address the entire spectrum of capabilities and resources across the United States. The resulting challenge is to integrate the different types of management systems and approaches utilized by all levels of government into a comprehensive and unified response to meet the unique needs and requirements of each incident.

1. Consequence Management

State and local consequence management organizations are generally structured to respond to an incident scene using a modular, functionally-oriented ICS that can be tailored to the kind, size and management needs of the incident. ICS is employed to organize and unify multiple disciplines with multi-jurisdictional responsibilities on-scene under one functional organization. State and local emergency operations plans generally establish direction and control procedures for their agencies' response to disaster situations. The organization's staff is built from a "top-down" approach with responsibility and authority placed initially with an Incident Commander who determines which local resources will be deployed. In many States, State law or local jurisdiction ordinances will identify by organizational position the person(s) that will be responsible for serving as the incident commander. In most cases, the incident commander will come from the State or local organization that has primary responsibility for managing the emergency situation.

When the magnitude of a crisis exceeds the capabilities and resources of the local incident commander or multiple jurisdictions become involved in order to resolve the crisis situation, the ICS command function can readily evolve into a Unified Command (see Figure 1). Under Unified Command, a multi-agency command post is established incorporating officials from agencies with jurisdictional responsibility at the incident scene. Multiple agency resources and personnel will then be integrated into the ICS as the single overall response management structure at the incident scene.

Multi-agency coordination to provide resources to support on-scene operations in complex or multiple incidents is the responsibility of emergency management. In the emergency management system, requests for resources are filled at the lowest possible level of government. Requests that exceed available capabilities are progressively forwarded until filled, from a local Emergency Operations Center (EOC), to a State EOC, to Federal operations centers at the regional or national level.

State assistance may be provided to local governments in responding to a terrorist threat or recovering from the consequences of a terrorist incident as in any natural or man-made disaster. The governor, by State law, is the chief executive officer of the State or commonwealth and has full authority to discharge the duties of his office and exercise all powers associated with the operational control of the State's emergency services during a declared emergency. State agencies are responsible for authorities and Incident Commander when requested. When State assistance is provided, the local government retains overall responsibility for command and control of the emergency operations, except in cases where State or Federal statutes transfer authority to a specific State or Federal agency. State and local governments have primary responsibility for consequence management. FEMA, using the FRP, directs and coordinates

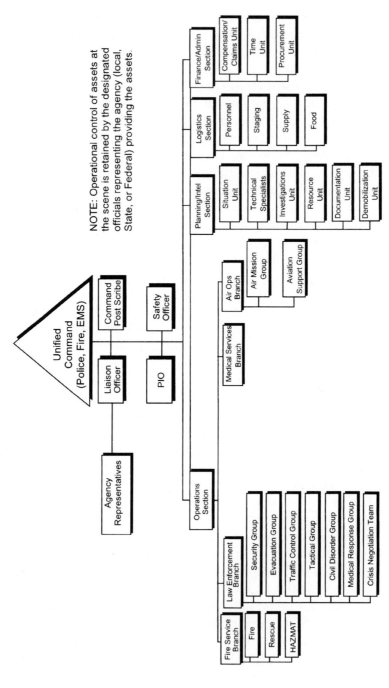

NOTE: Operational control of assets at the scene is retained by the designated officials representing the agency (local, State, or Federal) providing the assets.

FIGURE Q-1 Incident Command System/Unified Command

all Federal response efforts to manage the consequences in domestic incidents, for which the President has declared, or expressed an intent to declare, an emergency.

2. **Crisis Management**

As the lead agency for crisis management, the FBI manages a crisis situation from an FBI command post or JOC, bringing the necessary assets to respond and resolve the threat or incident. These activities primarily coordinate the law enforcement actions responding to the cause of the incident with State and local agencies.

During a crisis situation, the FBI Special Agent In Charge (SAC) of the local Field Division will establish a command post to manage the threat based upon a graduated and flexible response. This command post structure generally consists of three functional groups, Command, Operations, and Support, and is designed to accommodate participation of other agencies, as appropriate (see Figure 2). When the threat or incident exceeds the capabilities and resources of the local FBI Field Division, the SAC can request additional resources from the FBI's Critical Incident Response Group, located at Quantico, VA, to augment existing crisis management capabilities. In a terrorist threat or incident that may involve a WMD, the traditional FBI command post is expanded into a JOC incorporating a fourth functional entity, the Consequence Management Group.

Requests for DOD assistance for crisis management during the incident come from the Attorney General to the Secretary of Defense through the DOD Executive Secretary. Once the Secretary has approved the request, the order will be transmitted either directly to the unit involved or through the chairman of the Joint Chiefs of Staff.

C. Unification of Federal, State and Local Response

1. **Introduction**

Throughout the management of the terrorist incident, crisis and consequence management components will operate concurrently (see Figure 3). The concept of operations for a Federal response to a terrorist threat or incident provides for the designation of an LFA to ensure multi-agency coordination and a tailored, time-phased deployment of specialized Federal assets. It is critical that all participating Federal, State, and local agencies interact in a seamless manner.

2. **National Level Coordination**

The complexity and potential catastrophic consequences of a terrorist event will require application of a multi-agency coordination system at the Federal agency headquarters level. Many critical on-scene

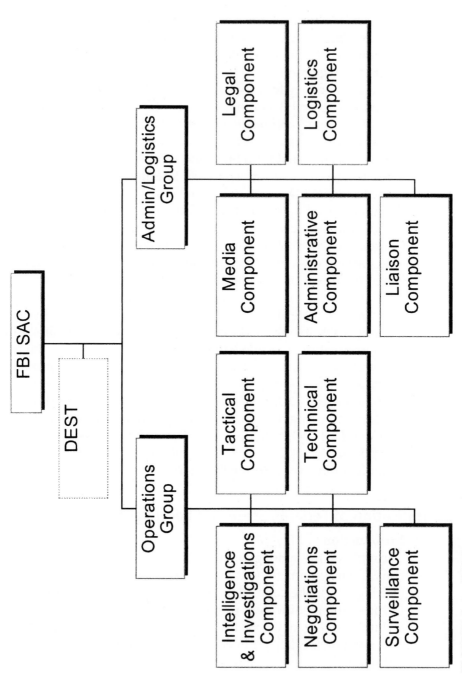

FIGURE Q-2 FBI Command Post

decisions may need to be made in consultation with higher authorities. In addition, the transfer of information between the headquarters and field levels is critical to the successful resolution of the crisis incident.

Upon determination of a credible threat, FBI Headquarters (FBIHQ) will activate its Strategic Information and Operations Center (SIOC) to coordinate and manage the national level support to a terrorism incident. At this level, the SIOC will generally mirror the JOC structure operating in the field. The SIOC is staffed by liaison officers from other Federal agencies that are required to provide direct support to the FBI, in accordance with PDD-39. The SIOC performs the critical functions of coordinating the Federal response and facilitating Federal agency headquarters connectivity. Affected Federal agencies will operate headquarter-level emergency operations centers, as necessary.

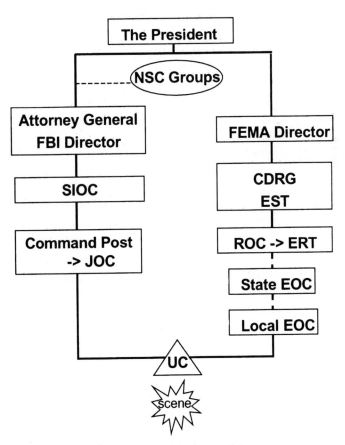

FIGURE Q-3 Coordinating Relationships

Upon notification by the FBI of a credible terrorist threat, FEMA may activate its Catastrophic Disaster Response Group. In addition, FEMA will activate the Regional Operations Center and Emergency Support Team, as required.

3. Field Level Coordination

During a terrorist incident, the organizational structure to implement the Federal response at the field level is the JOC. The JOC is established by the FBI under the operational control of the Federal OSC, and acts as the focal point for the strategic management and direction of on-site activities, identification of State and local requirements and priorities, and coordination of the Federal response. The local FBI field office will activate a Crisis Management Team to establish the JOC, which will be in the affected area, possibly collocated with an existing emergency operations facility. Additionally, the JOC will be augmented by outside agencies, including representatives from the DEST (if deployed), who provide interagency technical expertise as well as inter-agency continuity during the transition from and FBI command post structure to the JOC structure.

Similar to the Area Command concept within the ICS, the JOC is established to ensure inter-incident coordination and to organize multiple agencies and jurisdictions within an overall command and coordination structure. The JOC includes the following functional groups: Command, Operations, Admin/Logistics, and Consequence Management (see Figure 4). Representation within the JOC includes officials from local, State and Federal agencies with specific roles in crisis and consequence management.

The Command Group of the JOC is responsible for providing recommendations and advice to the Federal OSC regarding the development and implementation of strategic decisions to resolve the crisis situation and for approving the deployment and employment of resources. In this scope, the members of the Command Group play an important role in ensuring the coordination of Federal crisis and consequence management functions. The Command Group is composed of the FBI Federal OSC and senior officials with decision making authority from local, State, and Federal agencies, as appropriate, based upon the circumstances of the threat or incident. Strategies, tactics and priorities are jointly determined within this group. While the FBI retains authority to make Federal crisis management decisions at all times, operational decisions are made cooperatively to the greatest extent possible. The FBI Federal OSC and the senior FEMA official at the JOC will provide, or obtain from higher authority, an immediate resolution of conflicts in priorities

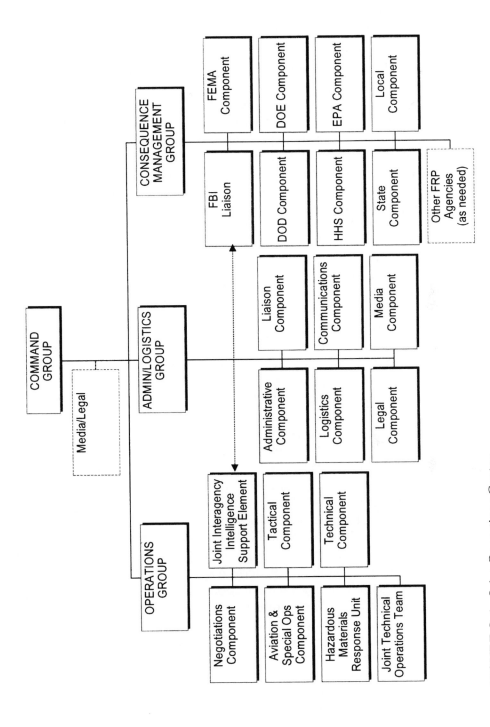

FIGURE Q-4 Joint Operations Center

for allocation of critical Federal resources between the crisis and consequence management responses.

A FEMA representative coordinates the actions of the JOC Consequence Management Group, and expedites activation of a Federal consequence management response should it become necessary. FBI and FEMA representatives will screen threat/incident intelligence for the Consequence Management Group. The JOC Consequence Management Group monitors the crisis management response in order to advise on decisions that may have implications for consequence management, and to provide continuity should a Federal consequence management response become necessary.

Should the threat of a terrorist incident become imminent, the JOC Consequence Management Group may forward recommendations to the ROC Director to initiate limited pre-deployment of assets under the Stafford Act. Authority to make decisions regarding FRP operations rests with the ROC Director until an FCO is appointed. The senior FEMA official in the JOC ensures appropriate coordination between FRP operations and the JOC Command Group.

4. **On-Scene Coordination**

Once a WMD incident has occurred (with or without a pre-release crisis period), local government emergency response organizations will respond to the incident scene and appropriate notifications to local, State, and Federal authorities will be made. Control of this incident scene will be established by local response authorities (likely a senior fire or law enforcement official). Command and control of the incident scene is vested with the Incident Commander/Unified Command. Operational control of assets at the scene is retained by the designated officials representing the agency (local, State, or Federal) providing the assets. These officials manage tactical operations at the scene in coordination with the UC as directed by their agency counterparts at field-level operational centers, if used. As mutual aid partners, State and Federal responders arrive to augment the local responders. The incident command structure that was initially established will likely transition into an Unified Command (UC). This UC structure will facilitate both crisis and consequence management activities. The UC structure used at the scene will expand as support units and agency representatives arrive to support crisis and consequence management operations. On-scene consequence management activities will be supported by the local and State EOC, which will be augmented by the ROC or Disaster Field Office, and the Emergency Support Team, as appropriate.

When Federal resources arrive at the scene, they will operate as a Forward Coordinating Team (FCT). The senior FBI representative will join the Unified Command group while the senior FEMA representative will coordinate activity of Federal consequence management liaisons to the Unified Command. On-scene Federal crisis management resources will be organized into a separate FBI Crisis Management Branch within the Operations Section, and an FBI representative will serve as Deputy to the Operations Section Chief. Federal consequence management resources will assist the appropriate ICS function, as directed (see Figure Q-5).

Throughout the incident, the actions and activities of the Unified Command at the incident scene and the Command Group of the JOC will be continuously and completely coordinated.

V. Phasing of the Federal Response

Phasing of the Federal response to a threat or act of terrorism includes Notification; Activation and Deployment; Response Operations; Response Deactivation; and Recovery. Phases may be abbreviated or bypassed when warranted.

A. Notification

Receipt of a terrorist threat or incident may be through any source or medium, may be articulated, or developed through intelligence sources. It is the responsibility of all local, State, and Federal agencies and departments to notify the FBI when such a threat is received.

Upon receipt of a threat of domestic terrorism, the FBI will conduct a formal threat credibility assessment of the information with assistance from select interagency experts. For a WMD threat, this includes three perspectives:

- Technical feasibility: An assessment of the capacity of the threatening individual or organization to obtain or produce the material at issue;
- Operational practicability: An assessment of the feasibility of delivering or employing the material in the manner threatened;
- Behavioral: A psychological assessment of the likelihood that the subject(s) will carry out the threat, including a review of any written or verbal statement by the subject(s)

The FBI manages a Terrorist Threat Waning System to ensure that vital information regarding terrorism reaches those in the U.S. counterterrorism and law enforcement community responsible for countering terrorist threats. This information is transmitted via secure teletype. Each message transmitted under this system is an alert, and advisory, or an assessment— an alert if the terrorist threat is credible and specific; an advisory, if the

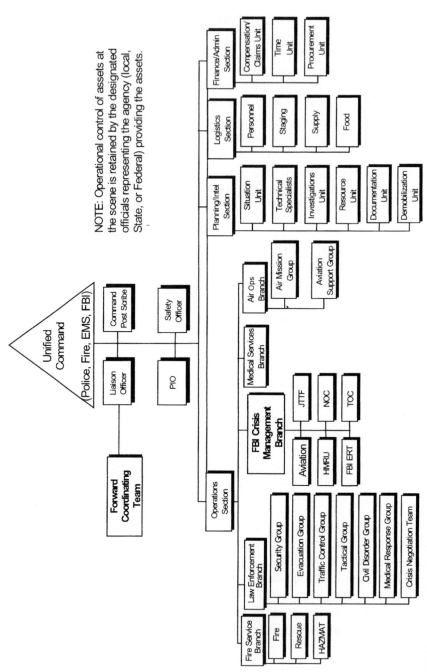

NOTE: Operational control of assets at the scene is retained by the designated officials representing the agency (local, State, or Federal) providing the assets.

FIGURE Q-5 On-Scene Coordination

threat is credible but general in both timing and target; or an assessment to impart facts and/or threat analysis concerning terrorism.

1. **The role of the FBI is to:**
   a. Verify the accuracy of the notification,
   b. Initiate the threat assessment process,
   c. Notify Domestic Emergency Support Team agencies, and
   d. Notify other Federal, State and local agencies, as appropriate.
2. **The role of FEMA is to:**
   a. Advise the FBI of consequence management considerations,
   b. Verify that the state and local governments have been notified, and
   c. Notify other Federal agencies under the FRP, as appropriate.

B. Activation and Deployment

Upon determination that the threat is credible, or an act of terrorism has occurred, FBIHQ will initiate appropriate liaison with other Federal agencies to activate their operations centers and provide liaison officers to the SIOC. In addition, FBIHQ will initiate communications with the SAC of the responsible Field Office apprising him/her of possible courses of action and discussing deployment of the DEST. The FBI SAC will establish initial operational priorities based upon the specific circumstances of the threat or incident. This information will then be forwarded to FBIHQ to coordinate identification and deployment of appropriate resources.

Based upon a credible threat assessment and a request by the SAC, the FBI Director, in consultation with the Attorney General, may request authorization through National Security Council groups to deploy the DEST to assist the SAC in mitigating the crisis situation. The DEST is rapidly deployable, inter-agency team responsible for providing the FBI expert advice and support concerning the U.S. Government's capabilities in resolving the terrorist threat or incident. This includes crisis and consequence management assistance, technical or scientific advice and contingency planning guidance tailored to situations involving chemical, biological, or nuclear/radiological weapons.

Upon arrival at the FBI Command Post or forward location, the DEST may act as a stand alone advisory team to the SAC providing recommended courses of action. While the DEST can operate as an advance element of the JOC, DEST deployment does not have to precede JOC activation. Upon JOC activation, the SAC is the Federal On-Scene Commander (OSC). The Federal OSC serves as the on-scene manager for the United States Government and coordinates the actions of the JOC Command Group. The DEST consequence management component merges into the JOC structure under the leadership of the Senior FEMA Official.

1. **The role of the FBI is to:**
   a. Designate a Federal OSC,
   b. Deploy the DEST if warranted and approved, and provide liaison to State and local authorities as appropriate,
   c. Establish multi-agency coordination structures, as appropriate, at the incident scene, area, and national level in order to:
      (1) Coordinate the determination of operational objectives, strategies, and priorities for the use of critical resources that have been allocated to the situation, and communicate multi-agency decisions back to individual agencies and incidents.
      (2) Coordinate the evaluation of emerging incidents, prioritization of incidents, and projection of future needs.
      (3) Establish a Joint Information Center and coordinate information dissemination.

2. **The role of FEMA is to:**
   a. Activate the appropriate FRP elements, as needed,
   b. Designate and deploy an individual to serve as the Senior FEMA Official to the JOC. Primary responsibilities include:
      (1) Managing the Consequence Management Group.
      (2) Serving as senior consequence management official on the Command Group.
      (3) Designate an individual to work with the FBI liaison to screen intelligence for consequence management related implications.
   c. Identify the appropriate agencies to staff the JOC Consequence Management Group and advise the FBI. With FBI concurrence, notify consequence management agencies to request they deploy representatives to the JOC.

C. Response Operations

These response operations phase involves those activities necessary for an actual Federal response to address the immediate and short-term effects of a terrorist threat or incident. These activities support an emergency response with a bilateral focus on the achievement of law enforcement goals and objectives, and the planning and execution of consequence management activities to address the effects of terrorist incident. Prior to the use or functioning of a WMD, crisis management activities will generally have priority. When an incident results in the use of WMD, consequence management activities will generally have priority. Activities may overlap and/or run concurrently during the emergency response, and are dependent on the threat and/or the strategies for responding to the incident. Events may preclude certain activities from occurring, particularly in an attack without prior warning.

D. Response Deactivation

Each Federal agency will discontinue emergency response operations under the CONPLAN when advised that their assistance is no longer required in support of the FBI, or when their statutory responsibilities have been fulfilled.

Upon determination that applicable law enforcement goals and objectives have been met, no further immediate threat exists, and that Federal crisis management actions are no longer required, the Attorney General, in consultation with the FBI Director and the FEMA Director, shall transfer the LFA role to FEMA. The Federal OSC will deactivate and discontinue emergency response operations under the CONPLAN. Prior to this activity, the Federal OSC will apprise the senior officials representing agencies in the JOC Command Group of the intent to deactivate in order to confirm agreement for this decision.

Consequence management support to the State and local government(s) impacted by the incident may continue for a very long period. Termination of consequence management assistance will be handled according to the procedures established in the FRP.

E. Recovery

The State and local governments share primary responsibility for planning the recovery of the affected area. Recovery efforts will be initiated at the request of the State or local governments following mutual agreement of the agencies involved and confirmation from the LFA that the incident has stabilized and that no further threat exists to public health and safety. The Federal government will assist the State and local governments in developing mitigation and recovery plans, with FEMA coordinating the overall activity of the Federal agencies involved in this phase.

Acronyms

CONPLAN	Concept of Operations Plan
DEST	Domestic Emergency Support Team
DOD	Department of Defense
DOE	Department of Energy
DOJ	Department of Justice
EM	Emergency Management
EMS	Emergency Medical Services
EOC	Emergency Operations Center
EPA	Environmental Protection Agency
ERT	Evidence Response Team (FBI)

FBI	Federal Bureau of Investigation
FCO	Federal Coordinating Officer
FEMA	Federal Emergency Management Agency
FRP	Federal Response Plan
HAZMAT	Hazardous Materials
HHS	Department of Health and Human Services
HMRU	Hazardous Materials Response Unit
JIC	Joint Information Center
JIISE	Joint Interagency Intelligence Support Element
JOC	Joint Operations Center
JTTF	Joint Terrorism Task Force
ICS	Incident Command System
LFA	Lead Federal Agency
NCP	National Oil and Hazardous Substances Pollution Contingency Plan
NOC	Negotiations Operations Center
OSC	On-Scene Commander (FBI) On-Scene Coordinator (EPA)
PIO	Public Information Officer
PDD-39	Presidential Decision directive 39
ROC	Regional Operations Center
SAC	Special Agent-in-Charge
SFO	Senior FEMA Official
SIOC	Strategic Information and Operations Center
STOC	Sniper Tactical Operations Center
TOC	Tactical Operations Center
UC	Unified Command
USCG	United States Coast Guard
WMD	Weapon of Mass Destruction

Definitions

Assessment—The evaluation and interpretation of measurements and other information to provide a basis for decision-making.

Combating Terrorism—The full range Federal programs and activities applied against terrorism, domestically and abroad, regardless of the source or motive.

Consequence Management—Consequence management is predominantly an emergency management function and includes measures to protect public health and safety, restore essential government services, and provide emergency relief to governments, businesses, and individuals affected by the consequences of terrorism. In an actual or potential terrorist incident, a consequence management response will be managed by FEMA using structures and resources of the Federal Response Plan (FRP). These efforts will include support missions as described in other Federal operations plans, such as predictive modeling, protective action recommendations, and mass decontamination.

Coordinate—To advance systematically an exchange of information among principals who have or may have a need to know certain information in order to carry out their role in a response.

Counterterrorism—The full range of activities directed against terrorism, including preventive, deterrent, response and crisis management efforts.

Crisis Management—Crisis management is predominantly a law enforcement function and includes measures to identify, acquire, and plan the use of resources needed to anticipate, prevent, and/or resolve a threat or act of terrorism. In a terrorist incident, a crisis management response may include traditional law enforcement missions, such as intelligence, surveillance, tactical operations, negotiations, forensics, and investigations, as well as technical support missions, such as agent identification, search, render safe procedures, transfer and disposal, and limited decontamination. In addition to the traditional law enforcement missions, crisis management also includes assurance of public health and safety.

Disaster Field Office (DFO)—The office established in or near the designated area to support Federal and State response and recovery operations. The Disaster Field Office houses the Federal Coordinating Officer (FCO), the Emergency Response Team, and, where possible, the State Coordinating Officer and support Staff.

Emergency—Any natural or man-caused situation that results in or may result in substantial injury or harm to the population or substantial damage to or loss of property.

Emergency Operations Center (EOC)—the site from which civil government officials (municipal, county, State and Federal) exercise direction and control in an emergency.

Emergency Public Information—Information which is disseminated primarily in anticipation of an emergency or at the actual time of an emergency and in addition to providing information, frequently directs actions, instructs, and transmits direct orders.

Emergency Response Team—(1) A team composed of Federal program and support personnel, which FEMA activates and deploys into an area affected by a major disaster or emergency. This team assists the FCO in carrying out his/her responsibilities under the Stafford Act, the declaration, applicable laws, regulations, and the FEMA-State agreement. (2) The team is an interagency team, consisting of the lead representative from each Federal department or agency assigned primary responsibility for an Emergency support Function and key members of the FCO's staff, formed to assist the FCO in carrying out his/her responsibilities. The team provides a forum for coordinating the overall Federal consequence management response requirements.

Emergency Support Function—A functional area of response activity established to facilitate coordinated Federal delivery of assistance required during the response phase to save lives, protect property and health, and maintain public safety. These functions represent those types of Federal assistance which the State likely will need most because of the overwhelming impact of a catastrophic event on local and State resources.

Evacuation—Organized, phased, and supervised dispersal of civilians from dangerous or potentially dangerous area, and their reception and care in safe areas.

Federal Coordinating Officer (FCO)—(1) The person appointed by the FEMA Director, or in his/her absence, the FEMA Deputy Director, or alternatively the FEMA Associate Director for Response and Recovery, following a declaration of a major disaster or of an emergency by the President, to coordinate Federal assistance. The FCO initiates action immediately to assure that Federal Assistance is provided in agreement. (2) The FCO is the senior Federal official appointed in accordance with the provisions of Public Law 93-288, as amended (the Stafford Act), to coordinate the overall consequence management response and recovery activities. The FCO represents the President as provided by Section 303 of the Stafford Act for the purpose of coordinating the administration of Federal relief activities in the designated area. Additionally, the FCO is delegated responsibilities and performs those for the FEMA Director as outlined in Executive Order 12148 and those responsibilities delegated to the FEMA Regional Director in the Code of Federal Regulations, Title 44, Part 205.

Federal On-Scene Commander (OSC)—The FBI official designated upon JOC activation to ensure appropriate coordination of the overall United States government response with Federal, State and local authorities, until such time as the Attorney General transfers the LFA role to FEMA.

Federal Response Plan (FRP)—(1) The plan designed to address the consequences of any disaster or emergency situation in which there is a need for Federal assistance under the authorities of the Robert T. Stafford Disaster Relief and Emergency Assistance Act, 42 U.S.C. 5 121 et seq. (2) the FRP is the Federal government's plan of action for assisting affected States and local jurisdiction in the event of a major disaster or emergency.

First Responder—Local police, fire, and emergency medical personnel who first arrive on the scene of an incident and take action to save lives, protect property, and meet basic human needs.

Joint Information Center (JIC)—A center established to coordinate the Federal public information activities on-scene. It is the central point of contact for all news media at the scene of the incident. Public information officials from all participating Federal agencies should collocate at the JIC. Public information officials from participating State and local agencies also may collocate at the JIC.

Joint Interagency Intelligence Support Element (JIISE)—The JIISE is an interagency intelligence component designed to fuse intelligence information from the various agencies participating in a response to a WMD threat or incident within an FBI JOC. The JIISE is an expanded version of the investigative/intelligence component which is part of the standardized FBI command post structure. The JIISE manages five functions including: security, collections management, current intelligence, exploitation, and dissemination.

Joint Operations Center (JOC)—Established by the LFA under the operational control of the Federal OSC, as the focal point for management and direction of onsite activities, coordination/establishment of State requirements/priorities, and coordination of the overall Federal response.

Lead Agency—The Federal department or agency assigned lead responsibility under U.S. law to manage and coordinate the Federal response in a specific functional area. For the purposes of the CONPLAN, there are two lead agencies, the FBI for Crisis Management and FEMA for Consequence Management. Lead agencies support the overall Lead Federal Agency (LFA) during all phases of the response.

Lead Federal Agency (LFA)—the agency designated by the President to lead and coordinate the overall Federal response is referred to as the LFA and is determined by the type of emergency. In general, an LFA establishes operational structures and procedures to assemble and work with agencies providing direct support to the LFA in order to provide an initial assessment of the situation; develop an action plan; monitor and update operational priorities; and ensure each agency exercises its concurrent and distinct authorities under US law and supports the LFA in carrying out the President's relevant policy. Specific responsibilities of an LFA vary according to the agency's unique statutory authorities.

Liaison—An agency official sent to another agency to facilitate interagency communications and coordination.

Local Government—Any county, city, village, town, district, or political subdivision of any State, and Indian tribe or authorized tribal organization, or Alaska Native village or organization, including any rural community or unincorporated town or village or any other public entity.

On-Scene Coordinator (OSC)—The Federal official pre-designated by the EPA and U.S. Coast Guard to coordinate and direct response and removals under the National Oil and Hazardous Substances Pollution Contingency Plan.

Public Information Officer—Official at head quarters or in the field responsible for preparing and coordinating the dissemination of public information in cooperation with other responding Federal, State, and local agencies.

Recovery—Recovery, in this document, includes all types of emergency actions dedicated to the continued protection of the public or to promoting the resumption of normal activities in the affected area.

Recovery Plan—A plan developed by each State, with assistance from the responding Federal agencies, to restore the affected area.

Regional Director—The Director of one of FEMA's ten regional offices and principal representative for working with other Federal regions, State and local governments, and the private sector in that jurisdiction.

Regional Operations Center (ROC)—The temporary operations facility for the coordination of Federal response and recovery activities, located at the FEMA Regional Office (or at the Federal Regional Center) and led by the FEMA Regional director or Deputy Regional Director until the Disaster Field Offices becomes operational.

Response—Those activities and programs designed to address the immediate and short-term effects of the onset of an emergency or disaster.

Senior FEMA Official (SFO)—The official appointed by the Director of FEMA, or his representative, that is responsible for deploying to the JOC to: (1) serve as the senior interagency consequence management representative on the Command Group, and (2) manage and coordinate activities taken by the Consequence Management Group.

State Coordinating Officer—An official designated by the Governor of the affected State, upon a declaration of a major disaster or emergency, to coordinate State and local disaster assistance efforts with those of the Federal government, and to act in cooperation with the FCO to administer disaster recovery efforts.

Terrorism—Terrorism includes the unlawful use of force or violence against persons or property to intimidate or coerce a government, the civilian population, or any segment thereof, in furtherance of political or social objectives.

Weapon of Mass Destruction (WMD)—A WMD is any device, material, or substance used in a manner, in a quantity or type, or under circumstances evidencing and intent to cause death or serious injury to persons or significant damage to property.

Special Bulletin

Provided by the National Domestic Preparedness Office (NDPO), in co-ordination with The Weapons of Mass Destruction Operations Unit of the FBI, The Hazardous Materials Response Unit of the FBI, The Centers for Disease Control and Prevention (COC), and the U.S. Army Medical Research Institute of Infectious Diseases (USAMRIID)

WMD Threats: Sample Guidelines Reissue

Because of the recent series of anthrax hoaxes, the National Domestic Preparedness Office is reissuing the sample guidelines for responding to a WMD threat.

1. **Anonymous caller indicating a WMD threat (including anthrax).**
   - Law enforcement response including local authorities and FBI agent.
   - Fire department/HazMat response not recommended unless device or substance is found.
   - Routine law enforcement investigation.
   - Investigative actions during this response may include:
     - Information gathering at the scene.
     - Building evacuation/search following local protocol.
     - Taking control of the building ventilation system may be warranted based upon investigative findings.
     - Attention should be focused on appliances or devices foreign to the surroundings.
     - Included should be an assessment of the building ventilation system to rule out forced entry and tampering.
       Protective equipment should not be required unless hazards or risks are indicated.
     - Investigation similar to a telephonic bomb threat.
   - Suspicious findings during investigation should initiate a public safety response including:

- Fire/EMS/HazMat.
- EOD team.
- Notifications per local plan which should include local and state health departments
- Notifications per FBI plan.

2. *Potential WMD device located*
   - Follow local protocols for risk assessment and evaluation of potential explosive devices. Included in the response should be:
     - Law enforcement including local authorities and FBI agent.
     - Fire/EMS/HazMat.
     - EOD team.
     - Local and state health departments.
   - If explosive device is not ruled out, coordinated efforts with local/regional EOD authority and notify FBI Bomb Data Center (BDC).
   - If explosive device is ruled out:
     - Evaluate for potential chemical, biological or radioactive filler.
     - If radioactive filler appears to be present, follow FBI plans for requesting additional assistance.
     - If no hazardous materials appear to be present, response continues as a law enforcement investigation.
   - Device with potential chemical or biological filler or supplement.
     - Follow FBI ERT protocols for documentation of the crime scene.
     - Contain the package following recommendations from a hazardous materials authority. Assure notification of FBI/HMRU, through local FBI.
     - Options include double bagging, steel cans, polly containment vessels, or utilization of a hazardous materials over-pack.
     - Control the material as evidence and follow FBI plan for laboratory analysis.
   - Potential release of WMD material from a device.
     - Control the ventilation system.
     - Follow protocols for a hazardous materials incident.
     - Evaluate the extent of contamination.
     - Evacuation of affected areas and decontamination procedures should be selected on the basis of an incident and risk assessment
     - Provide medial attention following the recommendations from the local/regional public health medical authority
     - Control and or isolate the hazard.
     - Treat as a FBI hazardous materials crime scene.
     - Request assistance from FBI/HMRU through local FBI.

3. **Specific situations—envelope with potential threat of anthrax, letter opened and material present.**
   - Public safety response including local authorities and FBI agent.
   - Contain the package following recommendations from a hazardous materials authority.
     - Options include double bagging, steel cans, poly containment vessels, or utilizatin of a hazardous materials over-pack.
     - Control the material as evidence and follow FBI plan for laboratory analysis.
   - Provide medical attention/decontamination following the recommendations from the local/regional public health medical authority.
     - Evaluate the extent of contamination.
     - Evacuation of the affected area and decontamination procedures should be selected on the basis of an incident hazard and risk assessment.
     - **Generally, medical prophylaxis and decontamination have not been indicated except for washing hands with soap and warm water.**

4. **Specific situations—envelope with potential threat of anthrax, letter opened and no material present.**
   - Law enforcement response including local authorities and FBI agent
     - Fire department/EMS/HazMat response not recommended unless suspicious material is found or individuals are presenting symptoms.
   - Handle the package following FBI ERT protocols
     - Double bag the material and place in a suitable container such as an evidence paint can
     - Control the material as evidence and follow FBI plan for laboratory analysis.
   - **No medical attention/decontamination is necessary unless symptoms are present, although local public health authorities should be notified.**
   - Handle as a law enforcement investigation.

5. **Specific situations—envelope with potential threat of anthrax, letter not opened.**
   - Law enforcement response including local authorities and FBI agent.
     - Fire department/HazMat response not recommended unless suspicious material is found.
   - Handle the package following FBI ERT protocols.

- Double bag the material and place in a suitable container such as evidence paint can.
- Control the material as evidence and follow FBI plan for laboratory analysis.
- **No medical attention/decontamination is necessary.**
- Handle as a law enforcement investigation.

<u>Please Note</u>: **According to the CDC, hand washing is sufficient for those who have touched the envelope and letter. Decontamination or prophylaxis is not warranted.**

These guidelines are reissued from the NDPO "WMD Threats: Sample Guidelines," which was issued with the Special Bulletin (SB) #4.

For additional information, please refer to the following NDPO Special Bulletins:
SB-1 "Anthrax Advisory"
SB-2 "Anthrax Threats"
SB-3 "Anthrax Threat Guidance for Law Enforcement"
SB-4 "Anthrax Facts"

Questions or comments can be sent to the NDPO at ndpo@leo.gov.

Glossary

absorption: The process of a substance being taken into the body through the skin (transdermal).

absorbed dose: The amount of energy deposited in any material by ionizing radiation. The unit of absorbed dose, the RAD, is a measure of energy absorbed per gram of material. The unit used in countries other than the United States is the Gray. One Gray equals 100 RAD.

acetylcholine (ACh): The neurotransmitter substance widely distributed throughout the tissues of the body at cholenergic synapses, which causes cardiac inhibition, vasodilation, gastrointestinal peristalsis, and other parasympathetic effects.

acetylcholinesterase: An enzyme that hydrolizes the neurotransmitter acetylcholine. Nerve agents inhibit the action of this enzyme.

acute dose: A radiation dose received over a short periof of time.

adenopathy: Swelling or morbid enlargement of the lymph.

adrenergic: *See* **sympathomimetic.**

adsorption: The process of a substance becoming chemically attached to a surface.

aerosol: Fine liquid or solid particles suspended in air (e.g., smoke/fog).

aerosols: A suspension or dispersion of small particles in a gaseous medium.

agency representative: An individual assigned to an incident by a responding or cooperating agency, who is delegated complete authority to execute decisions for all dealings affecting that agency's incident participation. Representatives report to the incident liaison officer.

agent dosage: Concentration of a toxic vapor in the air multiplied by the time that the concentration was present.

agonist: A substance that causes a physiological response.

air purification devices: Respirators or other filtration equipment that remove gases, particulate matter, or vapors from the atmosphere.

airborne pathogen: Pathological microorganisms spread by droplets expelled or dispersed in the air. While this occurs typically through productive sneezing and/or coughing, a deliberate release of pathogens can be accomplished via the use of a variety of aerosol delivery systems.

ALARA: The guiding principle behind radiation protection is that radiation exposures should be kept "as low as reasonably achievable (ALARA)," economic and social factors being taken into account. This common sense approach means that radiation doses for both workers and the public are typically kept lower than their regulatory limits.

alpha particle: Alpha particles are composed of two protons and two neutrons. Alpha particles do not travel very far from their radioactive source. They cannot pass through a piece of paper, clothes, or even the layer of dead cells that normally protects the skin. Because alpha particles cannot penetrate human skin, they are not considered an "external exposure hazard." This means that if the alpha particles stay outside the human body, they cannot harm it. However, alpha particle sources located within the body may pose an "internal" health hazard if they are present in great enough quantities. The risk from indoor radon is due to inhaled alpha particle sources that irradiate lung tissue.

alkali: Basic compound that possesses the ability to neutralize acids and form a salt.

all hazards planning: A fundamental planning approach that takes into account all hazards that a community may need to confront. This is the basic premise of planning and management presently being advocated by FEMA.

ALS (Advanced Life Support): Allowable procedures and techniques used by paramedics and EMT-Intermediate personnel to stabilize patients who exceed basic life support procedures.

ALS responder: Certified or licensed paramedic or EMT/Intermediate.

alveoli: Microscopic air sacs of the lungs where gas exchange occurs with the circulatory system.

ambulance: A ground vehicle providing patient care transportation capability, specified equipment capability, and qualified personnel (EMT, EMT/I, and Paramedic).

analgesic: A compound capable of producing pain relief by altering perception of nociceptive stimuli without producing anesthesia or loss of consciousness.

anaphylaxis: An acute, sometimes violent immunological (allergic) reaction characterized by contraction of smooth muscle and dilation of capillaries due to release of pharmacologically active substances (histamines, serotonin). Can be fatal.

anoxia: Absence of oxygen.

antagonist: A substance that inhibits (blocks) a physiological response or inhibits the response of another drug or substance.

antiadrenergic: *See* **sympatholytic.**

antibiotic: A drug that inhibits the growth of, or kills microorganisms.

antibody: A component of the immune system. A protein that eliminates or counteracts a foreign substance macromolecule or antigen in the body.

anticholinergic: *See* **parasympatholytic.**

anticholinestrase: A drug or substance that blocks the action of cholinesterase.

anticonvulsant: A substance that prevents or arrests seizures.

antidote: A drug or substance that neutralizes a poison or the effects of a poison.

antigen: Any substance that is capable of inducing an immune response.

antisera: A liquid component of blood that contains antibodies.

APR: Acronym for air purifying respirators.

apnea: Not breathing.

arsenical: Pertaining to or containing arsenic; a reference to the vesicant lewisite.

asphyxia: Condition in which cells experience oxygen deprivation.

AST: Abbreviation for aspartate aminotransferase, a liver enzyme.

assessment: (1) Patient assessment—evaluation of patient medical condition. (2) Scene assessment—evaluation of the emergency scene that occurs immediately upon arrival and periodically throughout the operation to ensure safety of members, status of activity, and to determine extent and implications of response operations.

assigned resources: Resources checked in and assigned work tasks on an incident.

assisting agency: An agency directly contributing emergency medical service, fire suppression, rescue, support, or service resources to another agency.

asthenia: Weakness or debility.

ataxia: Inability to coordinate muscle activity during voluntary movement so that smooth movements occur.

atelectasis: Absence of gas from part or all of the lungs due to failure of expansion or resorption of gas from the alveoli.

atom: Atoms are the smallest part of any material, which cannot be broken up by chemical means. Each atom has a center (the nucleus), which contains protons and neutrons. Electrons orbit around the nucleus. In an uncharged atom the number of electrons orbiting the nucleus equals the number of protons in the nucleus. The atom is primarily empty space. If the nucleus of an atom was the size of the button on a baseball pitcher's cap, the electrons would be like dust particles revolving around the outside of the baseball stadium at nearly the speed of light.

atropine: An anticholinergic medication used as an antidote for nerve agents to counteract excessive amounts of acetylcholine.

autoignition temperature: Lowest possible temperature at which a flammable gas or vapor/air mixture will ignite from its own heat source or a contacted heated surface without the necessity of flame or spark.

autonomic nervous system: Part of the nervous system controlling involuntary bodily functions; separated into the sympathetic and parasympathetic nervous systems.

bacillus: A genus of bacteria belonging to the family Bacillaceae. All specimens are rod shaped, sometimes occurring in chains. They are spore bearing, aerobic, motile or nonmotile, most are gram positive.

background radiation: Radiation is a part of our natural world. People have always been exposed to radiation that originates from within the Earth (terrestrial sources) and from outer space (cosmogenic or galactic sources).

bacteria: Single-celled organisms that multiply by cell division and that can cause disease in humans, animals, and plants.

base: *See* **alkali.**

beta particle: Beta particles are similar to electrons, except they come from the atomic nucleus and are not bound to any atom. Beta particles cannot travel very far from their radioactive source. For example, they can travel only about one-half inch in human tissue, and they may travel a few yards in air. They are not capable of penetrating something as thin as a book or a pad of paper.

bio: Abbreviation for biological.

bio-chemicals: Chemicals that make up or are produced by living things.

bio-terrorism agent: Living organisms, or materials derived from them, that cause disease in or injury to humans, animals or plants or cause deterioration of material. Agents can be used in liquid droplets, aerosols, or dry powders.

bio-terrorism: The deliberate use of biological agents/substances as weapons to kill or harm humans, animals, or plants, or to incapacitate equipment.

bio-regulators: Biochemicals that regulate bodily functions and are produced naturally in the body. Inappropriate levels can cause harmful effects.

biotechnology: Applied biological science (e.g., biofermentation processes or genetic engineering).

binary munition: A chemical munition divided into two sections, each containing precursor chemicals, which, when combined, release a chemical agent.

BL/P: There are four (4) biosafety levels (BLs) that conform to specified conditions. These conditions consist of a combination of laboratory practices and techniques, safety equipment, and laboratory facilities appropriate for the operations performed and the hazards posed by the infectious agents. Previously decribed as physical containment (P) levels.

blister agent: A chemical agent that produces local irritation and damage to the skin (vesicant) and mucous membranes, pain and injury to the eyes, reddening and blistering of the skin, and damage to the respiratory system when inhaled. Examples are lewisite, nitrogen mustard, and sulfur mustard.

blood agent: A chemical agent that is inhaled and absorbed into the blood, acting upon hemoglobin in blood cells. The blood carries the agent to all body tissues where it interferes with the tissue oxygenation process. Examples are cyanogen chloride and hydrogen cyanide.

blood-borne pathogen: Pathological microorganisms that are present in human blood and that can cause disease in humans (OSHA definition). Note: The term blood includes blood, blood components, and products developed from human blood.

BLS (basic life support): Basic noninvasive, pre-hospital care used by emergency medical technicians and the lesser trained certified first responders to stabilize critically sick and injured patients.

BLS responder: Certified or licensed emergency medical technician; basic or certified first responder.

B-NICE: An acronym for biological, nuclear, incendiary, chemical, or explosives.

body substance isolation (BSI): An infection control strategy that considers all body substances potentially infectious, requiring the use of universal precautions.

boiling point: Temperature at which a liquid changes its matter state to a gas/vapor. Also the temperature at which the pressure of the liquid equals atmospheric pressure.

branch: The organizational level having functional or geographic responsibility for major segments of incident operations. The branch level is organizationally between the section and the group/division.

breakthrough time: The time required for a given chemical to permeate a protective barrier material. This is usually defined as the time elapsed between the application of a chemical to a protective materials exterior surface and its initial apperance on the inner surface.

briefing: An organized face-to-face meeting between IMS managers/officers during an MCI or disaster.

bronchi: The two large sets of branches that come off the trachea and enter the lungs. There are right and left bronchi.

bronchitis: Inflammation of the mucous membranes of the bronchial tubes.

bronchioles: The finer subdivisions of the bronchus.

bronchiolitis: Inflammation of the bronchioles, often associated with bronchopneumonia.

brucella: A genus of encapsulated, nonmotile bacteria (family Brucellaceae) containing short, rod-shaped to coccoid, gram-negative cells. These organisms do not produce gas from carbohydrates; are parasitic, invading all animal tissues and causing infection of the genital organs, mammary gland, and the respiratory and intestinal tracts; and are pathogenic for humans and various species of domestic animals.

bubo: Inflammatory swelling of one or more lymph nodes, usually in the groin; the confluent mass of nodes usually suppurates and drains pus.

carbuncle: Deep-seated pyogenic infection of the skin and subcutaneous tissues, usually arising in several contiguous hair follicles with formation of connecting sinuses; often preceded or accompanied by fever, malaise, and prostration.

catecholamine: A hormone that acts on the autonomic nervous system including epinephrine (adrenaline), norepinephrine, and dopamine.

CAS registry number: A number assigned to a material by the Chemical Abstract Service to provide a single unique identifier.

causative agent: The organism or toxin that is responsible for causing a specific disease or harmful effects.

caustic: A substance that strongly burns, corrodes, irritates, or destroys living tissue.

CBR: An acronym for chemical, biological, and radiological.

C-cubed I (C3I): A military command term meaning command, control, communications, and intelligence.

ceiling exposure value: The maximum airborne concentration of a biological or chemical agent to which a worker may be exposed at a given time.

central nervous system (CNS): Pertaining to the body's central nervous system.

cerebrospinal: Relating to the brain and the spinal cord.

chain of command: The flow of orders/information from the command/management level to other levels in the IMS. In effective systems, information must flow upward as well as downward.

chemical agent symbol: A designation code assigned to a chemical agent, which is usually two letters. For example HD mustard, GB sarin, CX phosgene oxime.

chemical/biological (chem-bio) terrorism: The use of chemical or biological weapons to create a high-impact EMS incident (sometimes written as CW/BW).

chemical degradation: The altering of the chemical structure of a hazardous material usually accomplished during decontamination.

chemical protective ensemble (CPE): Garments specifically designed to protect the eyes and skin from direct chemical contact There are encapsulating and nonecapsulating versions available for use depending on the operation and the substance. Garments are usually worn with additional respiratory protection as required.

chemical resistance: Ability of the CPE to maintain its protective qualities when it has been contacted by a hazardous substance.

chemoprophylaxis: Prevention of disease through the use of chemicals or drugs.

choking agent: Substances that cause physical injury to the lungs. Exposure is through inhalation. In extreme cases, membranes swell, lungs fill with fluid (edema) and death results from lack of oxygen. The victim is choked. Examples are chlorine and phosgene.

cholinergic: *See* **parasympathomimetic.**

chronic dose: A radiation dose received over a long period of time.

chronic exposure: Repeated low dose exposures to a hazardous substance over an extended time frame.

clear text/clear speak: The use of plain English in radio communication transmissions. No "10" codes, agency-specific codes, or jargon is used when using clear speak/text communications.

CNS depressants: Compounds that have the predominate effect of depressing or obstructing activity of the central nervous system. The primary mental status effects yielded include the disruption of cognitive (thinking) ability, sedation, and lack of motivation (lethargy).

CNS stimulants: Compounds that have the predominant effect of flooding the brain with too much information (stimulus). The primary mental status effects yielded include loss of concentration, indecisiveness, and loss of the ability to act in a sustained, purposeful manner.

coagulated necrosis: Destructive process where acids cause proteins to precipitate as dense, coagulum (clotting) over an injured area.

coagulopathy: A disease affecting the coagulability of the blood.

cocobacillus: A short, thick bacterial rod in the shape of an oval or slightly elongated coccus.

coccus: A type of bacteria that is spherical or ovoid in form. Many are pathogenic, causing diseases such as septic sore throats, scarlet fever, rheumatic fever, pneumonia, and meningitis.

colorimetric tubes: Testing mediums that are used to identify the presence and approximate concentration of a substance in the atmosphere.

combustible gas indicator (CGI): Assessment equipment that measures the ambient concentration of flammable vapors or gases it has been set (calibrated) to monitor.

combustible liquid: Any liquid that has a flash point at or above 100°F (37.7°C) and below 200°F (93.3°C).

command: The act of directing, ordering, and/or controlling resources by virtue of explicit legal, agency, or delegated authority.

command cache: A kit of administrative materials and supplies necessary to operate the EMS incident management system; includes IMS forms, vests, checklists, and protocols.

command staff: The command staff consists of the safety officer, liaison officer, and public information officer, who all report directly to the incident commander.

communicable disease: A disease that can be transmitted from one person to another. Also know as contagious disease.

communications failure protocol: A protocol that dictates agency/unit operations when there is a failure of the telephone and/or EMS radio system.

communication order model: The process of briefly restating an order received to allow for verification and confirmation. This permits all involved to ensure that what was communicated and what was heard coincide, thus ensuring that the correct action is executed.

community disaster plan: A formal document that specifies who is in charge of various incidents, and what agencies/resources will be committed.

concentration: The amount of a chemical agent present in a unit volume of air, usually expressed in milligrams per cubic meter (mg/m3).

concentration time: The amount of a chemical agent present in a unit volume of air, multiplied by the time an individual is exposed to that concentration.

conjunctiva: A fine mucous membrane that lines the eyelids and covers the exposed surface of the eyeball.

contagious: Capable of being transmitted from one person to another.

consequence management: The measures to alleviate the damage, loss, hardship, or suffering caused by emergencies. Consequence management includes measures to protect public health and safety, restore essential government services, and provide emergency relief to affected governments, businesses, and individuals. Consequence management is implemented under the primary jurisdiction of the affected state and local governments. As directed by PDD 39, FEMA is designated the lead federal agency for consequence management, and as such provides support to the state when required.

contaminant/contaminated: A substance or process that poses a threat to life, health, or the environment (definition from NFPA 472).

cooperating agency: An agency that provides indirect support or service functions such as the American Red Cross, Salvation Army, EPA, etc.

corrosives: Substances that destroy the texture or substance of a tissue.

crisis management: The measures to identify, acquire, and plan the use of resources needed to anticipate, prevent, and/or resolve a terrorist threat or incident. As directed by PDD 39, the FBI is designated the lead federal agency for crisis management, which is implemented under its direction.

cryogenics: Materials that exist at extremely low temperatures, such as nitrogen.

CSF: Abbreviation for cerebrospinal fluid.

culture: A population of microorganisms grown in a medium.

cumulative: Additional exposure rather than repeated exposure. The collective effect of having an HD exposure for 30 minutes and several hours later being exposed again for 60 minutes would yield the same effect as a single 90-minute exposure.

cutaneous: Pertaining to the skin.

cyanosis: A dark bluish or purplish coloration of the skin and mucous membranes due to deficient oxygen levels in the blood (hypoxia). Cyanosis is evident when reduced hemoglobin in the blood exceeds 5g per 100 ml.

decontamination (decon): The removal of contamination from responders, patients, vehicles, and equipment. Patients must be decontaminated before treatment/transport; usually accomplished through a physical or chemical process.

delayed patient: A patient that is stable, but will require medical care; could deteriorate to the immediate category; triage color is yellow.

dermal: Relating to the skin or derma.

dermis: The inner layer of the skin, beneath the epidermis, which contains blood vessels, nerves, and structures of the skin.

desiccation: Violent mechanical dehydration of the cells.

desorption: The reverse process of absorption. The agent is removed from the surface, outgassing.

diplopia: A condition in which a single object is perceived as two objects (double vision).

distal: Away from the center of the body or point of origin.

dilution factor: Dilution of contaminated air with uncontaminated air in a general area, room, or building for the purpose of health hazard or nuisance control, and/or for heating and cooling.

disaster cache: A store of predetermined supplies/equipment that is immediately transported to an MCI or disaster.

disaster-catastrophic incidents/events: An MCI that overwhelms both local and regional EMS response capabilities, typically involves multiple over-lapping jurisdictional boundaries, and requires significant multi-jurisdictional response and coordination.

disaster committee: A formal committee of response agencies, planners, support agencies, and volunteer organizations that serves as a vehicle for threat assessment, emergency response planning, disaster exercises, and post-incident analysis.

disease: An alteration of health, with a characteristic set of symptoms, which may affect the entire body or specific organs. Diseases have a variety of causes and are known as infectious diseases when due to a pathogenic microorganism such as a bacteria, virus, or fungus.

disinfection: A procedure that inactivates virtually all recognized pathogenic microorganisms, but not necessarily all microbial forms (e.g., bacterial endospores) on inanimate objects (OSHA definition).

DMAT: Disaster Medical Assistance Team—a deployable team (usually 35 people) of medical personnel and support units under the command of the Office of Emergency Preparedness, U.S. Public Health Service.

DNA: Deoxyribonucleic acid—the genetic material of all organisms and viruses (except for a small class of RNA-containing viruses), which code structures and materials used in normal metabolism.

domestic terrorism: The unlawful use of force or violence, committed by a group(s) of two (2) or more individuals against persons or property to intimidate or coerce a government, the civilian population, or any segment thereof, in furtherance of criminal, political, or social objectives.

dorsal: Toward the back.

dosage: (1) The proper theraputic amount of a drug to be adminstered to a patient. (2) The concentration of a chemical agent in the atmosphere (C), multiplied by the time (t) the concentration remains, expressed as mgmin/m. The dosage (Ct) received by a person depends on how long they are exposed to the concentration. That is, the respiratory dosage in mgmin/m is equal to the time in minutes an individual is unmasked in an agent cloud, multiplied by the concentration of the cloud. The dosage is equal to the time of exposure in minutes of an individual's unprotected skin, multiplied by the concentration of the agent cloud.

DOT.: *See* **U.S. DOT.**

DOT Hazard Classifications: *See* **U.S. DOT hazard classifications.**

downwind distance: The distance a toxic agent vapor cloud will travel from its point of origin, with the wind.

dysphagia: Difficulty swallowing.

dysphonia: Altered voice production.

dyspnea: Shortness of breath, breathing distress.

ecchymosis: A purplish patch caused by extravasation of blood into the skin, differing from petechiae only in size (larger than 3 mm).

edema: An accumulation of an excessive amount of watery fluid in cells, tissues or serous cavities.

electron: Electrons are very small particles with a single negative charge. They are a part of the atom and orbit around the nucleus. Electrons are much smaller than protons or neutrons. The mass of an electron is only about one two-thousandth of a proton or neutron.

emergency medical operations: Delivery of emergency medical care and transportation prior to the arrival at a hospital or other health care facility (according to NFPA 1581).

Emergency Operations Center (EOC): A central disaster management center staffed by representatives from response and support agencies.

emergency operations plan (EOP): An operational document that has resulted in the delineation of response plans for a community or organization. This plan is usually the result of issues that have been identified through a threat assessment survey.

emergency support functions (ESF): Support functions outlined in the Federal Response Plan. ESFs identify lead and secondary agencies, and are not a management system. The ESFs are grouped by 12 identified functional tasks. ESF 1 Transportation, 2 Communications, 3 Public Works and Engineering, 4 Firefighting, 5 Information and Planning, 6 Mass Care, 7 Resource Support (Logistics), 8 Health and medical services, 9 USAR, 10 Hazardous Materials, 11 Food, 12 Energy.

EMS: Emergency medical service generally referring to pre-hospital care resources.

EMS branch: The organization level having functional responsibility for conducting emergency medical operations at a multiple casualty incident.

EMT-Basic: An individual trained in basic life support according to the standards set forth by the authority having jurisdiction (local, regional, or state EMS authority).

EMT-Intermediate: An individual trained in basic life support having received additional training in advanced life support according to the standards set forth by the authority having jurisdiction (local, regional, or state EMS authority).

EMT-Paramedic: An individual trained in advanced life support according to the standards set forth by the authority having jurisdiction (local, regional, or state EMS authority).

endogenous: Oriniating or produced from within.

endotoxin: Endotoxin is composed of compounds called lipopolysaccharides found in bacteria such as E. coli. The presence of endotoxin from a blood-borne infection (sepsis) of a gram-negative bacteria can cause clotting, organ failure, and subsequent death.

endotoxemia: Presence in the blood of endotoxins.

endotracheal intubation: The introduction of a tube through the oral or nasal cavities into the trachea for maintenance of a patent airway.

enterotoxin: A cytotoxin specific for the cells of the intestinal mucosa.

enzyme: A protein-like substance that acts as an organic catalyst in chemical reactions.

enzyme poisons: Chemicals that inhibit (block) specific cellular reactions by competing with or modifying the enzymes needed to catalyze those reactions.

epidermis: The outer layer of skin.

epistasis: Profuse bleeding from the nose.

equivalent dose: The equivalent dose is a measure of the effect that radiation has on humans. The concept of equivalent dose involves the impact that different types of radiation have on humans. Not all types of radiation produce the same effects in humans. The equivalent dose takes into account the type of radiation and the absorbed dose. For example, when considering beta, x-ray, and gamma ray radiation, the equivalent dose (expressed in REMS) is equal to the absorbed dose (expressed in RADS). For alpha radiation, the equivalent dose is assumed to be 20 times the absorbed dose.

erythema: Redness of the skin due to capillary dilatation.

erythema multiforme: An acute eruption of macules, papules, or subdermal vesicles presenting multiform appearance, the characteristic lesion being the target or iris lesion over the dorsal aspect of the hands and forearms.

erythrocyte: A mature red blood cell.

erythropoiesis: The formation of red blood cells.

etiological agent: A living organism that may cause human disease (according to NFPA 472).

evaporation rate: The rate at which a liquid changes to vapor at normal room temperature.

exanthema: Skin eruption occurring as a symptom of an acute viral or coccal disease.

exogenous: Originating or produced externally.

exothermic reaction: A chemical reation that produces heat.

explosives: Compounds that are unstable and break down with the sudden release of large amounts of energy.

explosive range: *See* **flammable range.**

extraocular: Adjacent but exterior to the eyeball.

extrication: The action of disentangling and freeing a person from entrapment.

extrication sector: The EMS IMS organizational component responsible for freeing and disentangling victims from wreckage.

facilities unit: Responsible for support facilities, including shelter, rehabilitation, sanitation, and auxiliary power.

fasciculation: Involuntary contractions or twitching of groups (fasciculi) of muscle fibers; a coarser form of muscular contraction than fibrillation. Commonly described as movement resembling a "bag of worms."

febrile: Having or referring to a fever.

Federal Bureau of Investigation (FBI): The FBI is the principal investigative arm of the U.S. Department of Justice. It has the authority and responsibility to investigate specific crimes assigned to it. The FBI also is authorized to provide other law enforcement agencies with cooperative services, such as fingerprint identification, laboratory examinations, and police training. For the purposes of this textbook, the FBI is the lead federal agency for the crisis management of terrorism incidents as directed by PDD 39 and the FRP.

federal coordinating officer (FCO): Is the president of the United States' representative at a disaster incident. For the purposes of this textbook, the FCO is the individual responsible for coordinating the consequence management response, and will usually be a representative of FEMA.

Federal Emergency Management Agency (FEMA): Is the lead federal agency for the consequence management response to a terrorism incident as directed by PDD 39 and the FRP.

Federal Response Plan (FRP): The FRP provides the system for the overall delivery of federal assistance in a disaster. Twenty-seven federal departments and agencies and the American Red Cross provide resources. Resources are grouped into 12 emergency support functions (ESFs), each headed by a primary or lead agency. For the purposes of this textbook, the FRP presents the federal government's consequence management response to terrorism incidents.

FIRESCOPE: Firefighting Resources of California Organized Against Potential Emergency; developed as a response to California wildfires and became the benchmark for the EMS incident management system.

finance/administration section: The section responsible for all costs and financial actions of the incident and administrative functions, which include the time unit, procurement unit, compensation/claims unit, and cost unit.

first responder: The first trained personnel to arrive at the scene of an emergency.

flammability: The inherent capacity of a substance to ignite and burn rapidly.

flammable range: Range of a gas or vapor concentration that will burn or explode if an ignition source is present. Usually expressed as a percentage by volume of air. Depending on range there are lower expolsive limits (LEL) and upper explosive limits (UEL). For an iginition to take place, the substance must be within the LEL and UEL. Any presences that do not meet or exceed these ranges should not ignite.

flash point: Minimum temperature at which a liquid gives off sufficient enough vapor to ignite and flashover, but not continue to burn without the availablity of more heat.

fomite: Objects such as clothing, towels, utensils that possibly harbor agents of disease and are capable of spreading it.

formalin: A 37 percent aqueous solution of formaldehyde.

fulminant hepatitis: Severe rapidly progressive loss of hepatic function due to viral infection or other cause of inflammatory destruction of liver tissue.

fungus: A group of microorganisms including molds and yeasts, similar to the cellular structure of plants. Some fungi are pathogenic.

G series nerve agents: Chemical agents developed in the 1930s with moderate to high toxicity that act by inhibiting a key nervous system enzyme (GA, GB, GD).

gamma rays: Gamma rays are an example of electromagnetic radiation, as is visible light. Gamma rays originate from the nucleus of an atom. They are capable of traveling long distances through air and most other materials. Gamma rays require more shielding material, such as lead or steel, to reduce their numbers than is required for alpha and beta particles.

general staff: The individuals responsible for incident management. These individuals include the IM, operations section chief, logistics section chief, plans section chief, and administration chief. This level is above the branch/division/group/sector level.

genetic effects: Effects seen in the offspring of the individual who received the agent. The agent must be encountered before conception.

genetic engineering: The directed alteration or manipulation of genetic material.

group/sector: The organizational level having responsibility for a specified functional assignment at an incident (triage, treatment, extrication, etc.).

half-life: The time required for the level of a substance in the blood to be reduced by 50 percent of its initial level.

hazardous materials: Substances that can cause harm to people or the environment upon release.

hemagglutination: The agglutination of red blood cells; may be immune as a result of specific antibody either for red blood cell antigens per se, or other antigens that coat the red blood cells; or may be nonimmune as in hemagglutination caused by viruses or other microbes.

hemagglutinin: A substance, antibody, or other that causes hemagglutination.

hematemesis: Vomiting blood.

hematuria: Urination with blood or red blood cells.

hemodynamic: Referring to the physical aspects of the circulation of blood.

hemoglobin: The iron-containing pigment of the red blood cells. Its function is to carry oxygen from the lungs to the tissues.

hemolysis: The destruction of red blood cells with the liberation of hemoglobin that diffuses into the fluid surrounding them.

hemoptysis: Bloody or blood-tinged sputum.

hepatic: Pertaining to the liver.

high impact incident/event: Any emergency that requires mutual aid resources in order to effectively manage the incident or to maintain community 911 operations.

histamine: A chemical released by mast cells and basophils on stimulation. One of the most powerful vasodilators known and a major mediator of anaphylaxis.

hormone: A chemical substance released by a gland that controls or influences other glands or body systems.

hospital alert system: A communications system between EMS personnel on-site of an MCI and a medical facility that provides available hospital patient receiving capability and/or medical control.

Hospital Emergency Incident Command System (HEICS): A system for the management of internal/external hospital emergencies based on the IMS model.

host: A person that can harbor or nourish a disease-producing organism. The host is infected.

hydration: The combining of a substance with water.

hydrolysis: The reaction of any chemical substance with water by which decomposition of the substance occurs, and one or more new substances are produced.

hyperemia: Presence of an increased amount of blood in a part or organ.

hyperesthesia: Abnormal acuteness of sensitivity to touch, pain, or other sensory stimuli.

hypertension: Abnormally high blood pressure.

hypocalcemia: Reduction of blood calcium below normal levels.

hypotension: Abnormally low blood pressure.

hypovolemia: Abnormally low amount of blood in the body. Generally a result of trauma or internal bleeding.

hypoxia: A condition when insufficient oxygen is available to meet the oxygen demands of the cells.

hypoxemia: The reduction of oxygen content in the arterial blood.

IDLH: Immediate danger to life and health.

idiopathic: Referring to a disease of unknown origin.

immediate patient: A patient who is critical and in need of immediate care; triage color is red.

immunization: The process of rendering a person immune or highly resistant to a disease. Usually accomplished through vaccination.

immunoassay: Detection and assay of substances by serological (immuno-logical) methods; in most applications the substance in question serves as antigen, both in antibodies production and in measurement of antibodies by the test substance.

incapacitating agents: Substances that produce temporary physiological and/or mental effects via action on the central nervous system. Effects may persist for hours or days. Victims usually do not require medical treatment, but treatment will assist in speeding recovery.

incident action plan: A plan consisting of the strategic goals, tactical objectives and support requirements for the incident. All incidents require an action plan. For simple/smaller incidents, that action plan is not usually in written form. Larger or complex responses require the action plan be documented in writing.

incident command post (ICP): The location from which command functions are executed.

incident management system (IMS): Originally known as the incident command system, IMS has evolved into a systematic management approach with a common organizational structure responsible for the management of assigned resources to effectively accomplish stated objectives pertaining to an incident.

incident manager (IM): The designated person with overall authority for management of the incident (varies by jurisdiction).

incident objectives: Statements of guidance and direction necessary for the selection of appropriate strategy(s) and the tactical direction of resources to accomplish the same. Incident objectives are based on realistic expectations of operational accomplishments when all anticipated resources have been deployed. Incident objectives must be achievable and measurable, yet flexible enough to allow for strategic and tactical realignment.

incident termination/securement: The conclusion of emergency operations at the scene of an incident, usually the departure of the last resource from the incident scene.

incubation period: The time from exposure to the disease until the first appearance of symptoms.

industrial agents: Chemicals developed or produced for use in industrial operations or research by academia, government, or industry. These substances are not primarily produced for the specific purpose of harming humans or incapacitating equipment, but if utilized by rogue individuals as a weapon, will yield results similar to a chemical agent.

infection: Growth of pathogenic organisms in the tissues of a host, with or without detectable signs of injury.

infectious: Capable of causing infection in a suitable host.

infectious disease: An illness or disease resulting from invasion of a host by disease-producing organisms such as bacteria, viruses, fungi, or parasites.

infectivity: (1) The ability of an organism to spread. (2) The number of organisms required to cause an infection to secondary hosts. (3) The capabilities of an organism to spread out from site of infection and cause disease in the host organism. Infectivity can also be defined as the number of organisms required to cause an infection.

ingestion: Exposure to a substance through the gastrointestinal tract.

inhalation: Exposure to a substance through the respiratory tract.

initial response: The resources initially committed to an incident.

injection: Exposure to a substance through a break in the skin.

inoculation: *See* **vaccine.**

insecticide: Chemicals that are used to kill insects.

international terrorism: The unlawful use of force or violence, committed by a group(s) of two (2) or more individual(s) who is foreign based, and/or directed by countries or groups outside the continental United States (CONUS), or whose activities transcend national boundaries, against persons or property to intimidate or coerce a government, the civilian population, or any segment thereof, in furtherance of criminal, political, or social objectives.

Integrated Emergency Management System (IEMS): A system of emergency planning and consequence response that integrates local response agencies, state agencies, and federal agencies into a comprehensive emergency management plan.

ions, ionization: Atoms that have the same number of electrons and protons have zero charge since the number of positively charged protons equals the number of negatively charged electrons. If an atom has more electrons than protons, it has a negative charge, and is called a negative ion. Atoms that have fewer electrons than protons are positively charged, and are called positive ions. Some forms of radiation can strip electrons from atoms. This type of radiation is appropriately called "ionizing radiation."

ionizing radiation: Ionizing radiation is radiation that has enough energy to cause atoms to lose electrons and become ions. Alpha and beta particles, as well as gamma and x-rays, are all examples of ionizing radiation. Ultraviolet, infrared, and visible light are examples of non-ionizing radiation.

in vitro: An artificial environment, as in a test tube or culture media.

in vivo: In the living body, referring to a reaction or process therein.

ischemia: The lack of blood supply to a part of the body, leading to deficient oxygen levels and subsequent damage to anatomical structures.

ischemic necrosis: Death of cells subsequent to the lack of blood flow to affected tissues or organs.

Law Enforcement Incident Command System (LEICS): A law enforcement incident management system based on the IMS model.

LD$_{50}$: Dose (LD is lethal dose) that will kill 50 percent of the exposed population.

Leader: The individual responsible for command of a task force, strike team, or unit.

LEL: Lower explosive limit. The minimum concentration of a substance (gas or vapor) required for a substance to burn.

Level A protection: The level of protective equipment required in situations where the substance is considered acutely vapor toxic to the skin and the hazards are unknown. Use of Level A is recommended when immediate identification of the substance is unavailable or unknown. Level A consists of a full encapsulating protective ensemble with SCBA or supplied air breathing apparatus (SABA).

Level B protection: The level of protective equipment in situations where the substance is considered acutely vapor toxic to the skin and the hazards may cause respiratory effects. Level B consists of a Level B encapsulating (non-airtight) protective ensemble or chemical splash suit with SCBA or SABA.

Level C protection: The level of protective equipment required to prevent respiratory exposure, but not to exclude possible skin contact. Chemical splash suits with cartridge respirators (APRs).

Level D protection: The level of protective equipment required when the atmosphere contains no known hazard, when splashes, immersions, inhalation, or contact with hazardous levels of any substance is precluded. Work uniform such as coveralls, boots, leather gloves, and hard hat.

liaison: The coordination of activities between agencies operating at the incident.

liaison officer: The point of contact (POC) for assisting or coordinating agencies and members of the management staff. The liaison officer is a member of the management staff.

liquid agent: Chemical agent that appears to be an oily film or droplet form, usually brownish in color.

liquifaction necrosis: The destructive process by which alkali causes cell death and turns solid tissue into a soapy liquid.

liver: The largest and one of the most complex internal organs of the body. The liver produces bile, secretes glucose, protein, vitamins, fat, and other compounds. It also processes hemoglobin to forage iron content, and is the body's primary detoxification center.

logistics section: The section responsible for providing facilities, services, and materials for the incident. Includes the communication unit, medical unit, and food unit within the service branch; and the supply unit, facilities unit, and ground support unit within the support branch.

low impact incident/event: An MCI that can be managed by local EMS resources and members without mutual aid resources from outside organizations.

lymph: A clear, transparent, colorless, alkaline fluid found in the lymphatic vessels.

lymph nodes: A rounded body consisting of lymphatic tissue found at intervals in the course of lymphatic vessels.

lymphadenopathy: Any disease process affecting a lymph node or lymph nodes.

macula: A small spot, perceptibly different in color from surrounding skin.

management staff: The incident manager's direct support staff, consisting of the public information officer, liaison officer, safety officer, and the stress management officer.

Mass Casualty Incident (MCI): An incident with several patients or an unusual event associated with minimal casualties (airplane crash, terrorism, haz mat, etc.); incident with negative impact on hospitals, EMS, and response resources.

mass decontamination: The decontamination of mass numbers of patients (pediatric, adult, and geriatric) from exposure to radiation or a chemical/biological agent.

mechanism of injury: A sudden and intense energy transmitted to the body that causes trauma, or exposure to a chemical or biological agent. A contaminated patient can transport a chem-bio mechanism of injury.

mediastinitis: Inflammation of the tissue of the mediastinum.

mediastinum: The median partition of the thoracic cavity, covered by the mediastinal pleura and containing all the thoracic viscera and structures except the lungs.

median incapacitating dosage (ID50): The amount of liquid chemical agent expected to incapacitate 50 percent of a group of exposed, unprotected individuals.

median lethal dosage (LCT50): The amount of liquid chemical agent expected to kill 50 percent of a group of exposed, unprotected individuals.

medical control: Central focus of all medical treatment and direction being rendered at the scene of an emergency; usually an M.D. on site or via radio.

medical supply coche: A cache consisting of standardized medical supplies and equipment stored in a predesignated location for dispatch to MCI incidents.

medical unit: The unit within the service branch of the logistics section responsible for providing emergency medical treatment to emergency responders. This unit does not provide treatment to civilians. Rehab is often a function assigned to the medical unit.

medium: Substance used to provide nutrients for the growth and multiplication of microorganisms.

melena: Passage of dark-colored, tarry stools due to the presence of blood altered by the intestinal juices.

meningococcemia: Presence of meningococci in the circulating blood.

meninges: Membranous coverings of the brain and spinal cord.

METTAG: A four-color system of tagging patients during the triage process. Each color is demonstrative of a differing medical priority; red—critical, yellow—less serious and delayed transport, green—minor injuries (walking wounded), and black—deceased or unsalvageable.

Metropolitan Medical Strike Team (MMST): The Metro Medical Strike Team shall, at the request of local and/or regional jurisdictions, respond to and assist with the medical treatment/management and public health consequences of chemical, biological, and nuclear incidents resulting from deliberate or accidental acts.

microcyst: A tiny cyst, frequently of such dimensions that a magnifying lens or microscope is required to visualize it.

microorganism: Any organism, such as bacteria, viruses, and some fungi, that can be seen only with a microscope.

microscopy: Observation/investigation of minute objects by means of a microscope.

minor patient: A patient with minor injuries that requires minimal treatment; triage color is green.

miosis: A condition where the pupil of the eye significantly constricts (pinpoint) impairing vision, especially night vision.

mitigation: Actions taken to prevent or reduce the likelihood of harm.

mists: Liquid droplets dispersed in the air.

M8 chemical agent detector paper: A paper used to detect and identify liquid V and G class nerve agents and H class blister agents.

M256 Kit: A kit that detects and identifies vapor concentrations of nerve, blister, and blood agents.

MMST: *See* **Metropolitan Medical Strike Team.**

morgue: A segregated area for deceased victims that is coordinated with the medical examiner and/or jurisdictional law enforcement authorities; triage color for deceased victims is black.

MSDS: Material Safety Data Sheets. A comprehensive document that delineates all pertinent information about a hazardous substance. This information sheet is provided by and available from the manufacturer of the product.

mucocutaneous: Referring to the mucous membrane and skin.

mustard agent: *See* **blister agent.**

myalgia: Muscular pain.

mycotoxin: A toxin produced by fungi.

mydriasis: Dilation of the pupil.

nasopharynx: The part of pharynx that lies above the level of the soft palate and directly posterior to the nose.

National Interagency Incident Management System (NIIMS): An adaptation of fire ICS for interagency disaster operations.

naturally occurring radioactive materials (NORM): The term NORM is used to identify naturally occuring radioactive materials that may have been technologically enhanced in some way. The enhancement occurs when a naturally occurring radioactive material has its composition, concentration, availability, or proximity to people altered by human activity. The term is usually applied when the naturally occuring radionuclide is present in sufficient quantities or concentrations to require control for purposes of radiological protection of the public or the environment. NORM does not include source, by-product, or special nuclear material (terms defined by law and referring primarily to uranium, thorium, and nuclear fuel cycle products); or commercial products containing small quantities of natural radioactive materials (e.g., phosphate fertilizer, potassium chloride for road deicing) or natural radon in buildings.

NBC: A military acronym for nuclear, biological, and chemical weapons.

necrosis: Pathological death of one or more cells or a portion of tissue or organ resulting in irreversible damage.

nerve agent: Chemical agent that acts by disrupting the normal function of the nervous system.

neutron: Neutrons are part of the nucleus of an atom. Neutrons are, as the name implies, neutral in their charge. That is, they have neither a positive nor a negative charge. Neutrons are about the same size as protons.

neurotransmitter: A substance that is released from the axon terminal of a presynaptic neuron on excitation, and travels across the synaptic cleft to influence (excite or inhibit) the target cell, such as epinephrine, dopamine, or norepinephrine.

NIOSH: National Institutes for Occupational Safety and Health.

nonlethal agents: Chemical agents that can incapacitate but which, by themselves, are not intended to cause death. Examples are tear gases, vomiting agents, and psychochemicals such as BZ .

non-persistent agent: An agent that upon release loses its ability to cause casualties after 10 to 15 minutes. It possesses a high evaporation rate, is lighter than air, and will disperse quickly. This type of agent is considered to be a short-term hazard, however, in small and unventilated areas the agent will be more persistent than out in the open.

non-stochastic effect: Effects that can be related directly to the dose received. The effect is more severe with a higher dose, (i.e., the burn gets worse as dose increases). It typically has a threshold, below which the effect will not occur. A skin burn from radiation is a non-stochastic effect.

ocular: Pertaining to the eye.

oliguria: Markedly reduced urine output.

operations section: The section responsible for all tactical operations at the incident.

organism: Any individual living thing whether human, animal, or plant.

organophosphate: A compound with a specific phosphate group that inhibits acetycholinesterase. Used in chemical warfare agents, insecticides, and pesticides.

organophosphorous compound: A compound containing the elements phosphorous and carbon, whose physiological effects include inhibition of acetycholinesterase. Most pesticides and virtually all nerve agents are organophosphate compounds.

oropharynx: Portion of the pharynx that lies posterior to the mouth.

OSHA: Occupational Safety and Health Administration. A part of the U.S. Department of Labor.

osteomyelitis: Inflammation of the bone marrow and adjacent bone.

oxygen meters: Device that measures or monitors the concentration of oxygen in a specific area.

pandemic: Denoting a disease affecting or attacking the population of an extensive region, country, or continent; extensively epidemic.

papule: A small, circumscribed, solid elevation on the skin.

parasite: Any organism that lives in or on another organism without providing benefit in return.

parasympathetic nervous system: A division of the autonomic nervous system that is responsible for controlling the body's vegetative functions.

parasympatholytic: A drug or other substance that blocks or inhibits the actions of the parasympathetic nervous system. *See also* **anticholinergic.**

parasympathomimetic: A drug or other substance that causes effects like those of the parasympathetic nervous system.

pathogen: A microorganism that can cause disease. Pathogens can be bacteria, fungi, parasites, or viruses.

pathogenic: Capable of causing disease.

penetration: The movement of a substance through a closure such as a flap, seam, zipper, or other vulnerable design feature of a chemical protective garment.

PEL: *See* **permissible exposure limits.**

percutaneous: Referring to the passage of substances through intact skin (e.g., needle puncture, etc.).

percutaneous agent: A substance able to be absorbed through the body.

percutaneous absorption: Substances absorbed through the skin.

perivascular: Surrounding a blood or lymph vessel.

permeation: The process by which a chemical moves through protective clothing.

permeation rate: The rate at which the challenge chemical permeates the protective fabric.

permissible exposure limit (PEL): An occupational health term used to describe exposure limits for employees. Usually described in time-weighted averages (TWA) or short-term exposure limits (STEL). The maximum average concentration (over eight continuous hours, aver-aged), to which 95 percent of otherwise healthy adults can be repeatedly and safely exposed for period 8-hour days and 40-hour weeks.

persistance: Measure of the duration for which a chemical agent is effective. This property is relative and varies by agent, method of dissemi-nation, and influencing environmental conditions such as weather and terrain.

persistent agent: A substance that remains in the target area for longer time frames. Hazards from both liquid and vapor may remain for hours, days, or in extreme cases weeks after distribution of an agent. As a rule of thumb, persistent agent duration will be greater than 12 hours.

personal protective equipment (PPE): Equipment for the protection of EMS personnel; includes gloves, masks, goggles, gowns, and biological disposal bags (red bags).

petechiae: Minute hemorrhagic spots in the skin, of pinpoint to pinhead size, that are not blanched by pressure.

pH: A scientific method of articulating the acid or base (alkali) content of a solution. pH is a logarithm of the hydrogen (H) ion concentration divided by 1. The higher the pH the greater the alkalinity, the lower the pH the greater the acidity.

pharyngeal: Relating to the pharynx.

phosgene: Carbonyl chloride; a colorless liquid below 8.2°C, but an extremely poisonous gas at ordinary temperatures. It is an insidious gas, and is not immediately irritating, even when fatal concentrations are inhaled.

photophobia: Significant dread and avoidance of light. Usually accompanied by severe pain with exposure to light.

physiological action: Most toxic chemical agents are used for their toxic effect. The effects are the production of harmful physiological reactions when the human body is exposed either through external, inhalation, or internal routes. The subsequent bodily response to the exposure is the physiological response.

planning meeting: Meetings, held as required throughout the duration of the incident, to select specific strategies and tactics for incident management and for service and support planning.

pleurisy: Inflammation of the pleura.

poison: Any substance which, taken into the body by absorption, ingestion, inhalation, or injection, interferes with normal physiological functions.

poisoning: The state of introduction of a poison into the body.

Poison Control Center: A toxicological information clearinghouse (usually staffed 24 hours a day/7 days a week) that serves as informational resource for toxic materials and the treatment of their exposures.

polyuria: Excessive urination.

post-incident analysis (PIA): A written review of major incidents for the purpose of implementing changes in operations, resources, logistics, and protocols, based on lessons learned.

ppm: Parts per million.

precursor: A chemical substance required for the manufacture of chemical agents.

Presidential Decision Directive (PDD): A presidential directive that establishes policy.

Presidential Decision Directive (PDD) 39: PDD 39 presents the United States policy on counterterrorism.

presynaptic: Pertaining to the area on the proximal side of a synaptic cleft.

prophylaxis: Prevention of disease or of a process that can lead to a disease.

proton: Protons, along with neutrons, make up the nucleus of an atom. Protons have a single positive charge. While protons and neutrons are about 2,000 times heavier than electrons, they are still very small particles. A grain of sand weighs about one hundred million trillion (100,000,000,000,000,000,000) times more than a proton or a neutron.

pruritus: Itching.

psychochemical agent: Chemical agent that incapacitates by distorting the perceptions and cognitive processes of the victim.

ptosis: Drooping of the eyelids.

public information officer (PIO): The person responsible for interface with the media and others requiring information direct from the incident scene. Information is only disseminated with the authorization of the incident commander. The PIO is a member of the command staff.

pull logistics: A process of ordering supplies by field units, via communications, as they are needed at an MCI or disaster.

pulmonary edema: Edema of the lungs. Left unattended in severe cases can be fatal.

pupil: Opening at the center of the iris of the eye for transmission of light.

push logistics: A process of forwarding predetermined supplies, usually as a disaster cache, to an MCI or disaster.

pyrogenic: Causing fever.

radiation: Radiation is energy in the form of waves or particles (see types of radiation). Radiation comes from sources such as radioactive material or from equipment such as x-ray machines or accelerators.

radiation dose: The effect of radiation on any material is determined by the dose of radiation that material receives. Radiation dose is simply the quantity of radiation energy deposited in a material. There are several terms used in radiation protection to precisely describe the various aspects associated with the concept of dose, and how radiation energy deposited in tissue affects humans.

radiation exposure: Radiation exposure is a measure of the amount of ionization produced by x-rays or gamma rays as they travel through air. The unit of radiation exposure is the roentgen (R), named for Wilhelm Roentgen, the German scientist who, in 1895, discovered x-rays.

radiation half-life: The time required for a population of atoms of a given radionuclide to decrease, by radioactive decay, to exactly one-half of its original number. No operation, either chemical or physical, can change the decay rate of a radioactive substance. Half-lives range from much less than a microsecond to more than a billion years. The longer the half-life the more stable the nuclide. After one half-life, half the original atoms will remain; after two half-lives, one-fourth (or 1/2 of 1/2) will remain; after three half-lives one-eighth of the original number (1/2 of 1/2 of 1/2) will remain; and so on.

radiation meters: Monitoring devices that detect, measure, and monitor for the presence of radiation.

radioactive contamination: Radioactive contamination is radioactive material distributed over some area, equipment, or person. It tends to be unwanted in the location where it is, and has to be cleaned up or decontaminated.

radioactive decay: Radioactive decay describes the process where an energetically unstable atom transforms itself to a more energetically favorable, or stable state. The unstable atom can emit ionizing radiation in order to become more stable. This atom is said to be radioactive, and the process of change is called radioactive decay.

rate of action: Rate at which the body reacts to or is affected by a chemical substance.

rate of detoxification: Rate at which the body can counteract the effects of a toxic chemical substance.

rate of hydrolysis: Rate at which the various chemical substances or compounds are decomposed by water.

reactivity: Ability of a substance to interact with other substances and/or body tissues.

real world: A phrase transmitted to all units when there is an injury or actual emergency during a disaster exercise.

recombinant DNA (rDNA): DNA prepared in the laboratory by splitting and splicing DNA from different species, with the resulting recombinant DNA having different properties than the original.

recombinant vaccine: A vaccine produced by genetic manipulation (gene splicing) usually in yeast.

reconnaissance: The primary survey to gather information.

rehabilitation (rehab): The function and location that includes medical evaluation and treatment, food and fluid replenishment, and relief from extreme environmental conditions for emergency responders, according to the circumstances of the incident.

resource status unit (RESTAT): The unit within the planning section responsible for recording the status of, and accounting for, resources committed to the incident, and for evaluation of (1) resources currently committed to the incident, (2) the impact that additional responding units will have on an incident, and (3) anticipated resource requirements. Note: RESTAT is normally utilized at actual or escalating high impact or long-term operations.

respiratory dosage: Equal to the time in minutes an individual is unmasked in an agent cloud multiplied by the concentration of the cloud.

restriction enzyme: Enzyme that splits DNA at a specific sequence.

retrosternal: Posterior to the sternum.

rhinorrhea: Runny nose.

rickettsia: Generic name applied to a group of microorganisms, family Rickettsiaceae, order Rickettsiales, which occupy a position intermediate between viruses and bacteria. They differ from bacteria in that they are obligate parasites requiring living cells for growth and differ from viruses in that the Berkefeld filter retains them. They are the causative agents of many diseases and are usually transmitted by lice, fleas, ticks, and mites (anthropods).

riot control agents: Substances usually having short-term effects that are typically used by governmental authorities for law enforcement purposes.

Robert T. Stafford Disaster Relief and Emergency Assistance Act: A federal law that assigns disaster responsibilities to FEMA and defines federal support to local communities.

routes of exposure: The mechaism by which a contaminant enters the body.

SABA: Supplied air breathing apparatus.

safety officer: The command staff member responsible for monitoring and assessing safety hazards, unsafe situations, and developing measures for ensuring member safety on-site.

sarin: A nerve poison which is an extremely potent, irreversible cholinesterase inhibitor.

SCBA: Self-contained breathing apparatus.

scarification: The making of a number of superficial incisions in the skin.

sclera: Tough white supporting tunic of the eyeball.

secondary device: An explosive device designed and placed to kill emergency responders.

section: The organizational level having functional responsibility for primary segments of incident operations such as operations, planning, logistics, and finance/administration. The section level is organizationally between branch and incident commander.

section chief: Title referring to a member of the general staff (operations section chief, planning section chief, logistics section chief, and finance/administration section chief).

sector/group officer: The individual responsible for supervising members who are performing a similar function or task (i.e., triage, treatment, transport, or extrication).

security unit: Responsible for personnel security, traffic control, and morgue security at an MCI or disaster.

seizure: A disorder of the nervous system owing to sudden, excessive, disorderly discharge of brain neurons.

sensitize: To become highly responsive (sensitive) or easily receptive to the effects of a toxic substance after initial exposure.

sequala (ae): A condition following as a consequence of a disease.

sequestration agents: Agents that bind specific salts and make them unavailable to the cells.

service branch: A branch within the logistics section responsible for service activities at an incident. Its components include the communications unit, medical unit, and food unit.

shigellosis: Bacillary dysentery caused by a bacteria of the genus Shigella, zoften occurring in epidemic patterns.

short-term exposure limits (STEL): A 15-minute, time weighted average (TWA) exposure that should not be exceeded at any time during a work day, even if the 8-hour TWA is within the threshold limit value (TLV). Exposures at the STEL should not be repeated more than four times a day and there should be at least 60 minutes between successive exposures at the STEL.

single resource: An individual ambulance or piece of equipment used to complete a task.

Situation Status Unit (SITSTAT): The unit within the planning section responsible for analysis of the situation as it progresses, reporting to the planning section chief. Note: SITSTAT is normally utilized at actual or escalating high-impact or long-term operations.

skin dosage: Equal to the time of exposure in minutes of an individual's unprotected skin, multiplied by the concentration of the agent cloud.

sloughing: Process by which necrotic cells separate from the tissues to which they have been attached.

solubility: (1) Ability of a material (solid, liquid, gas, or vapor) to dissolve in a solvent. (2) Ability of one material to blend uniformly with another.

solvent: Material that is capable of dissolving another chemical.

soman: An extremely potent cholinesterase inhibitor.

somatic effects: Effects from some agent, like radiation, that are seen in the individual who receives the agent.

span of control: The number of subordinates supervised by a superior; ideal span varies from three to five people.

Special Agent in Charge (SAIC): The senior individual appointed by a federal law enforcement agency to manage and coordinate all activities. For the purposes of this textbook, the SAIC is the FBI representative at the terrorism incident in charge of crisis management response. The SAIC is usually the FBI's on-site commander.

specific gravity: Weight of a liquid compared to the weight of an equal volume of water.

spore: A reproductive form some microorganisms can take to become resistant to environmental conditions such as cold or heat. This is referred to as the "resting phase."

Stafford Act: *See* **Robert T. Stafford Disaster Relief and Emergency Assistance Act.**

staging: A specific status where resources are assembled in an area at or near the incident scene to await deployment or assignment.

staging area: The location where incident personnel and equipment are assigned on an immediately available status.

standard operating procedures (SOPs): An organizational directive that establishes a standard course of action.

standing orders: Medical treatment policies, protocols, and procedures approved by a local, regional, or state EMS authority for use by EMS personnel without having to first make direct medical control contact for authorization.

START: Acronym for "simple treatment and rapid triage." This is an initial triage system utilized for triaging large numbers of patients at an emergency incident. This system was developed in Newport Beach, California, in the early 1980s.

status epilepticus: Two or more seizures in succession without a lucid interval.

stochastic effects: Effects that occur on a random basis, and are independent of the size of dose. The effect typically has no threshold and is based on probabilities, with the chances of seeing the effect increasing with dose. Cancer is thought to be a stochastic effect.

strategic goals: The overall plan that will be used to control the incident. Strategic goals are broad in nature and are achieved by the completion of tactical objectives.

stridor: High pitched, noisy respirations. Usually indicative of an upper airway obstruction, either foreign or anatomical.

strike team: Up to five of the same kind or type of resource with common communications and an assigned leader.

subdermal: Below the skin.

superinfection: A new infection in addition to one already present.

supply unit: The unit within the support branch of the logistics section responsible for providing the personnel, equipment, and supplies to support incident operations.

sympathetic nervous system: A division of the autonomic nervous system that prepares the body for stressful stimuli (fight or flight).

sympatholytic: A substance that produces effects that inhibit (block) the actions of the sympathetic nervous system, also referred to as antiadrenergic.

sympathomimetic: A substance that produces effects that mimic those of the sympathetic nervous system, also referred to adrenergic.

syncope: A transient loss of consciousness caused by inadequate blood flow.

tactical objectives: The specific operations that must be accomplished to achieve strategic goals. Tactical objectives must be specific and measurable, and are usually accomplished at the division or group level.

tactical ultraviolence: Maximum violence used to accomplish a criminal goal or objective.

task force: A group of any type or kind of resource, with common communications and a leader, temporarily assembled for a specific mission (not to exceed five resources).

technical advisor: Any individual with specialized expertise useful to the management/general staff.

technical specialists: Personnel with special skills who are activated only when needed. Technical specialists may be needed in the areas of rescue, water resources, and training. Technical specialists report initially to the planning section, but may be assigned anywhere within the IMS organizational structure as needed.

teratogenic effects: Effects seen in the offspring of the individual who received the agent. The agent must be encountered during the gestation period.

teratogenicity: Capacity of a substance to produce fetal malformation.

threat assessment: An assessment of a community's vulnerability and potential for natural, technological, and terrorist risks.

time unit: A unit within the finance section. Responsible for record keeping of time for personnel working at incident.

time-weighted averages (TWA): Average concentration for a normal 8-hour work day and a 40-hour work week, to which nearly all workers may be repeatedly exposed without adversity.

toxicity: Property a substance possesses that enables it to injure the physiological mechanism of an organism by chemical means with the maximum effect being incapacitation or death. The relative toxicity of an agent can be articulated in milligrams of toxin needed per kilogram of body weight to kill experimental animals.

toxoid: A toxin that has been treated so as to destroy its toxic property, but retain its antigenicity. Its capability of stimulating the production of antitoxin antibodies is retained, thus producing an active immunity.

toxins: Poisonous substances produced by living organisms.

tracking officer: The EMS IMS organizational position, usually a sub-component of the transportation sector/group, responsible for tracking all patients removed from the scene or treated and released.

transfer of command: A process of transferring command responsibilities from one individual to another. Commonly a formal procedure conducted in a face-to-face interaction with a event synopsis briefing and completed by a radio transmission announcing that a certain individual is now assuming command responsibility of an incident. A similar transition occurs when a sector/group or division/branch transfers responsibilities.

trauma intervention program (TIPS): A program to manage traumatic stress in emergency responders and disaster victims/families.

transportation sector/group: The EMS IMS organizational component responsible for acquisition and coordination of all patient transport resources. Most times this position is also responsible for coordinating the destination hospital for patients being removed from the scene.

treatment sector/group: The EMS IMS organizational component responsible for collecting and treating patients in a centralized location.

triage: The act of sorting patients by the severity of their medical conditions.

triage sector/group: The EMS IMS organizational component responsible for conducting triage of all patients at an MCI or high impact incident.

UEL: Upper explosive limits.

unified command: A standard method to coordinate command of an incident when multiple agencies have either functional or geographical jurisdiction. This results in a command system with shared responsibility.

unit: The organizational element having functional responsibility for a specific incident planning, logistics, or finance activity.

unity of command: The concept of an individual being a supervisor at each level of the IMS, beginning at the unit level, and extending upward to the incident manager.

universal precautions: System of infectious disease control which assumes that direct contact with body fluids is infectious (OSHA definition). Centers for Disease Control and Injury Prevention have published a series of procedures and precaution guidelines to assist the rescuer in fully understanding threat potential and protective measures required.

upwind: In or toward the direction from which the wind blows. Place yourself with the wind blowing toward the suspected release site.

urticaria: Skin condition characterized by intensely itching red, raised patches of skin.

USAR: Urban search and rescue.

U.S. DOT: The United States Department of Transportation. Federal agency responsible for regulating the transportation of hazardous materials.

U.S. DOT hazard classifications: Hazard class designations for specific hazardous materials as delineated in the U.S. DOT Regulations.

V series nerve agents: Generally persistent chemical agents of moderate to high toxicity developed in the 1950s that act by inhibiting a key nervous system enzyme. Examples are VX, VE, VG, VM, and VS.

vaccine: A preparation of a killed or weakened microorganism used to artificially induce immunity against a disease.

vapors: Gaseous form of a substance that is normally in a liquid or a solid state at room temperature and pressure.

vapor agent: A gaseous form of a chemical agent. If heavier than air, the cloud will be down to the ground; if lighter than air, the cloud will rise and dissipate more quickly.

vapor density: A comparison of any gas or vapor to the weight of an equal amount of air.

vesicant agent: *See* **blister agent.**

vesicles: Blisters on the skin.

vesiculation: Formation or presence of vessicles (blisters).

viremia: The presence of a virus in the blood.

virulence: The disease-evoking power of a microorganism in a given host.

virus: A microorganism usually only visible with an electron microscope. Viruses normally reside within other living (host) cells, and cannot reproduce outside of a living cell. It is an infectious microorganism that exists as a particle rather than as a complete cell. Particle sizes range from 200 to 400 nanometers (one-billionth of a meter).

viscosity: Degree to which a fluid resists flow.

volatility: Measure of how readily a substance will vaporize.

vomiting agent: Substance that produces nausea and vomiting effects; can also cause coughing, sneezing, pain in the nose and throat, nasal discharge and tears.

water reactive: Any substance that readily reacts with or decomposes in the presence of water with a significant energy release.

water solubility: Quantity of a chemical substance that will dissolve or mix with water.

weapons of mass destruction (WMD): Weaponization of nuclear, radiological, biological, or chemical substances.

weapons of mass effect (WME): Same definition as WMD, but reflects a more accurate description of the events surrounding use of these type weapons. Destruction is not guaranteed when utilized, but societal effects in many ways can be assured.

x-rays: X-rays are an example of electromagnetic radiation that arises as electrons are deflected from their original paths or inner orbital electrons change their orbital levels around the atomic nucleus. X-rays, like gamma rays, are capable of traveling long distances through air and most other materials. Like gamma rays, X-rays require more shielding to reduce their intensity than do beta or alpha particles. X-rays and gamma rays differ primarily in their origin: X-rays originate in the electronic shell; gamma rays originate in the nucleus.

Index

A

Ability-to-survive-operation (ATSO), 7
Administration, 68–69
  in IMS, 27
Advanced trauma life support (ATLS), 17
After-action analysis, 277–278
Alfred P. Murrah Federal Building (Oklahoma
    City), 5
Ammonium nitrate, 225
Anthrax, 163–167
Auf der Heide, Erik, 58
Aum Shinrikyo, 112, 115, 147
Awareness, 47

B

Background exposure, 212–213
Bagwan Sri Rajneesh Sect, 147
Biological agents
  decontamination and, 247–248
  symptoms and treatment table, 177–182
Biological IEDs, 227–228
Biological terrorism, 143–144
  biological agents: symptoms and treatments
      table, 177–182
  biotoxins
    botulinum toxins, 154–156
    clostridium toxins, 156–157
    ricin, 157–158
    saxitoxin, 158–159
    staphylococcal enterotoxin, 159–160
    tetrodotoxin, 160–161
    trichothecene myocotoxins (T2), 161–162
  history, 145–146
  live bacteriologic warfare agents, 162
    anthrax, 163–167
    brucellosis, 167–168
    cholera, 168–170
    ebola virus, 170
    plague, 171–172
    Q fever, 172–173
    smallpox, 173–174
    tularemia, 174–176
  threat assessment (reasons for unprepared-
      ness), 147

control of supplies, 149
deniability, 152
detection, 148–149
intelligence, 148
interagency coordination and public per-
    ception, 151–152
personal protective equipment, 149–150
prophylaxis, 150
realities and costs, 153–154
response time, 152–153
training, 151
treaties, 146–147
Biotoxins. *See* Biological terrorism
Black powder, 225
Blast overpressure, 224
Blood agents, decontamination require-
    ments, 247
Bohr Model of the atom, 327
Botulinum toxins, 154–156
Boyd's closed-loop decision cycles, 20–21
Brucellosis, 167–168

C

C4 explosive, 225
Care under fire, 18
Casualty care in the combat environment,
    17–18
Chain of custody of crime scene evidence,
    264–265
Change, in information age, 189
Chemical agents: symptoms and treatment
    table, 138–139
Chemical/biological/radiological (CBR) hot
    zone, 48–49
Chemical IEDs, 227–228
Chemical terrorism and warfare agents, 111–112
  cyanide, 124–126
  delivery/dissemination, 113–114
  different from hazardous materials and acci-
      dents, 113
  early recognition, 137
  lewisite, 131–132
  nerve agents
    antidotes, 120–122